Updates on Titanium Dioxide

Edited by Bochra Bejaoui

Published in London, United Kingdom

Updates on Titanium Dioxide
http://dx.doi.org/10.5772/intechopen.104130
Edited by Bochra Bejaoui

Scientific Contributor: Sharafat Ali

Contributors
Fawad Ur Rehman, Rida e Maria Qazi, Zahra Sajid, Chunqiu Zhao, Afsar Ali Mian, Javid Khan, Lei Han, Ganeshraja Ayyakannu Sundaram, Rajkumar Kanniah, Vaithinathan Karthikeyan, Bilal Ahmad Khan, Bilal Akram, Raieesa Batool, Jingyi Wang, Hui Xiao, Huaxin Wang, Mohsen Mhadhbi, Houyem Abderazzak, Barış Avar, Anna Zielinska-Jurek, Szymon Dudziak, Marta Kowalkińska, Patrick Martin, Naceur M'Hamdi, Bochra Bejaoui, Imen Bouchmila, Khaoula Nefzi, Imen Belhadj Slimen, Sidrine Koumbad, Nicolas Joly, Helena Cristina Vasconcelos, Telmo Eleutério, Maria Gabriela Meirelles, Susana Sério, Jaspal Singh, Ashwani Kumar Verma, Aymen Bourezgui, Imen Kacem

First published in London, United Kingdom, 2023 by IntechOpen
IntechOpen is the global imprint of INTECHOPEN LIMITED, registered in England and Wales, registration number: 11086078, 5 Princes Gate Court, London, SW7 2QJ, United Kingdom

British Library Cataloguing-in-Publication Data
A catalogue record for this book is available from the British Library

Additional hard and PDF copies can be obtained from orders@intechopen.com

Updates on Titanium Dioxide
Edited by Bochra Bejaoui
p. cm.
Print ISBN 978-1-80355-672-7
Online ISBN 978-1-80355-673-4
eBook (PDF) ISBN 978-1-80355-674-1

For EU product safety concerns:
IN TECH d.o.o., Prolaz Marije Krucifikse Kozulić 3, 51000 Rijeka, Croatia,
info@intechopen.com or visit our website at intechopen.com.

Meet the editor

Bochra Bejaoui, Ph.D., is an Associate Professor of Analytical Chemistry in the Department of Chemistry, Faculty of Sciences of Bizerte, Tunisia, and a researcher at the National Institute of Research and Physico-chemical Analysis (INRAP), Laboratory of Useful Materials, Sidi Thabet Technopark, Tunisia. She received a master's degree from the Faculty of Science at the University Tunis El Manar, Tunisia, and a Ph.D. in Analytical Chemistry from the same university in 2005. She is the author of several original papers, guest editor of a special issue of *Materials*, and a reviewer for numerous journals. Dr. Bejaoui is a member of many international projects. She is currently working on analytical chemistry, bioactive molecules, separation, and purification.

Contents

Preface

I am aware that this book will spark a lot of debate. It has never been easy to dispute the consensus because the "system"—of any kind, in any setting—will use whatever means possible to maintain the status quo. I feel obligated to share my knowledge, analyses, and conclusions after spending 20 years in the field of analytical chemistry, sample preparation, and separation techniques of bioactive molecules. I have published twenty papers, including four reviews and book chapters on various aspects of analytical chemistry. The beneficiaries will be all of us – ourselves, scientists, researchers, students, and society. One of the most notable aspects of this book is that it lacks a textbook format where the chapters must be read in the order provided to be comprehended. In reality, you can begin the adventure at any point, depending on your interests, tastes, and preferences.

I do not expect everybody in the scientific community to agree with the content and ideas put forth in this book. But I do hope that the information and knowledge presented will be useful for the students, researchers, and scientists. The book comprises eleven chapters.

Chapter 1 describes the processes for synthesizing titanium dioxide (TiO_2) with vanadium oxide V_0, its properties, and some of its applications for photocatalysis. It also discusses the reductive, adsorption, optical, and structural properties of TiO_2 nanoparticles containing V_0. This chapter is useful for engineers who seek to produce extremely effective photocatalysts and expand the functional applications of photocatalysis by developing defective semiconductors. Chapter 2 discusses the synthesis of black titania and its applications. It outlines various synthetic approaches employed to obtain black titania and describes the structural features of the black titania nanomaterials, along with their photocatalytic performances towards various applications. Chapter 3 focuses on the comparison of possible surface-related parameters and photocatalytic activity of anatase, rutile, and brookite polymorphs with exposed different crystal facets. It also summarizes computational data on their different possible surface structures, focusing on the geometry, energy, and possible reconstructions. Chapter 4 briefly describes the synthesis process of the different types of nanostructured (NS) TiO_2, their chemical and surface modifications, and their applications. It also describes the preparation of NS TiO_2, including nanoparticles, nanorods, nanowires, nanosheets, nanofibers, and nanotubes. This chapter discusses the effects of precursor properties and synthesis conditions on the structure, crystallinity, surface specificity, and morphology of titanium dioxide nanoparticles. Chapter 5 explains the recently reported methods that are used to synthesize TiO_2 nanoparticles, such as sol–gel, hydrothermal, precipitation, and so on. The chapter also highlights the different properties of TiO_2. Chapter 6 describes the main features of TiO_2, including processes, structure, and final properties, and reports and discusses different surface coating methods for TiO_2 with inorganic oxides and organic matter. Chapter 7 deals with tuning the magnetic and photocatalytic properties of wide-bandgap metal oxide semiconductors for environmental remediation. Chapter 8 describes the factors that influence the crystalline structure, morphology, and adhesion of TiO_2 films to ginger lily fibers. The surface properties of TiO_2 films are determined using analytical techniques such as X-ray diffraction (XRD) and X-ray photoelectron spectroscopy (XPS), and their morphology and elemental composition are determined by scanning electron microscopy/energy dispersive X-ray spectroscopy (SEM/EDS). Chapter 9 discusses the complex impact of gamma rays on materials of TiO_2 and their practical use in photovoltaic contexts. Chapter 10 discusses plasmonic–TiO_2 nanohybrids for environmental and energy applications. Chapter 11 reviews the design, fabrication, and theranostic biomedical applications of nano titania.

I hope this book is widely read. If we are to avoid the blunders of the past, then we need to change direction and start benefiting from the knowledge base created by the scientists. We did not have this chance a decade ago. Now is the right time.

Bochra Bejaoui
National Institute of Research and Pysico-chemical Analysis (INRAP),
Laboratory of Useful Materials,
Technopark of Sidi Thabet,
Ariana, Tunisia

Faculty of Sciences of Bizerte,
Department of Chemistry,
The University of Carthage,
Zarzouna, Tunisia

Section 1

Synthesis and Properties of TiO_2

Chapter 1

Oxygen Vacancy in TiO_2: Production Methods and Properties

Javid Khan and Lei Han

Abstract

Titanium dioxide (TiO_2) is a versatile material used in a variety of applications, including photocatalysis, photovoltaics, sensing, and environmental remediation. The properties of TiO_2 are influenced by its defect disorder, with oxygen vacancy (V_O) being a prominent defect that has been widely studied. Defective TiO_2 materials, particularly those containing V_O defects, are of interest for the development of next-generation semiconducting nanomaterials. Several methods, including high-temperature calcination, ion implantation, and chemical doping, are used to produce defective TiO_2 with varying degrees of V_O defects. The properties of defective TiO_2, including optical, electronic, and structural characteristics, are essential for determining the material's suitability for various applications. Modification of the defect structure of TiO_2 through doping with impurities can enhance the photocatalytic activity of the material. Researchers continue to investigate the impact of factors such as crystal structure and the presence of other defects on the properties of TiO_2-based materials, further enhancing their potential for various applications. Overall, a deeper understanding of defect disorder and the development of production methods for defective TiO_2 will play a crucial role in the design and production of next-generation semiconducting nanomaterials.

Keywords: TiO_2, defects, oxygen vacancies, structural, electronic, optical properties

1. Introduction

As a wide bandgap semiconductor material with industrially significant applications, titanium dioxide (TiO_2) is commonly used in catalysts [1–3], ointments [4–6], paints [7–9], sunscreens [10–12], and toothpaste [13–15]. Intense study has been put into TiO_2 materials since Honda and Fujishima [16] discovered the phenomenon that TiO_2 can be used for photocatalytic water splitting. This has allowed TiO_2 to be used in photoelectrochemical cells [17–19], photovoltaics [20–22], and photocatalysis [23–25]. TiO_2 offers several benefits over other semiconductor materials, including its low toxicity, resistance to photocorrosion, abundance on Earth, and chemical and thermal stability [26]. However, due to its significant recombination rate and broad band gap (3.2 eV), poor quantum efficiency as well as inadequate exploitation of visible light during photocatalytic reaction TiO_2's applications is severely limited [27]. Therefore, a variety of methods have been used to alter the TiO_2 in an effort to narrow the band gap and lengthen the lifetime of photogenerated charge carriers [28]. These methods include co-doping with metal ion/nonmetal ions, coupling TiO_2 with a semiconductor with a small band gap, encasing noble metal cores in a TiO_2 shell to create metal core@TiO_2 shell composite photocatalysts, noble metal deposition, and surface sensitization by organic dyes [29–32]. Recent research has shown that the defect disorder of TiO_2 may influence several of its physical and chemical characteristics, including selectivity, photocatalytic reactivity, and light absorption, among others [33–35].

One of the most prominent defects observed in TiO_2 is oxygen vacancies (V_O), which are also considered to be common defects in metal oxides and have been studied extensively by using both experimental and theoretical characterizations [36–38]. The V_O has the potential to function as active sites and adsorption points during heterogeneous catalysis [39–41]. The electrical structure, charge transport, surface properties, and other photocatalytic characteristics of metal oxides based on TiO_2 have also been demonstrated to be intimately connected to V_O [42–44]. It is theoretically possible that Ti^{3+} centers or unpaired electrons (e^-), which could lead to the creation of donor levels in TiO_2's electronic structure, are produced as a result of the production of V_O on TiO_2 [45–47]. Additionally, it is thought that V_O alters the recombination rate of electron-hole pairs during photocatalysis, which alters the chemical processes that rely on charge transfer from either hole (h^+) or e^- [48–50]. According to theoretical and experimental findings, the excess e^- on V_O states impacts the reactivity and surface adsorption of important adsorbates like H_2O or O_2 on TiO_2. In order to take use of their special features for photocatalytic applications, controlled synthesis of V_O incorporating TiO_2 is of utmost importance [51–53].

In this instructional chapter, we list the methods for producing TiO_2 with V_O, go over their characteristics, and touch on some of the uses for photocatalysis. The preparation technique is the main division used to classify the syntheses of TiO_2 nanomaterials with V_O. The readers may consult the relevant literature for comprehensive directions for each synthesis. In Section 3, the reductive, adsorption, optical, and structural characteristics of the TiO_2 nanomaterials containing V_O are discussed. To create highly effective photocatalysts and increase the functional applications of photocatalysis, it is intended that this chapter would be a beneficial resource for engineers who want to create defective semiconductors.

2. Synthetic methods

2.1 Treatment under hydrogen

A common technique to alter the photoelectrochemical and surface characteristics of TiO_2 is the hydrogen treatment [54–56]. When TiO_2 is heated, the oxygen (O) atoms in the lattice are interacted with by hydrogen (H) atoms to generate V_O and alter the material's surface characteristics [57]. Three stages may be distinguished between the interaction among H and TiO_2 throughout this process: The elimination of the adsorbed oxygen ESR signals in **Figure 1** is evidence that hydrogen physically interacts with the gas at temperatures below 300°C. Additionally, at temperatures above 300°C, O atoms in the TiO_2 lattice get e^- that were previously held by H atoms. Then, the surface of TiO_2 has its lattice O extracted, causing the O atom to separate from the H atom and create H_2O. As a result, TiO_2's surface develops V_O, as seen in **Figure 1**. Thirdly, as the temperature reaches 450°C, the contact between the two substances happens more significantly. In order to produce Ti^{3+} defects, the e^- in the H atoms are transported to the Ti^{4+} ions of the TiO_2. The V_O states' e^- are forced away

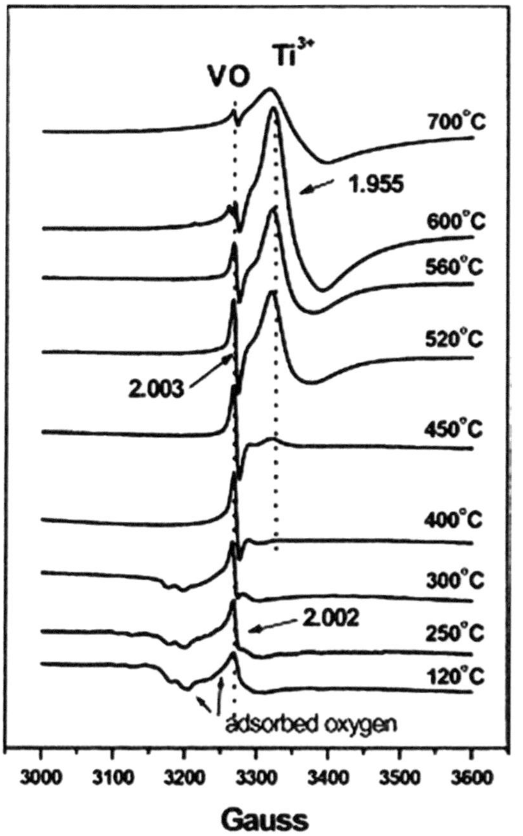

Figure 1.
ESR spectra of Ti^{3+} and V_o in TiO_2 after treatment with H_2. Reproduced with permission from ref. [57] Copyright Elsevier.

and moved to Ti^{4+} as the temperature rises to 560°C, where they remain until 600°C. As a consequence, the V_O states' ESR signal strength decreases and that of Ti^{3+} rises.

Notably, the hydrogen treatment used to reduce TiO_2 results in the formation of Ti interstitials as well as V_O in the matrix of TiO_2 [58]. The optical band gap of TiO_2 reduces when the amount of Ti exceeds that of O. According to Morgan and Watson [59], V_O creation is to some extent more favorable in rutile compared to anatase, whereas Ti interstitials formation occurs more in rutile. However, V_O is the preferred defect type in oxygen-rich environments. Still, the formation energies of both defect types are high. However, both defect types are stabilized in O-poor environments. Additionally, it is proposed that vacuum annealing and high-temperature type harsh conditions are needed for Ti ions interstitials formation than V_O [60]. Moreover, V_O are common defects in many oxides and not just significant defects in TiO_2, which has a significant impact on the physicochemical properties of such oxides. As a result, V_O has received a lot of attention and may be more interesting than Ti interstitials.

2.2 Bombardment with high-energy particles

Numerous studies have demonstrated that oxygen ions and neutral atoms can be selectively desorbed from TiO_2 surfaces, leading to the creation of vacancies [61–63].

Knotek and Feibelman [64] discovered that an interatomic Auger recombination practice enables e^- having energies of more than 34 eV to knockout surface oxygen. In their research, a potential mechanism for the formation of V_O when irradiated with e^- was also put forth. They suggested that the formation of V_O is caused by the removal of O^+ from the surface of TiO_2 due to e^- induced desorption. The benefit of using this technique for defect production is that e^- with moderate energies are bombarded to cause slight surface destruction and utterly create V_O. Even exposing the electron-irradiated surfaces at low temperatures to molecular oxygen can result in the creation of V_O [65].

Ion sputtering, specifically argon ion (Ar^+) sputtering, produces V_O on the surface of TiO_2, much like electron bombardment does [66]. When exposed to oxygen at low temperatures, the defects at the surface due to Ar^+ sputtering do not go away. This shows surface bridging V_O on Ar^+ sputtered surfaces along with other subsurface defects are suggested to be more highly reduced surface species. However, only by treating under oxygen at low temperatures, these kinds of defects cannot be repaired.

Additionally, reducing gas atmosphere plasma treatment at low temperatures is frequently used to produce V_O on the surface of metal oxides [67, 68]. Species with high energies like radicles, atoms, and e^- are used under low-temperature plasma. Due to moderate reaction conditions, the outer layer of metal oxides is changed, while the bulk materials are unaffected.

2.3 Doping

In the lattice of TiO_2, V_O often occur when they are doped with a nonmetal or metal ions. For instance, Krol and Wu have shown that the production of V_O in the TiO_2 lattice may occur when Fe^{3+} ions are substituted for Ti^{4+} ions in the lattice [69]. Additionally, Domen's group [70] revealed that aliovalent cations can be used to successfully dope and improve the defects of the photocatalyst. They proved that the extrinsic V_O are introduced with the help of a cation that has a lower valence as compared to the parent cation (Ti), preventing the creation of Ti^{3+}, as seen in **Figure 2a**. As a lower valence cation, trivalent cation (M^{3+}) doping occupies the Ti^{4+}

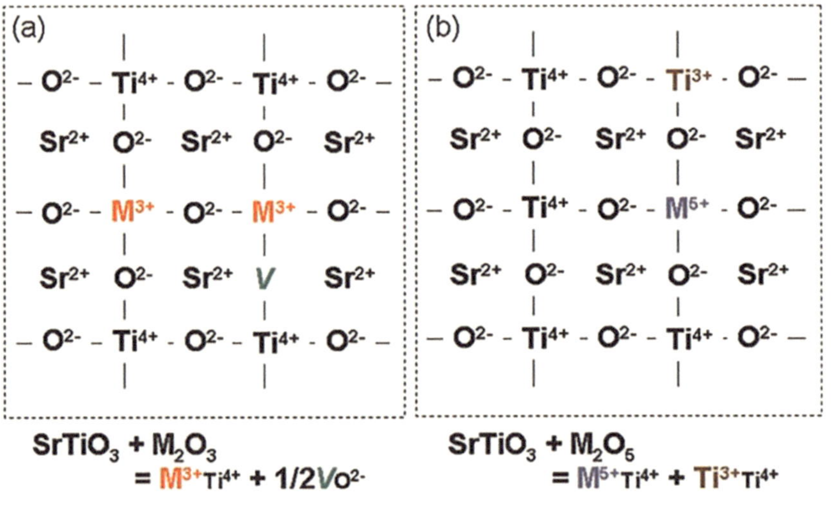

Figure 2.
Schematic diagram of doping of (a) trivalent cations and (b) pentavalent cations in $SrTiO_3$. Replicated with consent from Ref. [70] Copyright American Chemical Society.

sites. As seen in the equation of **Figure 2a**, this causes the production of V_O to be aided without producing Ti^{3+} species. In contrast, as seen in **Figure 2b**, the Ti^{3+} is stabilized by the higher valence cations without developing V_O. Similarly, the equation in **Figure 2b** shows that Ti^{4+} sites have been occupied by the pentavalent cation (M^{5+}) so Ti^{3+} sites would be created and the development of V_O would be prevented when a higher valence cation is doped.

Similar to how doping with metal ions may produce V_O in the TiO_2 lattice, doping with nonmetal ions like fluorine or nitrogen can also do so [51, 71–74]. According to calculations using density functional theory (DFT), adding N to bulk TiO_2 results in a significant decrease in the energy required to generate V_O. This shows that N doping makes V_O more likely to occur. Additionally, N doping is often performed in a decreasing environment. TiO_2 may be partially reduced by this reducing environment, which will lead to the development of V_O.

2.4 Through different reaction conditions

Another possible effect of the lattice oxygen participation in the thermally driven catalytic reaction of organic molecules is oxygen removal from the surface of TiO_2 [75]. A surface vacancy is created as a consequence of this process, which involves oxidizing organic materials on the surface of oxides while losing oxygen atoms from the surface lattice. For example, as shown in **Figure 3**, Morris and Panayotov [76] showed that methoxyl groups could be burnt thermally by activated lattice oxygen, leading to shallow donor states (Ti^{3+} and V_O) below the conduction band of TiO_2.

On the surface of several semiconductors, photochemically induced oxidation reactions, the mechanism of reaction-driven V_O production is also at work [77–81]. For instance, under the circumstances of a photocatalytic process, Xu et al. [82] have only recently discovered that V_O are photoinduced formed on TiO_2. According to the

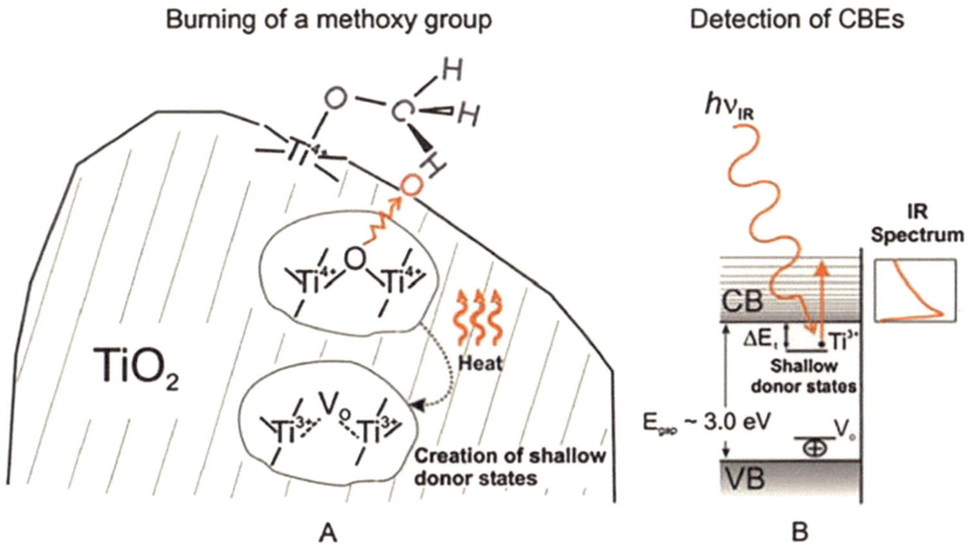

Figure 3.
Schematic illustration of (A) thermally activated oxygen leaving Ti^{3+}-V_o-Ti^{3+} donors in the bridge lattice. (B) Methoxyl groups attached to CUS Ti^{4+} Lewis acid sites burn on the particle surface where the oxygen atoms diffuse to. Reprinted with permission from ref. [76]. Copyright American Chemical Society.

detailed production method, when UV light is applied, molecular oxygen absorbs the photo-generated e^-, while the h^+ diffuses to the surface of the TiO_2 and are trapped by the lattice oxygen. Consequently, the lattice oxygen and Ti atom's binding link become weaker due to the trapped h^+, and this bond is broken by the adsorbed molecule benzyl alcohol. The oxygen from the lattice is subsequently removed from the surface of TiO_2, creating V_O defects on the catalyst's surface.

2.5 Thermal treatment under oxygen deficient conditions

Another method for producing V_O is to anneal TiO_2 in the pure form above 400°C under Ar, N_2, or He gas atmosphere, or in a vacuum [83]. The following equilibrium may be used to explain how V_O arises at high temperatures, using the common Kroger-Vink notation:

$$O_O \overset{TiO_2}{\leftrightarrow} V_O + \frac{1}{2}O_2(g) + 2e^- \tag{1}$$

Following is an expression for the equilibrium constant of this reaction:

$$K = [V_O]n^2 p(O_2)^{\frac{1}{2}} \tag{2}$$

As a function of $p(O_2)$, Eq. (2) may be transformed to indicate the V_O concentration:

$$[V_O] = Kn^{-2}p(O_2)^{-\frac{1}{2}} \tag{3}$$

Where O_O signifies the lattice oxygen; $[V_O]$ signifies the concentration of V_O, V_O the number of oxygen vacancies, and $p(O_2)$ the oxygen pressure. From Eq. (3), we may infer that the concentration of V_O rises as O_2 pressure falls, i.e., the oxygen deprived state during thermal annealing would promote the production of V_O.

Reaction (1) is reversible even at room temperature, and thus, the V_O as they have generated will gradually vanish when the TiO_2 is exposed to air [69]. The TiO_2 nanoparticles may be doped with foreign ions working as accepter-like Fe dopants, to stabilize the V_O.

$$Fe_2O_3 \overset{TiO_2}{\rightarrow} 2Fe_{Ti} + 3O_O + V_O \tag{4}$$

There is no way to undo this breakdown process. Positively charged V_O in the TiO_2 lattice would be made up for by the inclusion of Fe^{3+} [52]. The amount of free e^- in TiO_2 is subsequently reduced as a consequence. Since Fe^{3+} ions are doped into TiO_2, V_O are so stabilized.

3. Properties

3.1 Structural properties

3.1.1 Electronic structure

One or two e^- localize in a V_O state when an oxygen atom from the bulk or surface of TiO_2 is absent. The highly ionic crystal's Madelung potential is the main factor behind the e-'s localization in the V_O state [60]. One or two "free" e^- in the flawed crystal take the O_2 anion's position from the normal lattice in this fashion, reducing the energy cost of vacancy creation. As illustrated in **Figure 4**, these e^- on the V_O states directly affect the electrical structure of TiO_2 by producing a donor level below the conduction band. Localized donor states generated from V_O have an energy level between 0.75 and 1.18 eV below TiO_2's conduction band [84]. Additionally, the elimination of oxygen atoms to create V_O may also result in the redistribution of extra e^- to the Ti atoms in the neighborhood of the V_O site, leading to the formation of shallow donor states below the conduction band derived from Ti 3d orbits [85]. These donor states are shown to rise with increasing V_O in both rutile and anatase TiO_2. For anatase TiO_2 with very low oxygen content, they may even overlap the conduction band [86]. These results indicate that the formation of V_O leads to a large shift of the Fermi level of TiO_2 toward higher energies.

3.1.2 Geometric structure

In addition to altering the electrical structure of TiO_2, V_O also changes the material's geometric structure [87–89]. The formation of V_O changes the surface structure of TiO_2, as shown by Park et al. [90] who also found that the generation of V_O resulted in an upshift of the Eg mode of the Ti-O bond in the Raman spectrum. Due to the existence of V_O, this may be ascribed to atomic rearrangement. In order to strengthen their connection with the remaining lattice, the three Ti atoms closest to an ejected O atom have a tendency to move away from the vacancy [91]. Similarly, Watson et al. reported that the bond length of the Ti-O bond is reduced as a result of outward relaxation, which also reduces the overlap between the three Ti dangling bonds. According to Dal Santo et al. [58] experimental electron diffraction data support this conclusion. They have seen a TiO_2 lattice shrinkage brought on by V_O.

As illustrated in **Figure 5**, Cheng et al. [89] recently revealed that the prominent surface reorganization is caused by V_O in TiO_2 sheets with surface-terminated

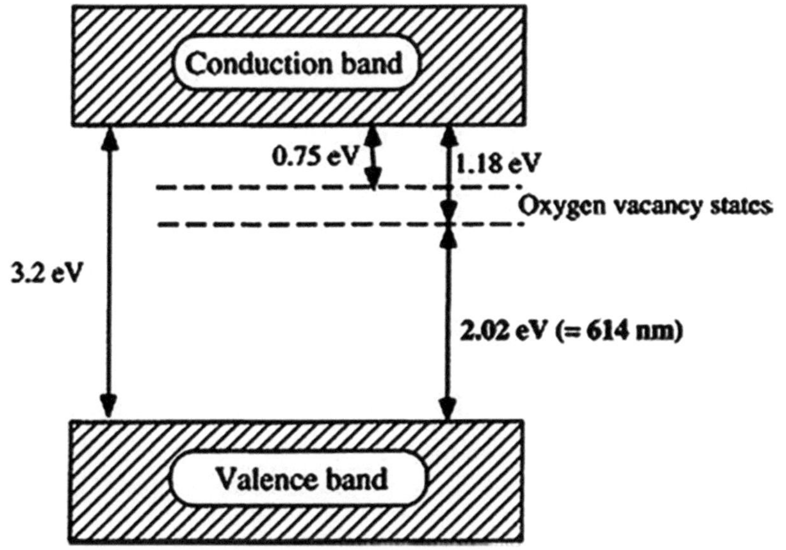

Figure 4.
Projected band structure model for V_o in anatase TiO_2. Copied with approval from Ref. [84] Copyright Elsevier.

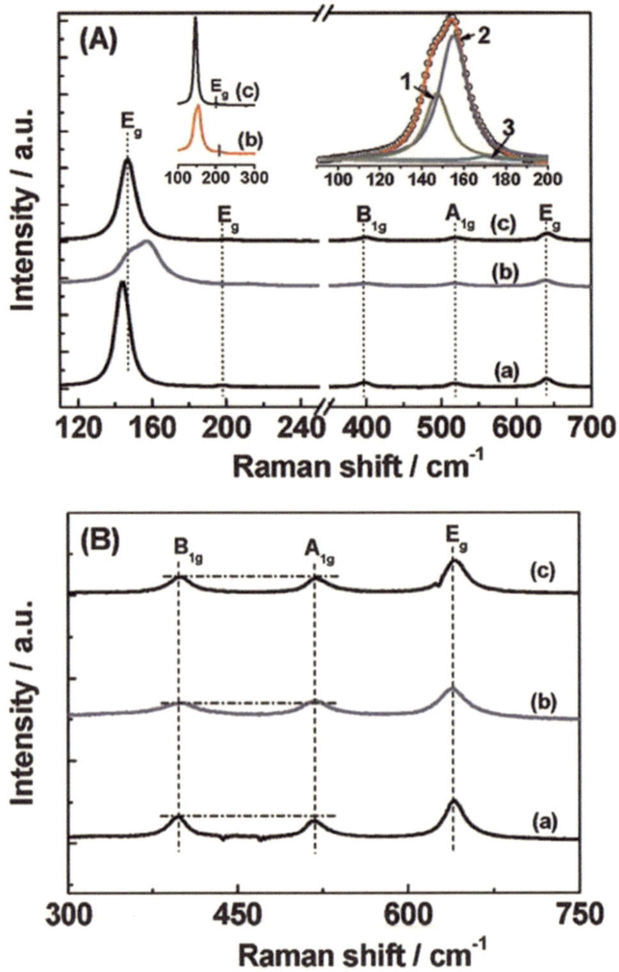

Figure 5.
Raman spectra of (a) anatase sheets with V_o, (b) anatase sheets without V_o (a), and (c) reference anatase TiO_2 in the ranges of (A) 110–700 cm^{-1} and (B) 300–750 cm^{-1}, respectively. Partial Raman spectra of curves b and c between 100 and 300 cm^{-1} are shown in the left inset of panel A, while the fitted Eg mode of curve b at 100–200 cm^{-1} is shown in the right inset of panel A. Copied with permission from Ref. [89] Copyright American Chemical Society.

fluorine. This is supported by two new Raman modes at 155 and 171 cm^{-1} and the weaker B1g mode at 397 cm^{-1}. By simple calcination of the sample in air, the two modes at 155 and 171 cm^{-1} entirely vanished when the surface fluorine and V_O were removed from the TiO_2 sheets. Additionally, only "fluorine-terminated anatase TiO_2 sheets" are capable of producing a novel active mode. This finding suggests that the two new Raman modes are separate from the single-surface fluorine. Additionally, it was shown that the lone V_O only produces a few additional weak modes above 300 cm^{-1}. The surface fluorine and V_O in oxygen-deficient anatase TiO_2 have synergistic effects that likely change the bonding length of the Ti—O—Ti network and the atomic coordination numbers. It is suggested that the rebuilt surface is made up of Ti atoms having smaller coordination numbers such as Ti with four coordination numbers, which may provide reactants in catalytic processes with more advantageous places to bind to. This is significant because, in contrast to TiO_2 sheets without V_O, a regenerated surface may boost the contact between the TiO_2 matrix and the loaded Pt through a unique e^- transfer mechanism induced by V_O that increases the photocatalytic hydrogen generation rate.

3.2 Optical properties

To generate e^--h^+ pairs for surface reactions, heterogeneous photocatalysis depends on the ability of photocatalysts to harvest light energy. However, TiO_2 can only absorb UV light because of its broad band gap. Fortunately, defect engineering allows for the manipulation of TiO_2's optical properties. As local states are created by V_O below the conduction band edge, the light harvesting ability of TiO_2 increases from UV to the visible light range. The V_O states that have been developed may participate in a fresh photoexcitation process. In other words, visible light's energy is used to excite the electron from the valence band to the V_O states, resulting in the usual excitations seen in the visible spectrum. Because of this, V_O are referred to as F centers, which comes from the German word for color, Farbe. Additionally, by interacting with nearby Ti^{4+}, the e^- remaining in the V_O may create the Ti^{3+} species [92]. Just below the conduction band, the Ti^{3+} defects may generate a shallow donor level that might also affect the sensitivity to visible light.

3.3 Dissociative adsorption properties

Understanding the active locations on TiO_2 has been aided by research into defects using adsorbing probe molecules. On TiO_2 single-crystal surfaces, small molecules including HCOOH, O_2, H_2O, N_2O, H_2, and CO have been employed to study the performance of such defective sites [93–95]. Some of these molecules' adsorption properties are discovered to change as a result of defects linked to V_O [96]. During the photocatalytic activity, h^+ and e^- produced in the crystals of TiO_2 under UV light may transfer to the surface and then be transported to the adsorbed species, where they take part in the redox reaction [97]. Even though the microscopic specifics of the mechanism of these e^- transfer are still not fully known, it is anticipated that the transfer will be more effective if the surface and adsorbate are closely connected, as well as when the adsorbed materials are detached.

3.3.1 Oxygen adsorption

In areas like gas sensing and heterogeneous catalysis, the interaction of oxygen with TiO_2 is essential. TiO_2 catalyzes a number of photooxidation processes, where molecular oxygen acts as the oxidizing agent [98]. One crucial stage of the photocatalytic reaction in these systems is the adsorption of molecular oxygen on the surface of TiO_2 [99]. O_2 does not, however, adsorb on a perfectly neutral TiO_2 surface [100]. Only when there is sufficient negative charge available to form O—Ti bonds can O_2 adsorb onto the TiO_2 surface; this charge may come from subsurface V_O or photogenerated e^-, with adsorption energies of 2.52 and 0.94 eV, respectively, as shown in **Figure 6**. Superoxide radical groups may be created simultaneously by the O_2 adsorbed on the surface of TiO_2 and the free e^- present on V_O states. Both the charge separation process and the oxidation of organic materials are actively promoted by the production of these radical groups [101].

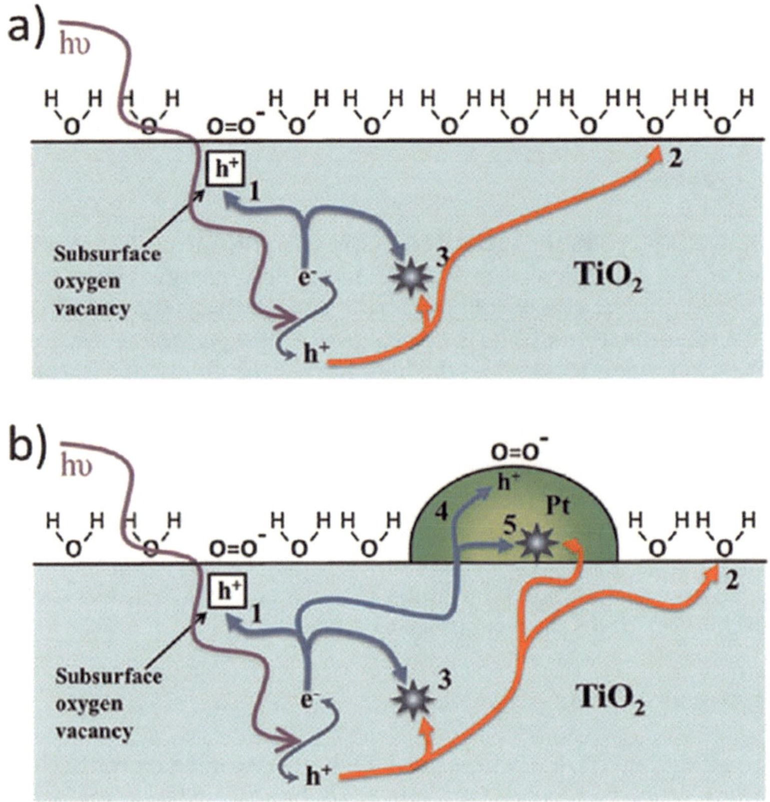

Figure 6.
Schematic illustration of photoexcited charge carriers in (a) TiO_2 and (b) Pt/TiO_2. Reprinted with approval from Ref. [101] Copyright American Chemical Society.

3.3.2 H_2O adsorption

As an example of a straightforward surface chemical reaction with considerable applications, water dissociation on TiO_2 is of basic importance. It has been thoroughly investigated how H_2O and TiO_2 interact with surfaces, significantly influencing the photocatalytic processes [102]. It is particularly well known that on the imperfect TiO_2 surface, H_2O molecules that have been chemically dissociated are energetically preferred, whereas H_2O molecules are only physically adsorbed on the ideal TiO_2 surface [103–105]. H_2O dissociation only occurs on defect sites linked to V_O at low coverage, according to research by Besenbacher et al. [106] that combines experimental and theoretical methods. They used scanning tunneling microscopy (STM) to show a direct correlation between V_O before water exposure and surface hydroxyl groups after exposure, and they used DFT to show that only the defect sites are energetically capable of supporting water dissociation. It is shown that V_O in the surface layer dissolves H_2O by transferring one proton to an oxygen atom nearby, resulting in the formation of two OH groups for every vacancy.

3.3.3 Adsorption of alcohol

The reactive sites on metal oxides, both in powder form and in single crystal, have been intensively probed using alcohols. Both experimental characterizations and theoretical calculations have been used to extensively study the adsorption of alcohols on TiO_2 [107]. Using theoretical calculations, Oviedo et al. [108] have shown that methanol dissociation is thermodynamically advantageous on the V_O states. According to Farfan-Arribas and Madix, temperature programmed reaction spectroscopy (TPRS) and X-ray photoelectron spectroscopy (XPS) are used to investigate the function of V_O in the adsorption of aliphatic alcohols on TiO_2. They discovered that the existence of V_O on the surface leads to greater alcohol adsorption on the surface. At room temperature, the adsorbed aliphatic alcohols spontaneously dissolve on the TiO_2-(110) surfaces containing V_O, generating hydroxide and alkoxide groups [109]. Particularly, it is discovered that the alkoxide species is more photocatalytically reactive than the physisorbed species. Additionally, chemically dissociated alcohols may swiftly scavenge the photogenerated h^+, substantially extending the lifespan of photogenerated e^- [110].

3.3.4 CO_2 adsorption

One of the potential options for lowering CO_2 emissions and utilizing CO_2 as a building block to produce valuable goods is the photochemical conversion of CO_2 into solar fuels by photocatalysts like TiO_2. The first stage in CO_2's photo-reduction is its adsorption on TiO_2 [39, 111]. According to theoretical research, the physisorption and the most stable chemisorption of CO_2 on the neutral charge of perfect anatase TiO_2 (001) have adsorption energies of 9.03 and 24.66 kcal mol1, respectively, on the spin-unpolarized TiO_2 with V_O. This suggests that CO_2 is tightly bound by V_O on a TiO_2 surface that is deficient. Additionally, it is shown that the CO_2 activation barrier on TiO_2 with V_O is lower than it is on TiO_2 with flawless anatase (001) [112]. Furthermore, it is discovered that the energetically favored conversion of CO_2 to CO occurs on the surface of flawed TiO_2, with V_O acting as the photocatalyst. Surprisingly, Li et al. [83] have shown that CO_2 spontaneously dissociates into CO on a Cu(I)/TiO_{2x} surface created by thermal annealing under an inert atmosphere even when it is dark. According to **Figure 7**, the surface V_O that provides the electrical charge as well as the locations for the adsorption of oxygen atoms from CO_2 is mostly responsible for the spontaneous dissociation of CO_2 in the dark. In addition, compared to those in the dark, CO_2 activation and dissociation may be markedly enhanced by photoirradiation.

In conclusion, the reactant molecule's dissociative adsorption would be facilitated by V_O on the TiO_2 surface. In the photocatalytic processes, it seems that the dissociative adsorption of the reactant molecule on the TiO_2 surface lowers its activation energy and influences the reaction mechanism at the molecular level. Additionally, compared to physisorbed species, chemically separated compounds have higher photocatalytic reactivity. It should be emphasized, nevertheless, that the surface of TiO_2 goes through re-oxidation often in conjunction with the dissociated adsorption of the adsorbates. Therefore, before investigating its dissociation adsorption capabilities for photocatalytic application, we first need to stabilize the V_O in reduced TiO_2. This may be accomplished by doping the Fe into the TiO_2 nanoparticles, a procedure we covered in Section 2.5.

3.4 Reductive properties

In addition to changing the properties of adsorbates, V_O on catalyst surfaces also plays a role in the reduction of a number of these adsorbates. As demonstrated by Lu et al., one can observe the reactivity of thermally created V_O sites for the reduction of

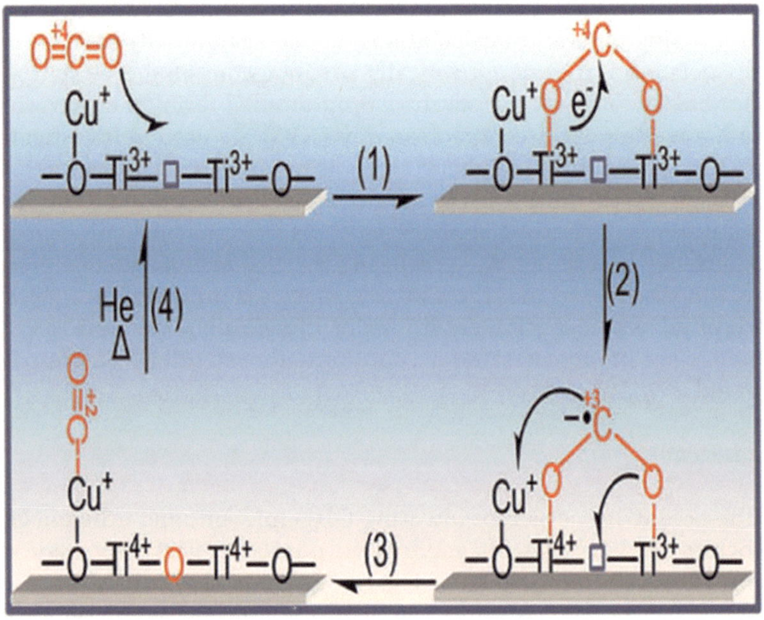

Figure 7.
Proposed mechanism for CO_2 dissociation on the surface of Cu/TiO_2 under ambient temperature in dark. Copied with permission from Ref. [83] Copyright American Chemical Society.

NO, CH_2O, or D_2O by adsorbing a test molecule on the defective surface as well as a fully stoichiometric surface and comparing the results of temperature programming desorption (TPD) [113]. By measuring the TPD, the reductive products (N_2O, C_2H_4, and D_2) are identified after adsorbing these adsorbates on the flawed surface. The oxidation of surface defect sites occurs concurrently with the deoxygenation of adsorbates. Therefore, the coverage of surface V_O is inversely correlated with the yield of reduction products. On the surface with no defects, there are no deoxygenation processes seen. It has also been shown that surface V_O sites are active in the reduction of metal ions. Our most recent study has shown that V_O plays a crucial role in the charge transfer from the damaged surface to gold ions. As a result, a very quick, direct development of metallic gold nanoparticles was accomplished on the surface of the semiconductor TiO_2 containing V_O. Ye and colleagues also saw the metal ions spontaneously reducing on damaged surfaces. They have shown that on the flawed surface of $WO_{2.72}$, the ions of noble metals are engaged in redox processes where the metal ions partly oxidize the reduced V_O states [114]. As a result, the metal ions are quickly reduced and nucleated on their surface, where they develop into clusters and then nanoparticles. The controlled synthesis of metal/semiconductor hybrid nanomaterials may be accomplished using this approach in a single step without the need for external reducing agents, stabilizing molecules, or pretreatment of the precursors. It is interesting to note that Li et al. [115] found that the sub-stoichiometric WO_{3-x}, which is produced by utilizing hydrogen treatment to create V_O in WO_3, is stable thermodynamically at room temperature and has a strong resistance to re-oxidation. So, in the absence of catalysts for oxygen evolution reaction, hydrogen-treated WO_3 is stabilized to be used for water oxidation in a neutral media. These ground-breaking findings imply that by carefully selecting the preparation techniques, the property of V_O states may be precisely regulated.

4. Conclusion

This chapter discusses various methods that can be employed to produce defective TiO_2 containing V_O. These methods include thermal processing in an oxygen-depleted atmosphere, doping with non-metal or metal ions, bombardment with high-energy particles, and even *in situ* catalytic processes. These processes remove lattice oxygen from the surface or bulk of TiO_2, resulting in a vacancy state. The presence of V_O defects provides defective TiO_2 with unique chemical and physical characteristics, including enhanced reductive and dissociative adsorption properties and visible light absorption capabilities. Although the function of V_O in photocatalytic processes is still not completely understood, defective TiO_2 has been shown to have benefits for a wide range of applications, such as selective charge separation and visible light response for photocatalytic activities. However, there are also conflicting claims about the role of V_O in the photocatalytic performance of semiconductors. Overall, the intentional introduction of V_O defects into TiO_2 has great potential for improving the material's properties and enhancing its performance in various applications. Further research is necessary to fully understand the impact of V_O defects on TiO_2 and to explore the potential of defective TiO_2 materials in emerging fields such as sensing, photoelectrochemical water splitting, and photocatalytic air purification.

Conflict of interest

The authors declare no conflict of interest.

Author details

Javid Khan* and Lei Han
College of Materials Science and Engineering, Hunan Joint International Laboratory of Advanced Materials and Technology for Clean Energy, Hunan University, Changsha, China

*Address all correspondence to: javidchemist@yahoo.com

References

[1] Potdar SB, Huang C-M, Praveen BVS, Manickam S, Sonawane SH. Highly photoactive titanium dioxide supported platinum catalyst: Synthesis using cleaner ultrasound approach. Catalysts. 2022;**12**(1):78

[2] Moradeeya PG, Sharma A, Kumar MA, Basha S. Titanium dioxide based nanocomposites–current trends and emerging strategies for the photocatalytic degradation of ruinous environmental pollutants. Environmental Research. 2022;**204**: 112384

[3] Kompa A, Kekuda D, Murari MS, Rao KM. Defect induced enhanced catalytic activity of Lu doped titanium dioxide (TiO_2) thin films. Surfaces and Interfaces. 2022;**31**:101988

[4] Szentmihályi K, Móricz K, Gigler G, May Z, Bódis E, Tóth J, et al. Ointment containing spray freeze-dried metronidazole effective against rosacea. Journal of Drug Delivery Science and Technology. 2022;**74**:103559

[5] Niessink T, Ringoot J, Otto C, Janssen M, Jansen TL. Clinical images: Detection of titanium dioxide particles by Raman spectroscopy in synovial fluid from a swollen ankle. Arthritis & Rheumatology. 2022;**74**(6):1069

[6] Javanmardi S, Ghojoghi A, Divband B, Ashrafi J. Titanium dioxide nanoparticle/gelatin: A potential burn wound healing biomaterial. Wounds. 2018;**30**(12):372-379

[7] Chen MC, Koh PW, Ponnusamy VK, Lee SL. Titanium dioxide and other nanomaterials based antimicrobial additives in functional paints and coatings. Progress in Organic Coatings. 2022;**163**:106660

[8] Bergamaschi E, Bellisario V, Macrì M, Buglisi M, Garzaro G, Squillacioti G, et al. A biomonitoring pilot study in workers from a paints production plant exposed to pigment-grade titanium dioxide (TiO_2). Toxics. 2022;**10**(4):171

[9] Mariappan T, Agarwal A, Ray S. Influence of titanium dioxide on the thermal insulation of waterborne intumescent fire protective paints to structural steel. Progress in Organic Coatings. 2017;**111**:67-74

[10] Newman MD, Stotland M, Ellis JI. The safety of nanosized particles in titanium dioxide–and zinc oxide–based sunscreens. Journal of the American Academy of Dermatology. 2009;**61**(4): 685-692

[11] Smijs TG, Pavel S. Titanium dioxide and zinc oxide nanoparticles in sunscreens: Focus on their safety and effectiveness. Nanotechnology, Science and Applications. 2011;**4**:95

[12] Trivedi M, Murase J. Titanium dioxide in sunscreen. Application of Titanium Dioxide. 2017:61-71

[13] Rompelberg C, Heringa MB, van Donkersgoed G, Drijvers J, Roos A, Westenbrink S, et al. Oral intake of added titanium dioxide and its nanofraction from food products, food supplements and toothpaste by the Dutch population. Nanotoxicology. 2016;**10**(10):1404-1414

[14] Hsu T-Y, Lin C-C, Lee M-D, Chang BP-H, Tsai J-D. Titanium dioxide in toothpaste causing yellow nail syndrome. Pediatrics. 2017;**139**(1)

[15] Fadheela A-S, Al AR, Hawraa A-S, Layla H, Safa T. Toxicity evaluation of TiO_2 nanoparticles embedded in

toothpaste products. GSC Biological and Pharmaceutical Sciences. 2020;**12**(1): 102-115

[16] Fujishima A, Honda K. Electrochemical photolysis of water at a semiconductor electrode. Nature. 1972; **238**(5358):37-38

[17] Pessoa RS, Fraga MA, Santos LV, Massi M, Maciel HS. Nanostructured thin films based on TiO_2 and/or SiC for use in photoelectrochemical cells: A review of the material characteristics, synthesis and recent applications. Materials Science in Semiconductor Processing. 2015;**29**:56-68

[18] Bashiri R, Samsudin MFR, Mohamed NM, Suhaimi NA, Ling LY, Sufian S, et al. Influence of growth time on photoelectrical characteristics and photocatalytic hydrogen production of decorated Fe_2O_3 on TiO_2 nanorod in photoelectrochemical cell. Applied Surface Science. 2020;**510**:145482

[19] Wannapop S, Somdee A. Enhanced visible light absorption of TiO_2 nanorod photoanode by $NiTiO_3$ decoration for high-performance photoelectrochemical cells. Ceramics International. 2020; **46**(16):25758-25765

[20] Kavan L. Conduction band engineering in semiconducting oxides (TiO_2, SnO_2): Applications in perovskite photovoltaics and beyond. Catalysis Today. 2019;**328**:50-56

[21] Lee K-M, Lin W-J, Chen S-H, Wu M-C. Control of TiO_2 electron transport layer properties to enhance perovskite photovoltaics performance and stability. Organic Electronics. 2020;**77**:105406

[22] Nguyen TT, Patel M, Kim S, Mir RA, Yi J, Dao V-A, et al. Transparent photovoltaic cells and self-powered photodetectors by TiO_2/NiO heterojunction. Journal of Power Sources. 2021;**481**:228865

[23] Guo Q, Zhou C, Ma Z, Yang X. Fundamentals of TiO_2 photocatalysis: Concepts, mechanisms, and challenges. Advanced Materials. 2019;**31**(50): 1901997

[24] Meng A, Zhang L, Cheng B, Yu J. Dual cocatalysts in TiO_2 photocatalysis. Advanced Materials. 2019;**31**(30):1807660

[25] Al-Mamun MR, Kader S, Islam MS, Khan MZH. Photocatalytic activity improvement and application of UV-TiO_2 photocatalysis in textile wastewater treatment: A review. Journal of Environmental Chemical Engineering. 2019;**7**(5):103248

[26] Peiris S, de Silva HB, Ranasinghe KN, Bandara SV, Perera IR. Recent development and future prospects of TiO_2 photocatalysis. Journal of the Chinese Chemical Society. 2021; **68**(5):738-769

[27] Kamat PV. TiO_2 Nanostructures: Recent Physical Chemistry Advances. Journal of Physical Chemistry C. 2012; **116**(22):11849-11851

[28] Ijaz M, Zafar M. Titanium dioxide nanostructures as efficient photocatalyst: Progress, challenges and perspective. International Journal of Energy Research. 2021;**45**(3):3569-3589

[29] Rosales M, Zoltan T, Yadarola C, Mosquera E, Gracia F, García A. The influence of the morphology of 1D TiO_2 nanostructures on photogeneration of reactive oxygen species and enhanced photocatalytic activity. Journal of Molecular Liquids. 2019;**281**:59-69

[30] Mirzaei M, Ahadi H, Shariaty-Niassar M, Akbari M. Fabrication and characterization of visible light active

Fe-TiO_2 nanocomposites as nanophotocatalyst. International Journal of Nanoscience and Nanotechnology. 2015;**11**(4):289-293

[31] Zhang T, Chen F, Ma Y, Qi L. The synthesis and photocatalytic performance of peapod-like one dimensional nanocomposites composed of Au nanoparticles and TiO2 nanofibers. Journal of Nanoscience and Nanotechnology. 2016;**16**(6):5843-5849

[32] Sun S, Song P, Cui J, Liang S. Amorphous TiO_2 nanostructures: Synthesis, fundamental properties and photocatalytic applications. Catalysis Science & Technology. 2019;**9**(16): 4198-4215

[33] Li W, Yang J, Wu Z, Wang J, Li B, Feng S, et al. A versatile kinetics-controlled coating method to construct uniform porous TiO_2 shells for multifunctional core–shell structures. Journal of the American Chemical Society. 2012;**134**(29):11864-11867

[34] Jun J, Jin C, Kim H, Park S, Lee C. Fabrication and characterization of CuO-core/TiO_2-shell one-dimensional nanostructures. Applied Surface Science. 2009;**255**(20):8544-8550

[35] Hong Noh J, Ding B, Soo Han H, Seong Kim J, Hoon Park J, Baek Park S, et al. Tin doped indium oxide core—TiO_2 shell nanowires on stainless steel mesh for flexible photoelectrochemical cells. Applied Physics Letters. 2012; **100**(8):084104

[36] Chen S, Xiao Y, Wang Y, Hu Z, Zhao H, Xie W. A facile approach to prepare black TiO_2 with oxygen vacancy for enhancing photocatalytic activity. Nanomaterials. 2018;**8**(4):245

[37] Schaub R, Wahlstrom E, Rønnau A, Lægsgaard E, Stensgaard I, Besenbacher F. Oxygen-mediated diffusion of oxygen vacancies on the TiO_2 (110) surface. Science. 2003; **299**(5605):377-379

[38] Jia R, Wang Y, Wang C, Ling Y, Yu Y, Zhang B. Boosting selective nitrate electroreduction to ammonium by constructing oxygen vacancies in TiO_2. Acs Catalysis. 2020;**10**(6):3533-3540

[39] Indrakanti VP, Kubicki JD, Schobert HH. Photoinduced activation of CO_2 on Ti-based heterogeneous catalysts: Current state, chemical physics-based insights and outlook. Energy & Environmental Science. 2009; **2**(7):745-758

[40] Zhang A-Y, Lin T, He Y-Y, Mou Y-X. Heterogeneous activation of H_2O_2 by defect-engineered TiO_{2-x} single crystals for refractory pollutants degradation: A Fenton-like mechanism. Journal of Hazardous Materials. 2016;**311**:81-90

[41] Fittipaldi M, Gatteschi D, Fornasiero P. The power of EPR techniques in revealing active sites in heterogeneous photocatalysis: The case of anion doped TiO_2. Catalysis Today. 2013;**206**:2-11

[42] Di Valentin C, Pacchioni G, Selloni A. Reduced and n-type doped TiO_2: Nature of Ti3+ species. The Journal of Physical Chemistry C. 2009;**113**(48): 20543-20552

[43] Khan H, Swati IK. Fe^{3+}-doped anatase TiO_2 with d–d transition, oxygen vacancies and Ti^{3+} centers: Synthesis, characterization, UV–vis photocatalytic and mechanistic studies. Industrial & Engineering Chemistry Research. 2016; **55**(23):6619-6633

[44] Zhou Y, Chen C, Wang N, Li Y, Ding H. Stable Ti^{3+} self-doped anatase-rutile mixed TiO_2 with enhanced visible

light utilization and durability. The Journal of Physical Chemistry C. 2016; **120**(11):6116-6124

[45] Shang H, Li M, Li H, Huang S, Mao C, Ai Z, et al. Oxygen vacancies promoted the selective photocatalytic removal of NO with blue TiO_2 via simultaneous molecular oxygen activation and photogenerated hole annihilation. Environmental Science & Technology. 2019;**53**(11): 6444-6453

[46] Jiang X, Zhang Y, Jiang J, Rong Y, Wang Y, Wu Y, et al. Characterization of oxygen vacancy associates within hydrogenated TiO_2: A positron annihilation study. The Journal of Physical Chemistry C. 2012;**116**(42): 22619-22624

[47] Wendt S, Schaub R, Matthiesen J, Vestergaard EK, Wahlström E, Rasmussen MD, et al. Oxygen vacancies on TiO_2 (110) and their interaction with H_2O and O_2: A combined high-resolution STM and DFT study. Surface Science. 2005;**598**(1–3):226-245

[48] Kim D, Hong J, Park YR, Kim KJ. The origin of oxygen vacancy induced ferromagnetism in undoped TiO_2. Journal of Physics: Condensed Matter. 2009;**21**(19):195405

[49] Jing L, Xin B, Yuan F, Xue L, Wang B, Fu H. Effects of surface oxygen vacancies on photophysical and photochemical processes of Zn-doped TiO_2 nanoparticles and their relationships. The Journal of Physical Chemistry B. 2006;**110**(36):17860-17865

[50] Ni Q, Dong R, Bai Y, Wang Z, Ren H, Sean S, et al. Superior sodium-storage behavior of flexible anatase TiO_2 promoted by oxygen vacancies. Energy Storage Materials. 2020;**25**:903-911

[51] Li D, Haneda H, Labhsetwar NK, Hishita S, Ohashi N. Visible-light-driven photocatalysis on fluorine-doped TiO_2 powders by the creation of surface oxygen vacancies. Chemical Physics Letters. 2005;**401** (4–6):579-584

[52] Wu Q, van de Krol R. Selective photoreduction of nitric oxide to nitrogen by nanostructured TiO_2 photocatalysts: Role of oxygen vacancies and iron dopant. Journal of the American Chemical Society. 2012;**134**(22): 9369-9375

[53] Carter E, Carley AF, Murphy DM. Evidence for O_2-radical stabilization at surface oxygen vacancies on polycrystalline TiO_2. The Journal of Physical Chemistry C. 2007;**111**(28): 10630-10638

[54] Bharti B, Kumar S, Lee H-N, Kumar R. Formation of oxygen vacancies and Ti^{3+} state in TiO_2 thin film and enhanced optical properties by air plasma treatment. Scientific Reports. 2016;**6**(1):1-12

[55] Li J-J, Weng B, Cai S-C, Chen J, Jia H-P, Xu Y-J. Efficient promotion of charge transfer and separation in hydrogenated TiO_2/WO_3 with rich surface-oxygen-vacancies for photodecomposition of gaseous toluene. Journal of Hazardous Materials. 2018; **342**:661-669

[56] Pan X, Yang M-Q, Fu X, Zhang N, Xu Y-J. Defective TiO_2 with oxygen vacancies: Synthesis, properties and photocatalytic applications. Nanoscale. 2013;**5**(9):3601-3614

[57] Liu H, Ma HT, Li XZ, Li WZ, Wu M, Bao XH. The enhancement of TiO_2 photocatalytic activity by hydrogen thermal treatment. Chemosphere. 2003; **50**(1):39-46

[58] Naldoni A, Allieta M, Santangelo S, Marelli M, Fabbri F, Cappelli S, et al. Effect of nature and location of defects on bandgap narrowing in black TiO_2 nanoparticles. Journal of the American Chemical Society. 2012;**134**(18): 7600-7603

[59] Morgan BJ, Watson GW. Intrinsic n-type defect formation in TiO_2: A comparison of rutile and anatase from GGA+ U calculations. The Journal of Physical Chemistry C. 2010;**114**(5): 2321-2328

[60] Pacchioni G. Oxygen vacancy: The invisible agent on oxide surfaces. ChemPhysChem. 2003;**4**(10):1041-1047

[61] Wang L-Q, Baer DR, Engelhard MH, Shultz AN. The adsorption of liquid and vapor water on TiO_2 (110) surfaces: The role of defects. Surface Science. 1995; **344**(3):237-250

[62] Eriksen S, Egdell RG. Electronic excitations at oxygen deficient TiO_2 (110) surfaces: A study by EELS. Surface Science. 1987;**180**(1):263-278

[63] Jiang N, Qiu J, Ellison A, Silcox J. Fundamentals of high-energy electron-irradiation-induced modifications of silicate glasses. Physical Review B. 2003; **68**(6):064207

[64] Knotek ML, Feibelman PJ. Ion desorption by core-hole auger decay. Physical Review Letters. 1978;**40**(14):964

[65] Zhao J, Pontius N, Winkelmann A, Sametoglu V, Kubo A, Borisov AG, et al. Electronic potential of a chemisorption interface. Physical Review B. 2008; **78**(8):085419

[66] Thompson TL, Yates JT. TiO_2-based photocatalysis: Surface defects, oxygen and charge transfer. Topics in Catalysis. 2005;**35**(3):197-210

[67] Takeuchi K, Nakamura I, Matsumoto O, Sugihara S, Ando M, Ihara T. Preparation of visible-light-responsive titanium oxide photocatalysts by plasma treatment. Chemistry Letters. 2000;**29**(12):1354-1355

[68] Ihara T, Miyoshi M, Ando M, Sugihara S, Iriyama Y. Preparation of a visible-light-active TiO_2 photocatalyst by RF plasma treatment. Journal of Materials Science. 2001;**36**(17): 4201-4207

[69] Wu Q, Zheng Q, van de Krol R. Creating oxygen vacancies as a novel strategy to form tetrahedrally coordinated Ti^{4+} in Fe/TiO_2 nanoparticles. The Journal of Physical Chemistry C. 2012;**116**(12):7219-7226

[70] Takata T, Domen K. Defect engineering of photocatalysts by doping of aliovalent metal cations for efficient water splitting. The Journal of Physical Chemistry C. 2009;**113**(45):19386-19388

[71] Czoska AM, Livraghi S, Chiesa M, Giamello E, Agnoli S, Granozzi G, et al. The nature of defects in fluorine-doped TiO_2. The Journal of Physical Chemistry C. 2008;**112**(24):8951-8956

[72] Prokes SM, Gole JL, Chen X, Burda C, Carlos WE. Defect-related optical behavior in surface modified TiO_2 nanostructures. Advanced Functional Materials. 2005;**15**(1):161-167

[73] Wang J, Tafen DN, Lewis JP, Hong Z, Manivannan A, Zhi M, et al. Origin of photocatalytic activity of nitrogen-doped TiO_2 nanobelts. Journal of the American Chemical Society. 2009; **131**(34):12290-12297

[74] Lin Z, Orlov A, Lambert RM, Payne MC. New insights into the origin of visible light photocatalytic activity of nitrogen-doped and oxygen-deficient

anatase TiO_2. The Journal of Physical Chemistry B. 2005;**109**(44): 20948-20952

[75] Batzill M, Morales EH, Diebold U. Influence of nitrogen doping on the defect formation and surface properties of TiO_2 rutile and anatase. Physical Review Letters. 2006;**96**(2):026103

[76] Panayotov DA, Morris JR. Thermal decomposition of a chemical warfare agent simulant (DMMP) on TiO_2: Adsorbate reactions with lattice oxygen as studied by infrared spectroscopy. The Journal of Physical Chemistry C. 2009; **113**(35):15684-15691

[77] Blount MC, Buchholz JA, Falconer JL. Photocatalytic decomposition of aliphatic alcohols, acids, and esters. Journal of Catalysis. 2001;**197**(2):303-314

[78] Chen M-T, Lien C-F, Liao L-F, Lin J-L. In-situ FTIR study of adsorption and photoreactions of CH_2Cl_2 on powdered TiO_2. The Journal of Physical Chemistry B. 2003;**107**(16):3837-3843

[79] Muggli DS, Falconer JL. Role of lattice oxygen in photocatalytic oxidation on TiO_2. Journal of Catalysis. 2000;**191**(2):318-325

[80] Sato S, Kadowaki T, Yamaguti K. Photocatalytic oxygen isotopic exchange between oxygen molecule and the lattice oxygen of titanium dioxide prepared from titanium hydroxide. The Journal of Physical Chemistry. 1984;**88**(14): 2930-2931

[81] Thompson TL, Panayotov DA, Yates JT, Martyanov I, Klabunde K. Photodecomposition of adsorbed 2-chloroethyl ethyl sulfide on TiO_2: Involvement of lattice oxygen. The Journal of Physical Chemistry B. 2004; **108**(46):17857-17865

[82] Pan X, Zhang N, Fu X, Xu Y-J. Selective oxidation of benzyl alcohol over TiO_2 nanosheets with exposed {001} facets: Catalyst deactivation and regeneration. Applied Catalysis A: General. 2013;**453**:181-187

[83] Liu L, Zhao C, Li Y. Spontaneous dissociation of CO_2 to CO on defective surface of Cu (I)/TiO_{2-x} nanoparticles at room temperature. The Journal of Physical Chemistry C. 2012;**116**(14): 7904-7912

[84] Nakamura I, Negishi N, Kutsuna S, Ihara T, Sugihara S, Takeuchi K. Role of oxygen vacancy in the plasma-treated TiO_2 photocatalyst with visible light activity for NO removal. Journal of Molecular Catalysis A: Chemical. 2000; **161**(1–2):205-212

[85] Liu G, Li F, Wang D-W, Tang D-M, Liu C, Ma X, et al. Electron field emission of a nitrogen-doped TiO_2 nanotube array. Nanotechnology. 2007;**19**(2):025606

[86] von Oertzen GU, Gerson AR. The effects of O deficiency on the electronic structure of rutile TiO_2. Journal of Physics and Chemistry of Solids. 2007; **68**(3):324-330

[87] Chen W, Fan Z, Zhang B, Ma G, Takanabe K, Zhang X, et al. Enhanced visible-light activity of titania via confinement inside carbon nanotubes. Journal of the American Chemical Society. 2011;**133**(38):14896-14899

[88] Ghosh AK, Wakim FG, Addiss RR Jr. Photoelectronic processes in rutile. Physical. Review. 1969;**184**(3):979

[89] Liu G, Yang HG, Wang X, Cheng L, Lu H, Wang L, et al. Enhanced photoactivity of oxygen-deficient anatase TiO_2 sheets with dominant {001} facets. The Journal of Physical Chemistry C. 2009;**113**(52):21784-21788

[90] Parker JC, Siegel RW. Raman microprobe study of nanophase TiO_2 and oxidation-induced spectral changes. Journal of Materials Research. 1990;**5**(6): 1246-1252

[91] Janotti A, Varley JB, Rinke P, Umezawa N, Kresse G, Van de Walle CG. Physical Review B: Condensed Matter. 2010;**81**:085212

[92] Serpone N. Is the Band Gap of Pristine TiO_2 Narrowed by Anion-and Cation-Doping of Titanium Dioxide in Second-Generation Photocatalysts? Journal of Physical Chemistry B. 2006; **110**(48):24287-24293

[93] Göpel W, Rocker G, Feierabend R. Intrinsic defects of TiO_2 (110): Interaction with chemisorbed O_2, H_2, CO, and CO_2. Physical Review B. 1983; **28**(6):3427

[94] Pan JM, Maschhoff BL, Diebold U, Madey TE. Interaction of water, oxygen, and hydrogen with TiO_2 (110) surfaces having different defect densities. Journal of Vacuum Science & Technology A: Vacuum, Surfaces, and Films. 1992; **10**(4):2470-2476

[95] Wang LQ, Shultz AN, Baer DR, Engelhard MH. Interactions of small molecules with TiO_2 (110) surfaces: The role of defects. Journal of Vacuum Science & Technology A: Vacuum, Surfaces, and Films. 1996;**14**(3): 1532-1538

[96] Farfan-Arribas E, Madix RJ. Role of defects in the adsorption of aliphatic alcohols on the TiO_2 (110) surface. The Journal of Physical Chemistry B. 2002; **106**(41):10680-10692

[97] Selloni A. Anatase shows its reactive side. Nature Materials. 2008;**7**(8): 613-615

[98] Thompson TL, Yates JT. Surface science studies of the photoactivation of TiO_2 new photochemical processes. Chemical Reviews. 2006;**106**(10): 4428-4453

[99] Wendt S, Sprunger PT, Lira E, Madsen GKH, Li Z, Hansen JØ, et al. The role of interstitial sites in the Ti 3d defect state in the band gap of titania. Science. 2008;**320**(5884):1755-1759

[100] Lira E, Wendt S, Huo P, Hansen JØ, Streber R, Porsgaard S, et al. The importance of bulk Ti3+ defects in the oxygen chemistry on titania surfaces. Journal of the American Chemical Society. 2011;**133**(17):6529-6532

[101] Muhich CL, Zhou Y, Holder AM, Weimer AW, Musgrave CB. Effect of surface deposited Pt on the photoactivity of TiO_2. The Journal of Physical Chemistry C. 2012;**116**(18):10138-10149

[102] Kajita S, Minato T, Kato HS, Kawai M, Nakayama T. First-principles calculations of hydrogen diffusion on rutile TiO_2 (110) surfaces. The Journal of Chemical Physics. 2007;**127**(10):104709

[103] Lindan PJD, Harrison NM, Holender JM, Gillan MJ. First-principles molecular dynamics simulation of water dissociation on TiO_2 (110). Chemical Physics Letters. 1996;**261**(3):246-252

[104] Hugenschmidt MB, Gamble L, Campbell CT. The interaction of H_2O with a TiO_2 (110) surface. Surface Science. 1994;**302**(3):329-340

[105] Henderson MA. An HREELS and TPD study of water on TiO_2 (110): The extent of molecular versus dissociative adsorption. Surface Science. 1996;**355** (1–3):151-166

[106] Schaub R, Thostrup P, Lopez N, Lægsgaard E, Stensgaard I, Nørskov JK,

et al. Oxygen vacancies as active sites for water dissociation on rutile TiO_2 (110). Physical Review Letters. 2001;**87**(26): 266104

[107] Zhou C, Ma Z, Ren Z, Mao X, Dai D, Yang X. Effect of defects on photocatalytic dissociation of methanol on TiO_2 (110). Chemical Science. 2011; **2**(10):1980-1983

[108] De Armas RS, Oviedo J, San Miguel MA, Sanz JF. Methanol adsorption and dissociation on TiO_2 (110) from first principles calculations. The Journal of Physical Chemistry C. 2007;**111**(27): 10023-10028

[109] Xu W, Raftery D. Photocatalytic oxidation of 2-propanol on TiO_2 powder and TiO2 monolayer catalysts studied by solid-state NMR. The Journal of Physical Chemistry B. 2001;**105**(19): 4343-4349

[110] Tachikawa T, Takai Y, Tojo S, Fujitsuka M, Majima T. Probing the surface adsorption and photocatalytic degradation of catechols on TiO_2 by solid-state NMR spectroscopy. Langmuir. 2006;**22**(3): 893-896

[111] Xi G, Ouyang S, Li P, Ye J, Ma Q, Su N, et al. Ultrathin $W_{18}O_{49}$ nanowires with diameters below 1 nm: Synthesis, near-infrared absorption, photoluminescence, and photochemical reduction of carbon dioxide. Angewandte Chemie International Edition. 2012;**51**(10):2395-2399

[112] Pipornpong W, Wanbayor R, Ruangpornvisuti V. Adsorption CO_2 on the perfect and oxygen vacancy defect surfaces of anatase TiO_2 and its photocatalytic mechanism of conversion to CO. Applied Surface Science. 2011; **257**(24):10322-10328

[113] Lu G, Linsebigler A, Yates JT Jr. Ti^{3+} defect sites on TiO_2 (110): Production and chemical detection of active sites. The Journal of Physical Chemistry. 1994; **98**(45):11733-11738

[114] Xi G, Ye J, Ma Q, Su N, Bai H, Wang C. In situ growth of metal particles on 3D urchin-like WO_3 nanostructures. Journal of the American Chemical Society. 2012;**134**(15): 6508-6511

[115] Wang G, Ling Y, Wang H, Yang X, Wang C, Zhang JZ, et al. Hydrogen-treated WO_3 nanoflakes show enhanced photostability. Energy & Environmental Science. 2012;**5**(3):6180-6187

Chapter 2

Perspective Chapter: Black Titania – From Synthesis to Applications

Bilal Akram, Bilal Ahmad Khan and Raieesa Batool

Abstract

Titanium dioxide (TiO_2) nanomaterials are very important for generating hydrogen through photoelectrochemical water splitting and remediation of environmental pollution. It has remained the focus of many researchers during last few decades due to their wide applications. Optical absorption properties of TiO_2 can lead to increase their photocatalytic activities. However, its overall solar performance is very restricted because of its large bandgap. The emergence of black titania nanomaterials has recently triggered worldwide research interest due to its significantly improved solar absorption and enhanced photocatalytic performances. In this chapter, various synthetic approaches employed to obtain black titania are outlined, and the structural features of the black titania nanomaterials are described in detail, along with their photocatalytic performances towards various applications.

Keywords: black titania, chemical synthesis, hydrogen treatment, structural disorder, photocatalysis

1. Introduction

By virtue of their large-scale applications in the removal of environmental pollutants and photocatalytic water splitting, TiO_2-based nanomaterials have fascinated massive interest [1–3]. TiO_2 exists in three main crystal phases that is brookite, anatase, and rutile. Electronic bandgaps of all these phases are above 3.0 eV, which is considered to be high bandgap. This limits their optical absorption in the ultraviolet (UV) region of the solar spectrum, which is below 5% of overall solar energy. If TiO_2 utilized this UV light very efficiently, its solar activity is still not better. It is the number of electrons and holes of the photocatalysts that determine its photocatalytic activity [4].

Upon appropriate light absorption, TiO_2 produces excited electrons in conduction band and excited holes in the valence band. For performing photocatalytic reactions, these excited charges, apart from each other, travel towards the surface. From all the excited charges, few of them are combined and vanished during the charge separation and migration processes. This absorption process leads to the generation of excited charges on the surface. If TiO_2 absorbs more light, then more excited charges

IntechOpen

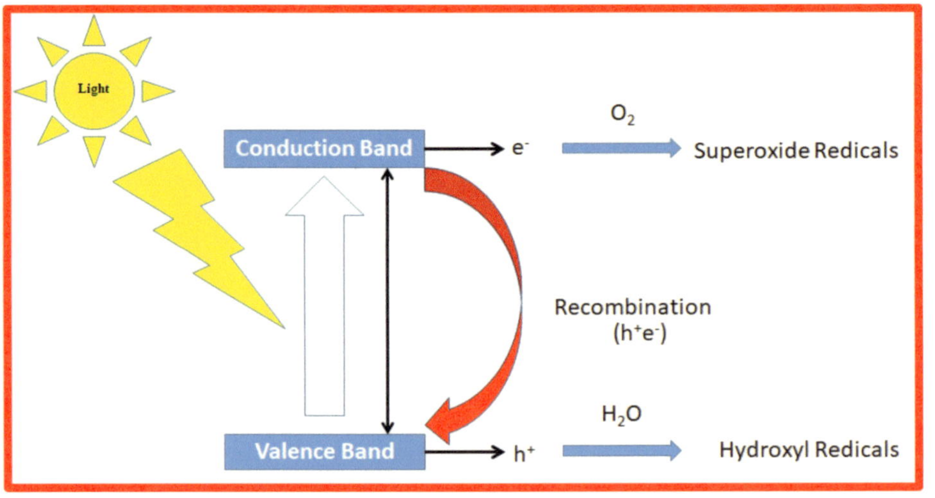

Figure 1.
Mechanistic illustration of the photocatalysis.

come on the surface. The process is schematically illustrated in **Figure 1**. Therefore, if we improve the optical absorption properties of TiO_2, its whole activity can be increased [5, 6].

During last decades, doping techniques have been extensively employed to make TiO_2 colourful for desirable optical absorption [7]. For instance, in the early 1990s, a number of metallic species were employed to substitute the Ti^{4+} ions in the TiO_2 lattice [8]. More efforts lead to the doping of several nonmetals till 2001. Currently, many metal and nonmetal elements have been used to substitute partial Ti^{4+} and O_2 ions in the TiO_2 lattice. All the aforementioned efforts lead to improved light absorption by TiO_2 and consequent photocatalytic performance.

Recently, black TiO_2 has become the focus of the research community because of much-enhanced photocatalytic performance. Materials appear black in colour if it absorbs 100% light from the overall visible light region. From the entire visible light region, if TiO_2 absorbs certain percentage evenly, then it will become partially black or grey. If no light absorbs from the overall visible light regions, then it will show white colour. In the other case in which light is not absorbed appropriately, different colours (e.g., green, yellow and brown) will be observed. The focus of the research about titania is to tune their colour from lighter to darker. The properties and performance of the TiO_2 nanomaterials are affected by the apparent colour. In the proceeding sections, different synthetic strategies to black titania as well as characteristic features and their properties related applications will be discussed.

2. Synthetic strategies used to obtain black Titania nanostructures

Black titania nanostructures have attracted extensive interest, and various reductive and oxidative approaches have been established to successfully fabricate the black or coloured titania [9]. A variety of structural and chemical modifications are in practice to impart unique features to black titania, like surface amorphousity, oxygen vacancy/Ti^{3+}, hydroxyl groups and Ti–H bonds. Various strategies have been briefly explained below.

2.1 Hydrogen treatment

To reduce TiO_2 nanocrystals using heat and hydrogen is an easy approach to obtaining black titania. Thermal hydrogen treatment changes TiO_2 (Ti^{4+}) into other chemical species, such as Ti^{3+} or any other reduction states. Consequently, their lattice structure and chemical/physical properties also change by changing reduction states. The chemistry involved in this reaction is illustrated in the following scheme.

$$TiO_2 + \times H_2 \xrightarrow{\Delta} TiO_2, TiO_{2-\times}, TiO_{2-\times}H_{2\times} \quad (1)$$

The above scheme illustrates that the chemical properties and concentration of starting TiO_2 nanomaterials, reaction temperature and time, pressure and concentration of hydrogen gas are the various factors that determine the final product of the reaction. When TiO_2 nanomaterials are treated with hydrogen, their final properties and the pathway through which the reaction proceeds will be different. This is because the final properties and the direction of reaction depend upon the conditions of hydrogen treatment. Many other factors that affect the chemical properties of nanomaterials are like morphology, shape, size, crystal facets and vacancies contents [10]. Reaction will be more complicated due to these variables. Preparation of these black TiO_2 nanomaterials is achieved normally by different research groups using different synthetic strategies. These alterations in structures lead to variations in their properties and functionalities. These structural and functional developments enable us to tune the structural features of a material and then consequent performances.

Hydrogenated environment used to obtain black titania may vary and it includes simply hydrogen thermal treatment, high-pressure pure hydrogen treatment, ambient or low-pressure pure hydrogen treatment, ambient hydrogen-argon treatment, ambient argon treatment and hydrogen plasma treatment.

Chen et al. synthesized black titania NPs *via* treatment of pure white titania NPs with 20.0-bar pure H_2 at 200°C for several days [11]. The precursor white titania NPs were synthesized following a solution-based route using titanium tetraisopropoxide

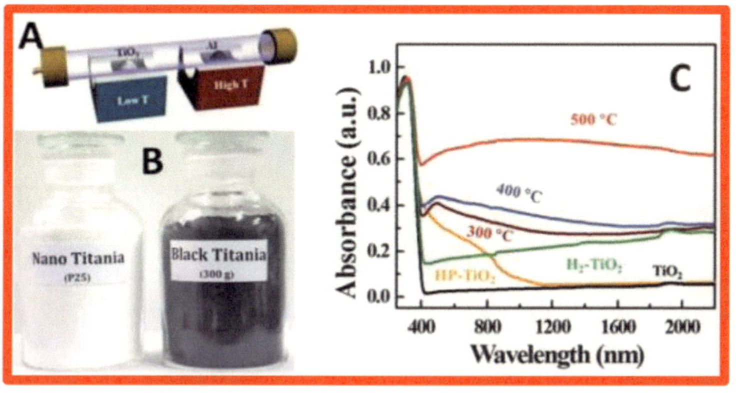

Figure 2.
(a) Schematic of the formation of black titania. (b) Digital photographs of white and black titania nanomaterials. High-resolution transmission electron microscopy images of (c) white and (d) black titania NPs, (e) UV-vis spectra of white and black titania NPs. Reprinted with permission from reference [11]. Copyright 2011 AAAS.

as precursor, Pluronic F127 surfactant, hydrochloric acid, deionized water, and ethanol solvents. **Figure 2** illustrates the schematic of the formation of black titania NPs (**Figure 2a**), along with digital photographs (**Figure 2b**) and electron microscopy images (**Figure 2c** and **d**) of white and black titania NPs. The obtained black titania NPs contained a well crystalline lattice core fenced by a disordered lattice shell from the hydrogen treatment. The amorphous boundary was expected to host the external hydrogen dopant and impart black colour to the hydrogenated titania NPs. The black titania NPs had broadband absorption as compared to corresponding untreated white titania as indicated in the UV-visible absorption spectrum (**Figure 2e**).

2.2 Chemical reduction

Chemical reduction is another route to obtain black titania where the Ti^{+4} species from the corresponding white titania precursors are reduced into low valent Ti species. The reducing agents involved in chemical reduction may include various natural products obtained from plants such as flavonoids, vitamins, phenolic acids, reducing sugars, polysaccharides, triterpenoids, tannins, and polysaccharides as all these are electron rich in nature. The resulting structure of the obtained materials can be tuned by controlling various reaction conditions. Besides these green reductants, some other chemical reducing agents such as aluminium, Zinc, $NaBH_4$, CaH_2 and imidazole have been extensively used to reduce TiO_2 to black titania.

A representative example of synthesis of black titania through chemical reduction is given here. Wang et al. reported the use of aluminium (Al) as a reducing agent to prepare black titania NPs in an evacuated two-zone vacuum furnace at elevated temperature. To obtain black titania, precursor white TiO_2 and Al were placed in separate zone of a two-zone tube furnace and pressure was set at 0.5 Pa through evacuation (**Figure 3a**). The Al was heated at 800°C, whereas precursor white titania was heated at 500°C for 7–18 hours. Thus, obtained reduced black titania NPs displayed intense absorption in the visible regions [12].

2.3 Chemical oxidation

Black titania can be obtained by chemically oxidizing the titanium hydride precursors. For instance, Liu et al. reported the preparation of black titania by oxidising TiH_2 powder with 25% H_2O_2 solution at elevated temperature. In this way, reduced TiO_{2-x} NPs obtained which possess characteristic blue colour. The obtained NPs are quite stable even in air atmosphere as indicated by the retention of their colour and

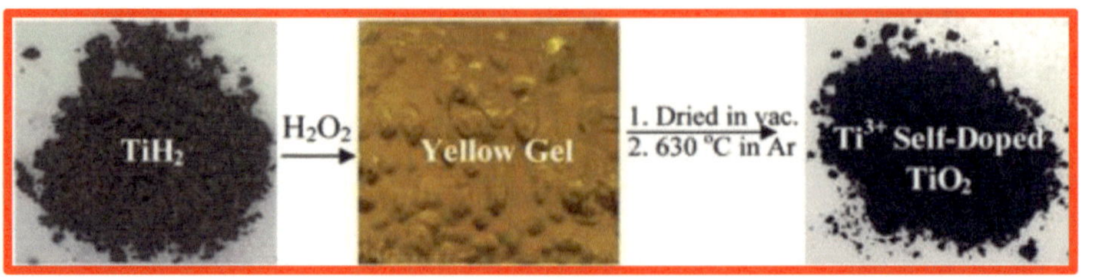

Figure 3.
(a) Schematic representation of the two-zone furnace. (b) Digital photograph of white and black titania NPs, (c) optical absorption spectra of titania NPs reduced at various temperatures. Reprinted with permission from reference [12]. Copyright 2013, The Royal Society of Chemistry.

significant absorption towards the UV to visible light. Similarly, in another report, Grabstanowicz et al. prepared black titania powders following a multistep approach, as indicated in **Figure 4**. First, H_2O_2 (15 mL) was added into TiH_2 powders (0.96 g) aqueous suspension (10 mL) and stirred for three hours at room temperature to obtain a miscible gel-like slurry, followed by additional H_2O_2 (12 mL and 15 mL) and stirring (4 hours and 16 hours) in forming a yellow gel. Second, the gel was vacuum-desiccated overnight, placed in an oven at 100°C for 12–20 hours to become a yellow powder, and then finally at 630°C for three hours in Ar. The black TiO_2 had a rutile phase and remarkably enhanced absorption in the visible-light and near-infrared regions [13].

2.4 Electrochemical reduction

Hydrogenated black titania nanotubes (NTs) were obtained through electrochemical reduction approach by Xu et al. [14]. The NTs were fabricated through two-step anodization at 150 V for an hour with a carbon rod serving as the cathode and Ti serving as the anode. Ethylene glycol, along with 0.3 wt% NH_4F and 10 vol% H_2O, was used as electrolytes. The NTs obtained through first-step anodization were removed using scotch tape and processed for the second anodization. The obtained NTs were then heated at 150°C for three hours and at 450°C for another five hours. The electrochemical doping that leads to reduced titania was achieved using a 5 V electric current for a very short period (5 to 40s) of time in 0.5 M Na_2SO_4 aqueous solution at room temperature. The NTs were used as the cathode, whereas Pt wire was used as the anode, respectively.

Likewise, Li et al. reported the fabrication of black titania NTs through anodization approach followed by electrochemical reduction as indicated schematically in **Figure 5** [15]. Precursor titania NTs were first obtained using titanium foil as anode and Pt gauze as the cathode. The voltage or applied was 80 V for 7200 s, or 4 mA for 5000 s. The electrolyte was an "aged" ethylene glycol with 0.2 M HF and 0.12 M H_2O_2 solution. The obtained NTs were further annealed at 450°C in air atmosphere for about 5 hours. The electrochemical reduction was achieved using conditions of 40 V voltage for 200 s in an ethylene glycol and 0.27 wt% NH_4F solution. The samples were treated at a higher voltage for activation before electrochemical doping.

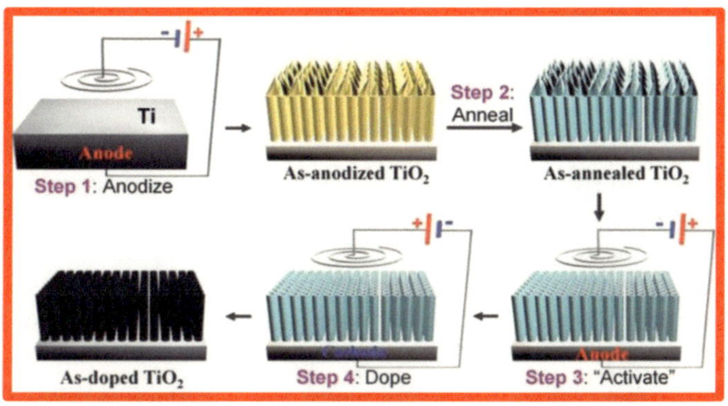

Figure 4.
Schematic illustration of the route from precursor to black titania NPs, along with their pictures. Reprinted with permission from reference [13]. Copyright 2013, American Chemical Society.

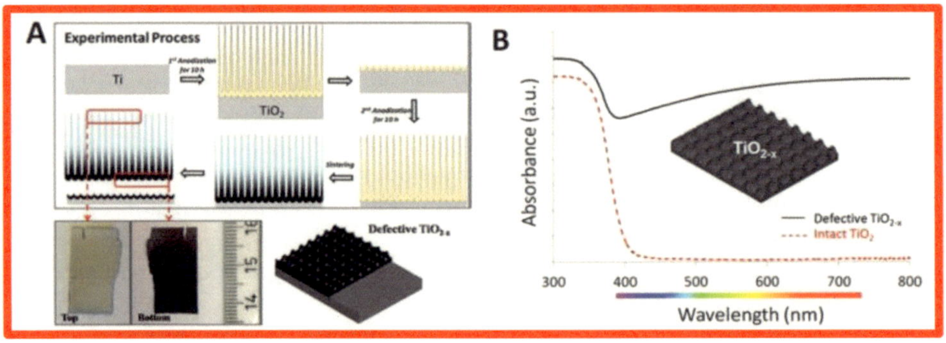

Figure 5.
Schematic of the formation of black titania NTs through an electrochemical reduction approach. Reprinted with permission from reference [15]. Copyright 2014, The Royal Society of Chemistry.

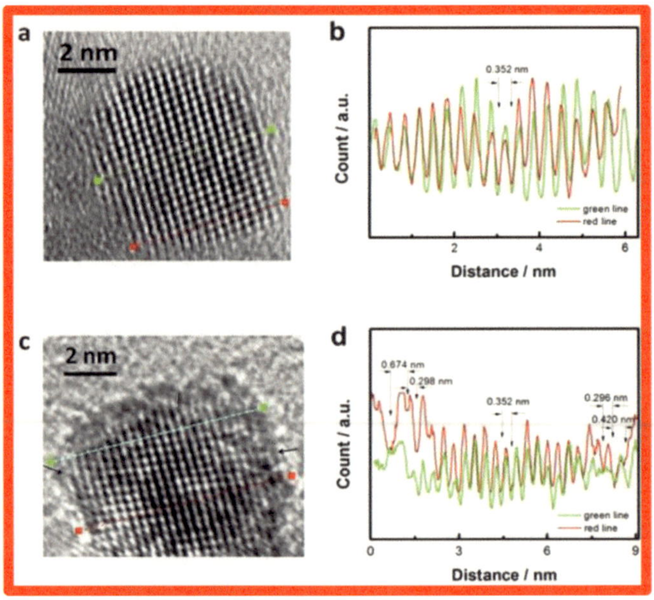

Figure 6.
(a) Representation of the experimental process and images of the stripped titania NTs. (b) Optical absorption spectra of titania NTs. Reprinted with permission from reference [16]. Copyright 2014, American Chemical Society.

2.5 Anodization-annealing

Dong et al. reported the fabrication of black titania NTs using an anodization approach followed by heat treatment at higher temperature as indicated schematically in **Figure 6** [16]. Precursor titania NTs were fabricated using 10-hour anodization on a Ti foil for two times at 60 V potential in an ethylene glycol solution having 0.25 wt.% NH_4F and 2 vol% deionized water. The titania NTs were removed from the reaction system after first anodization and then subjected to second anodization. The repeated washing and cleaning of anodized Ti foil was achieved using ethanol and distilled water. The pure cleaned analytes were then dried at 120°C, and then annealed at 450°C for an hour under ambient environment. The removal of top oxide layer from the substrate leads to the development of layer of black titania.

3. Characteristic features of black Titania nanostructures

It has been frequently stated that the various black titania nanostructures showed distinct chemical and physical characteristics, as briefly explained in the following sections, since the fabrication techniques and synthesis conditions of black titania nanostructures differ from one another in the literature. The observed black colour of the titania NPs has been attributed to some of these characteristics. The unique features responsible to impart colour to titania are given below.

3.1 Structural disorderness near the surface

According to certain investigations, the surface of black titania nanostructures with a crystalline/disordered core-shell structure shows disordered structural features. However, literature also contains different reports. For instance, Chen et al. reported the existence of a disordered surface layer surrounding the crystalline core in hydrogenated black titania NPs obtained in the conditions of 20 bar hydrogen pressure and 200°C [11]; Lu et al. have observed that hydrogenated titania nanocrystals made by treating commercial Degussa P25 under 35 bar hydrogen pressure and room temperature for up to 20 days also contain disorderness in the structure near surface [17]. Wang and Xu, observed the same structural features in the hydrogenated black titania nanosheets [18]. Therefore, the hydrogenation treatment also suggested a modest lattice expansion. Some groups reported lattice shrinkage in the disordered layer [19]. The surface of the hydrogenated titania NTs, on the other hand, was extremely transparent, according to Lu and Zhou et al. [20].

The disordered phase of the black titania nanostructures may be distinguished from the crystalline phase using high-resolution transmission electron microscopy (HRTEM). For instance, **Figure 6** illustrates the comparison of the structure of black and white titania NPs using HRTEM and line analysis [21]. Even at the surface of the nanocrystal, the white titania nanoparticle revealed sharply defined, and well-resolved lattice fringes (**Figure 7a**), and the spacing between the adjacent lattice planes was uniform throughout the whole nanocrystal and typical for anatase (0.352 nm) (**Figure 7b**). The crystalline-disordered core-shell structure of the black titania nanoparticle (**Figure 7c**) revealed a structural divergence from the typical crystalline anatase at the outer layer, where the straight lattice line was curved at the nanoparticle's edge, and the plane distance was no longer uniform (**Figure 7d**). The distinction between the amorphous structure and the crystalline phases has occasionally also been made using electron diffraction (ED).

3.2 The presence of Ti^{3+} ion

Ti^{3+} ions can be seen experimentally or not, depending on the synthetic approaches adopted to obtain black titania NPs. With conventional X-ray photoelectron spectroscopy (XPS), synchrotron X-ray absorption, emission, photoelectron spectroscopies, and electron spin resonance spectroscopy, Ti^{3+} ions were not detected in hydrogenated black titania nanocrystals obtained *via* hydrogen reduction or hydrogen plasma-derived black titania nanostructures. For instance, based on the almost identical Ti 2p XPS spectra of pure and hydrogenated titania nanowires (NWs), Wang et al. proposed the absence of Ti^{3+} in the black hydrogenated titania nanowires processed at 450°C [22]. However, in certain studies, Ti^{3+} ions in the black titania NPs were indicated after hydrogen treatment, chemical reduction, chemical oxidation, and electrochemical reduction,

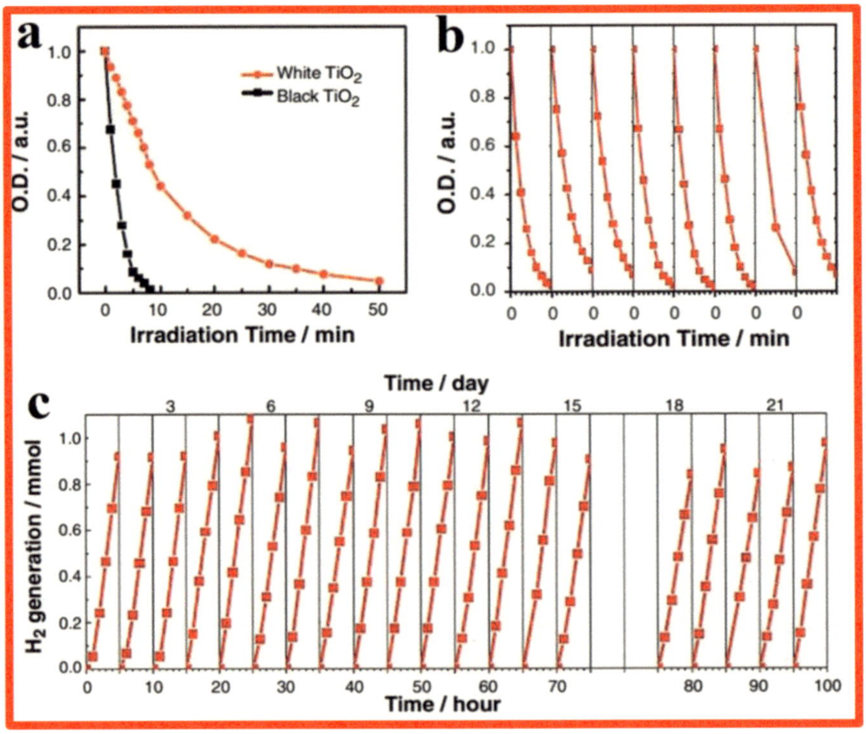

Figure 7.
HRTEM and line analyses of (a & b) one white titania nanoparticle, (c & d) one black titania nanoparticle. The zeros of the axis in b and d correspond to the left ends of the lines in a and c. The red and green curves in b and d correspond to the red and green lines in a and c. Reprinted with permission from reference [21]. Copyright 2013, Nature Publishing Group.

respectively [19]. The existence of Ti^{3+} ions can be detected even using XPS in case of the black titania NTs produced by electrochemical reduction and oxidation of TiH_2 [19].

3.3 Oxygen vacancies

Oxygen vacancies have been continuously documented in black titania nanostructures obtained *via* hydrogen thermal treatment, electrochemical reduction, chemical reduction, and chemical oxidation [19]. For instance, oxygen vacancies were detected by ESR spectroscopy in the black titania NTs formed through thermal hydrogen treatment [23], and electrochemical reduction [16], as well as in the black titania NPs synthesized through Al reduction approach [24]. Like Ti^{3+} ions, oxygen vacancies can also not always be detected. For instance, in the report of Xia et al. [25] no oxygen vacancy was found with ESR in the black titania NPs synthesized with thermal treatment.

3.4 The existence of Ti: OH groups

The hydrogenation treatment leads to a change in the OH content in the black titania nanostructures. A satellite peak characteristic of Ti–OH in the O 1 s XPS spectrum was observed in black titania NPs obtained *via* hydrogen treatment at 200°C for five days [11], in the hydrogen-treated titania NWs [22] and in the hydrogenated titania NTs obtained using ultrapure H_2 atmosphere and 200–600°C temperature for one hour [26]. However, the hydrogenated titania NTs arrays treated at 450°C for one hour in a reducing atmosphere of 5% H_2 and 95% argon did not manifest any alteration in the Ti-OH peak in the O 1 s XPS spectra, and the hydrogenated black titania NPs treated at 450°C for four hours under five bar H_2 displayed a decreased OH signal in the O 1 s XPS spectrum [25].

Black hydrogenated titania NPs demonstrated a change in the strength of the OH vibrational band in the Fourier transform infrared (FTIR) spectrum [25]. In hydrogenated titania nanosheets treated at 400°C for two hours in a pure H_2 environment, more surface OH groups were seen. Titania NPs that had been hydrogenated and subjected to a hydrogen plasma at 500°C for 4–8 hours showed additional peaks at wave numbers of 3685, 3670, and 3645 cm^{-1} [27]. The intensity of the OH peak was significantly lower in hydrogenated titania microspheres than in pure titania after being treated at 500°C for 4 hours with a flow of H_2 (5% in N_2, 300 sccm). Based on the weaker bands at 3446 and 1645 cm^{-1}, less water and/or hydroxyl groups were adsorbed onto the hydrogenated titania NTs when they were heated to 450°C for one hour in a reducing environment of 5% H_2 and 95% Ar [20]. Black hydrogenated titania treated at 200°C for five days showed a drop in O-H intensity, and hydrogenated titania NPs treated at 450°C for 4 hours under 5 bar H_2 showed no OH absorption bands [25]. When hydrogenated titania NPs were exposed to hydrogen plasma at 500°C for 4 to 8 hours, the proton nuclear magnetic resonance (NMR) spectra revealed a higher peak at 5.5 ppm from bridging hydroxyl groups and additional signals at 0.01 and 0.4 ppm from the internal and terminal hydroxyl groups [27]. In the hydrogenated black titania NPs treated at 200°C for five days, a peak at chemical shift +5.7 ppm with two additional tiny, narrow peaks at chemical shifts 0.03 and 0.73 ppm were observed [21]. The hydrogenated titania NPs, on the other hand, showed much smaller OH signals after being heated at 450°C for 4 hours under 5 bar H_2 [25].

3.5 Hydrogen doping

It has occasionally been noted that hydrogenated black titania NPs possess Ti-H groups. 106,117,134,137 Hydrogenated titania nanowire microspheres obtained *via* treatment at 500°C for four hours under a flow of H_2 (5% in N_2, 300 sccm) and hydrogenated titania NPs treated with hydrogen plasma at 500°C for 4–8 hours, a satellite peak at about 457.3 eV in the Ti 2p XPS spectrum was observed, which refers to the presence of surface Ti-H bonds [27]. The surface of hydrogenated titania nanosheets produced at 400°C for two hours under a 10-bar pure H_2 atmosphere was hypothesized to be converted to Ti-H bonds [18]. The diffraction peak at about 59.281 in the X-ray diffraction (XRD) pattern of hydrogenated titania NTs obtained *via* treatment in ultrapure H_2 atmosphere at 200–600°C for one hour also referred to the presence of the Ti-H bond [28].

3.6 Valence band edge modifications

The black titania nanostructures sometime exhibit a shift in the valence band. For instance, when hydrogenated titania was obtained *via* treatment at 200°C for four days under 20 bar H_2 [11] or heated at 500°C for one hour under a hydrogen atmosphere [29], a redshift of the valence band was observed as indicated from the valence band XPS spectrum. A similar shift was observed in the black brookite titania NPs, rutile titania NPs, and titania NTs prepared with the Al reduction method, in the black anatase titania NTs obtained through $NaBH_4$ reduction, in the black titania NPs fabricated *via* electron beam treatment, in the hydrogenated titania NTs treated at 450°C for an hour in a reducing 5% H_2 and 95% Ar environment and in the electrochemically hydrogenated black titania NTs [19]. It was discovered that Ti^{3+} ions did not contribute to the extra bandgap states of the hydrogenated TiO_2 nanocrystals because when Ti^{3+} appeared, these mid-gap states vanished [21]. On the other hand, pure and hydrogenated TiO_2 NWs treated for three hours at 200–550°C in an ultra-pure hydrogen atmosphere (ambient pressure) showed similar valence-band structures. Hydrogenated titania nanosheets exhibit similar valence band shifts.

4. Photocatalytic applications of black titania nanostructures

There are three main steps that are involved in photocatalysis. These are absorption of light in the form of photons, electrons, and holes excitation, separation of charges and migration to the surface of photocatalysts, and then the transfer of charges between the photogenerated carriers and the reactant. Prior to the discovery of hydrogenated black titania NPs, it had been demonstrated that hydrogen thermal treatment of titania could improve its photocatalytic properties. Harris and Schumacher discovered that hydrogen reduction at high temperatures decreased recombination canters and prolonged the lifespan of the holes [30]. Oxygen vacancies, Ti^{3+} species, and hydroxyl groups produced by this process are probably what contributed to the increased photoactivity. Black titania nanostructures exhibit superior performance towards various photocatalytic applications because of the distinct optical and charge transport features as compared to the pristine white titania. Various photocatalytic applications include photocatalytic removal of contaminations, photocatalytic hydrogen production through photoelectrochemical water splitting, photoelectrochemical sensor and photocatalytic CO_2 reduction.

4.1 Photocatalytic degradation of environmental pollutants

It has been estimated that about 20% of dye was wasted while the dyeing process in the textile industry and released as effluent in the water. The release of such a large number of coloured dyes into water causes it to be polluted and become a major source of environmental pollution [31, 32]. This polluted water can be photocatalytically detoxified using high-intensity solar energy. Titania is usually thought to be the best photocatalyst and has the tendency to treat wastewater, but only restricted to the ultraviolet region. However, this limitation can be overcome by using coloured titania.

Chen et al. first time prepared the hydrogenated black titania NPs from white titania. The detailed synthetic strategy involves the treatment of white titania with 20 bar of pure H_2 at 200°C for five days [11]. The obtained black titania had core-shell structure in which the core is well crystalline, and shell is amorphous. The much-enhanced photocatalytic performance towards methylene blue (MB) degradation was observed under simulated sunlight. For instance, the complete photodegradation of MB was achieved in a very short period of eight minutes under simulated solar light by using black titania as compared to white titania as indicated in **Figure 8a** and **b**.

4.2 Photocatalytic hydrogen generation

The recent energy and environmental situation suggests that hydrogen will be the ultimate source of clean and green energy. The photocatalytic water splitting facilitated by solar energy in which natural sunlight and water are employed as the

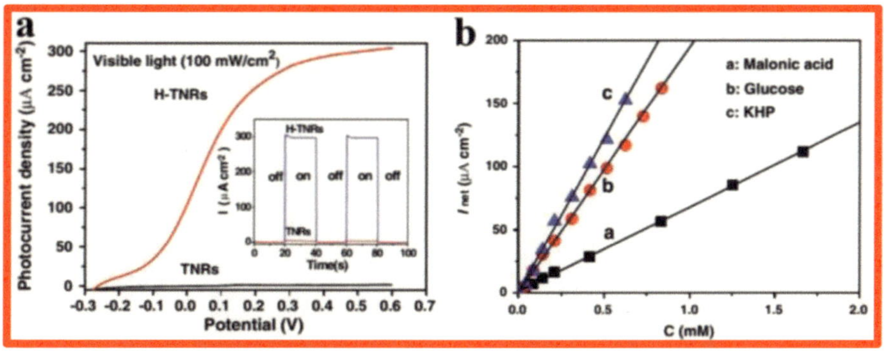

Figure 8.
a) Photocatalytic removal of MB from aqueous medium by white and black hydrogenated titania under stimulated solar light illumination, (b) cycling tests of solar-driven photocatalytic removal of aqueous MB of the black hydrogenated titania, c) stability tests of photocatalytic H_2 evolution of the black hydrogenated titania, and the H_2 evolution rate was calculated to be 10 mmol h^{-1} g^{-1}. Reproduced with permission from reference [11]. Copyright 2011, American Association for the Advancement of Science.

hydrogen source is considered an important source of hydrogen. Black titania-based photocatalysts have been extensively used for generating hydrogen through a water-splitting reaction [33]. For instance, black hydrogenated titania obtained using the high-pressure H_2 can generate H_2 from water at a rate of 10 mmol h^{-1} g^{-1} with exceptional stability under sunlight illumination, as displayed in **Figure 8c** [11]. The H_2-plasma assisted black titania also exhibited enhanced photocatalytic H_2-production rate of 8.2 mmol h^{-1} g^{-1}, about 13.5 times greater than the white titania.

4.3 Photoelectrochemical water splitting

Photoelectrochemical (PEC) water splitting is an important strategy to generate hydrogen following a green solar-to-hydrogen route. Significant research is being done to enhance its efficiency. The black titania is thought to be an emerging candidate for PEC water splitting because of its ideal band structure [33]. For instance, black titania NTs fabricated *via* Al reduction approach exhibited much-enhanced photocurrent as compared to unreduced white titania NTs [15]. The applied bias photon-to-current efficiency (ABPE) of black titania NTs attained 1.20% at a greater bias of 0.68 V (vs. Pt), remarkably greater than that of white titania NTs, 0.25% at 0.49 V (vs. Pt). The incident-photon-to-current-conversion efficiency (IPCE) of black titania NTs was also impressively improved as compared to white titania NTs, along both UV and visible light regions. The superior photocatalytic performance of black titania NTs was credited to the greater electron density and the subsequent better electric conductivity, as indicated from the Mott-Schottky plot. On the basis of this effect, enhanced PEC water-splitting activities were extensively observed in black titania NTs arrays synthesized *via* other reduction approaches. Moreover, Kim et al. reported that the black titania NTs displayed considerably unique electrocatalytic performance in producing OHs and Cl_2 in comparison with white anatase titania NTs [34].

4.4 Photoelectrochemical sensors

Titania can be employed as a photochemical sensor to measure the concentration and type of different organics found in an aqueous medium. It can be done by

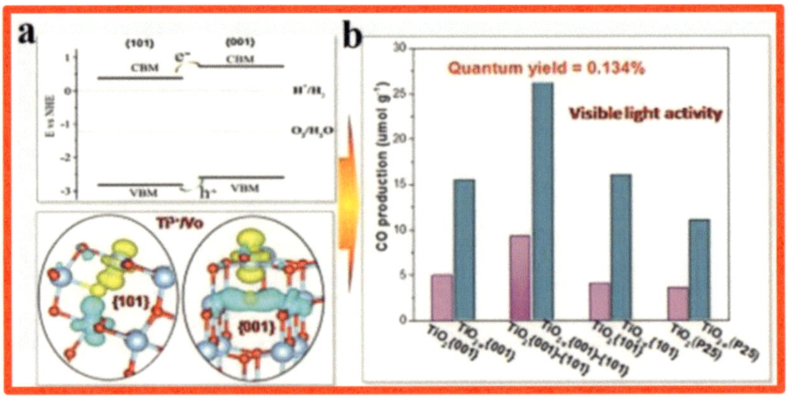

Figure 9.
a) The voltammograms of TNR and H-TNR photoanodes obtained at a scan rate of 5 mV s^{-1} under visible light. Inset shows the photocurrent responses for the TNR and H-TNR photoanodes. b) Relationships between the photocurrent related to the oxidation of the organics net and the concentrations of the organic at the H-TNR electrode. Reproduced with permission from reference [35]. Copyright 2014, Elsevier.

estimating the photocurrents generated from the dissociation process in PEC cells. The arrays of hydrogenated black titania NTs or nanorod fabricated through annealing in the H_2 atmosphere were used as a photoelectrochemical sensor to detect and quantify various organic compounds in solar light. For instance, the hydrogenated black titania nanorods arrays (H-TNRs) exhibited a much more sensitive and steady photocurrent response (~100-folds greater than the white titania nanorods (TNR)) in the $NaNO_3$ solution under solar light (**Figure 9a**). Under solar light illumination, the estimated photocurrents of the H-TNRs exhibited better linear correlations with the concentrations of most organics such as glucose, malonic acid and potassium hydrogen phthalate (**Figure 9b**), indicating that H-TNRs can sensitively and steadily quantify the given organic compounds. The enhancement in the photocurrent response was credited to the enhancement of conductivity [35].

4.5 Photocatalytic CO_2 reduction

Liu et al. reported the fabrication of oxygen-deficient blue titania NPs with two exposed (101) (001) facets. The obtained coloured titania showed enhanced photoreduction towards CO_2 under visible light illumination. A comparatively higher quantum efficiency (0.31% under full spectrum solar light and 0.134% under visible light region) for CO_2 reduction to CO was achieved using water vapour. This quantum yield was almost four times greater than titania having single exposed facets and P25 (**Figure 10b**). The superior performance of titania was attributed to exposure of more active sites, the facilitated electron transfer between (001) and (101) planes, and Ti^{3+} induced mid-gap states to extend the visible light response (**Figure 10a**) [36].

5. Conclusions

Black titania nanostructures have attracted extensive interest, and several reductive and oxidative approaches have been established to successfully fabricate the black or coloured titania. A variety of structural and chemical modifications are in practice to impart unique features to black titania, like surface amorphousity, oxygen vacancy/Ti3+,

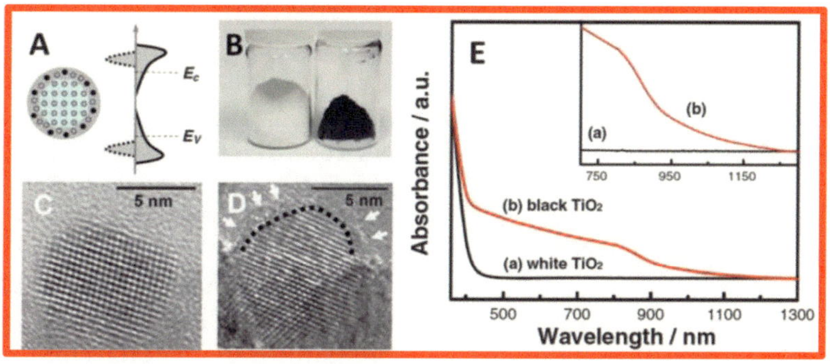

Figure 10.
a) Upper: Estimated relative band edges of the (101) and (001) facets of the anatase phase of titania, lower: Spin densities around the oxygen vacancy in (101) and (001) slab model of anatase TiO2 under antiferromagnetic alignment. The yellow and cyan isosurfaces refer to up-spin and down-spin densities, respectively. b) CO production over TiO_2 and TiO_{2-x} nanocrystals with different exposed facets under visible light. Reproduced with permission [36]. Copyright 2016, American Chemical Society.

hydroxyl groups and Ti–H bonds. These modifications lead to the change in the electronic structure of titania, for example, decreasing bandgap, which is responsible for the improved visible-light absorption features of black titania. Furthermore, the accompanying charge transport features are enhanced because of the decreased electron-hole recombination. Additionally, the most defective structure of black titania assists the adsorption and dissociation of the reactant on its surface, as well as the consequent charge transfer process. Such a novel optical and electrical features endow the material with enhanced photocatalytic activities as compared to white titania.

Conflict of interest

The authors declare no conflict of interest.

Author details

Bilal Akram[1], Bilal Ahmad Khan[2]* and Raieesa Batool[1]

1 Department of Chemistry, Women University of Azad Jammu and Kashmir, Bagh, Pakistan

2 Department of Chemistry, University of Azad Jammu and Kashmir, Muzaffarabad, Pakistan

*Address all correspondence to: bkhan@ajku.edu.pk

References

[1] Liu L, Chen X. Titanium dioxide nanomaterials: Self-structural modifications. Chemical Reviews. 2014;**114**(19):9890-9918

[2] Zhang J, Wang H, Wang L, Ali S, Wang C, Wang L, et al. Wet-chemistry strong metal–support interactions in titania-supported Au catalysts. Journal of the American Chemical Society. 2019;**141**(7):2975-2983

[3] Ali S, Li Z, Ali W, Zhang Z, Wei M, Qu Y, et al. Synthesis of Au-decorated three-phase-mixed TiO 2/phosphate modified active carbon nanocomposites as easily-recycled efficient photocatalysts for degrading high-concentration 2, 4-DCP. RSC Advances. 2019;**9**(66):38414-38421

[4] Chen X, Li C, Grätzel M, Kostecki R, Mao SS. Nanomaterials for renewable energy production and storage. Chemical Society Reviews. 2012;**41**(23):7909-7937

[5] Akram B, Wang X. Self-assembly of ultrathin nanocrystals to multidimensional superstructures. Langmuir. 2019;**35**(32):10246-10266

[6] Akram B, Ahmad K, Khan J, Khan BA, Akhtar J. Low-temperature solution-phase route to sub-10 nm titanium oxide nanocrystals having super-enhanced photoreactivity. New Journal of Chemistry. 2018;**42**(13):10947-10952

[7] Ali S, Li Z, Chen S, Zada A, Khan I, Khan I, et al. Synthesis of activated carbon-supported TiO2-based nano-photocatalysts with well recycling for efficiently degrading high-concentration pollutants. Catalysis Today. 2019;**335**:557-564

[8] Liu B, Chen HM, Liu C, Andrews SC, Hahn C, Yang P. Large-scale synthesis of transition-metal-doped TiO2 nanowires with controllable overpotential. Journal of the American Chemical Society. 2013;**135**(27):9995-9998

[9] Akram B, Butt AF, Akhtar J. Synthesis Approaches of Inorganic Materials: Inorganic Nanomaterials for Supercapacitor Design. Boca Raton: CRC Press; 2019. pp. 155-186

[10] Burda C, Chen X, Narayanan R, El-Sayed MA. Chemistry and properties of nanocrystals of different shapes. Chemical Reviews. 2005;**105**(4):1025-1102

[11] Chen X, Liu L, Yu PY, Mao SS. Increasing solar absorption for photocatalysis with black hydrogenated titanium dioxide nanocrystals. Science. 2011;**331**(6018):746-750

[12] Wang Z, Yang C, Lin T, Yin H, Chen P, Wan D, et al. Visible-light photocatalytic, solar thermal and photoelectrochemical properties of aluminium-reduced black titania. Energy & Environmental Science. 2013;**6**(10):3007-3014

[13] Grabstanowicz LR, Gao S, Li T, Rickard RM, Rajh T, Liu D-J, et al. Facile oxidative conversion of TiH2 to high-concentration Ti3+–self-doped rutile TiO2 with visible-light photoactivity. Inorganic Chemistry. 2013;**52**(7):3884-3890

[14] Xu C, Song Y, Lu L, Cheng C, Liu D, Fang X, et al. Electrochemically hydrogenated TiO2 nanotubes with improved photoelectrochemical water splitting performance. Nanoscale Research Letters. 2013;**8**(1):1-7

[15] Cui H, Zhao W, Yang C, Yin H, Lin T, Shan Y, et al. Black TiO2 nanotube arrays

for high-efficiency photoelectrochemical water-splitting. Journal of Materials Chemistry A. 2014;**2**(23):8612-8616

[16] Dong J, Han J, Liu Y, Nakajima A, Matsushita S, Wei S, et al. Defective black TiO2 synthesized via anodization for visible-light photocatalysis. ACS Applied Materials & Interfaces. 2014;**6**(3):1385-1388

[17] Lu H, Zhao B, Pan R, Yao J, Qiu J, Luo L, et al. Safe and facile hydrogenation of commercial Degussa P25 at room temperature with enhanced photocatalytic activity. RSC Advances. 2014;**4**:1128-1132

[18] Wei W, Yaru N, Chunhua L, Zhongzi X. Hydrogenation of TiO2 nanosheets with exposed {001} facets for enhanced photocatalytc activity. RSC Advances. 2012;**2**(22):8286-8288

[19] Chen X, Liu L, Huang F. Black titanium dioxide (TiO 2) nanomaterials. Chemical Society Reviews. 2015;**44**(7):1861-1885

[20] Lu Z, Yip CT, Wang L, Huang H, Zhou L. Hydrogenated TiO2 nanotube arrays as high-rate anodes for lithium-ion microbatteries. ChemPlusChem. 2012;**77**(11):991-1000

[21] Chen X, Liu L, Liu Z, Marcus MA, Wang W-C, Oyler NA, et al. Properties of disorder-engineered black titanium dioxide nanoparticles through hydrogenation. Scientific Reports. 2013;**3**(1):1-7

[22] Wang G, Wang H, Ling Y, Tang Y, Yang X, Fitzmorris RC, et al. Hydrogen-treated TiO2 nanowire arrays for photoelectrochemical water splitting. Nano Letters. 2011;**11**(7):3026-3033

[23] Liu N, Schneider C, Freitag D, Hartmann M, Venkatesan U, Müller J, et al. Black TiO2 nanotubes: Cocatalyst-free open-circuit hydrogen generation. Nano Letters. 2014;**14**(6):3309-3313

[24] Zhu G, Lin T, Lü X, Zhao W, Yang C, Wang Z, et al. Black brookite titania with high solar absorption and excellent photocatalytic performance. Journal of Materials Chemistry A. 2013;**1**(34):9650-9653

[25] Xia T, Zhang C, Oyler NA, Chen X. Hydrogenated TiO2 nanocrystals: A novel microwave absorbing material. Advanced Materials. 2013;**25**(47):6905-6910

[26] Lu X, Wang G, Zhai T, Yu M, Gan J, Tong Y, et al. Hydrogenated TiO2 nanotube arrays for supercapacitors. Nano Letters. 2012;**12**(3):1690-1696

[27] Wang Z, Yang C, Lin T, Yin H, Chen P, Wan D, et al. H-doped black titania with very high solar absorption and excellent photocatalysis enhanced by localized surface plasmon resonance. Advanced Functional Materials. 2013;**23**(43):5444-5450

[28] Zhang C, Yu H, Li Y, Gao Y, Zhao Y, Song W, et al. Supported Noble metals on hydrogen-treated TiO2 nanotube arrays as highly ordered electrodes for fuel cells. ChemSusChem. 2013;**6**(4):659-666

[29] Naldoni A, Allieta M, Santangelo S, Marelli M, Fabbri F, Cappelli S, et al. Effect of nature and location of defects on bandgap narrowing in black TiO2 nanoparticles. Journal of the American Chemical Society. 2012;**134**(18):7600-7603

[30] Harris L, Schumacher R. The influence of preparation on semiconducting rutile (TiO2). Journal of The Electrochemical Society. 1980;**127**(5):1186

[31] Akram B, Akhtar J. Nanomaterials for Textile Waste Treatment. Water

Pollution and Remediation: Organic Pollutants. Switzerland AG: Springer; 2021. pp. 663-684

[32] Ali S, Ali S, Ismail PM, Shen H, Zada A, Ali A, et al. Synthesis and bader analyzed cobalt-phthalocyanine modified solar UV-blind β-Ga2O3 quadrilateral nanorods photocatalysts for wide-visible-light driven H_2 evolution. Applied Catalysis B: Environmental. 2022;**307**:121149

[33] Liu X, Zhu G, Wang X, Yuan X, Lin T, Huang F. Progress in black titania: A new material for advanced photocatalysis. Advanced Energy Materials. 2016;**6**(17):1600452

[34] Kim C, Kim S, Lee J, Kim J, Yoon J. Capacitive and oxidant generating properties of black-colored TiO2 nanotube array fabricated by electrochemical self-doping. ACS applied materials & interfaces. 2015;**7**(14):7486-7491

[35] Zhang S, Zhang S, Peng B, Wang H, Yu H, Wang H, et al. High performance hydrogenated TiO2 nanorod arrays as a photoelectrochemical sensor for organic compounds under visible light. Electrochemistry Communications. 2014;**40**:24-27

[36] Liu L, Jiang Y, Zhao H, Chen J, Cheng J, Yang K, et al. Engineering coexposed {001} and {101} facets in oxygen-deficient TiO2 nanocrystals for enhanced CO_2 photoreduction under visible light. Acs Catalysis. 2016;**6**(2):1097-1108

Chapter 3

Crystal Facet Engineering of TiO_2 from Theory to Application

Szymon Dudziak, Marta Kowalkińska and Anna Zielińska-Jurek

Abstract

Recently, the surface structure effect on photocatalytic activity has gathered increasing attention due to its reported influence on the charge carrier trapping and separation. Detailed control over the surface structure can be achieved by exposing the specific crystal facets. As a result, the photogenerated electrons and holes can be effectively separated between the different facets of semiconductor crystals. TiO_2 is the most studied photocatalyst, with the particles exposing {0 0 1}, {1 0 0}, {1 0 1}, {1 1 0}, {1 1 1}, and {1 0 5} crystal facets. The performed studies have shown that the efficiency of the photocatalytic process strongly depends on the nature of the crystal facet exposed at the photocatalyst surface. In this regard, this chapter focuses on the comparison of possible surface-related parameters and photocatalytic activity of anatase, rutile, and brookite polymorphs with exposed different crystal facets. Particularly, computational data on their different possible surface structures are summarized, focusing on the geometry, energy, and possible reconstructions. This is followed by the general description of the hypothetical Wulff constructions and existing stabilization/synthesis strategies. Such an approach could help to further design, simulate, and optimize photocatalyst surface for efficient photoreduction and photooxidation processes.

Keywords: crystal facets, computational, titanium dioxide, photocatalysis, crystal growth, surface engineering

1. Introduction

Titanium dioxide (TiO_2) is one of the most studied photocatalysts, especially considering its application in the photocatalytic degradation of micropollutants. TiO_2 can induce specific redox reactions through photogenerated charge carriers in photocatalysis. Such process can be divided into subsequent steps, including (1) excitation of electrons in the TiO_2 structure; (2) dissociation of the generated excitons to free electrons and holes; (3) migration of the charge carriers to the surface; and (4) transfer of the e^- or h^+ to substrate present at the surface. All of these steps are common for every photocatalytic reaction, and each deals with limitations that influence the overall process efficiency. However, steps (3) and (4) occur at the surface. Therefore, any change at the interface between the photocatalyst and a substrate can induce significant changes in the involved elementary reactions. The importance of this interface and photocatalyst surface was realized very early in the photocatalytic studies, discussing problems like surface polarization with excess electrons, modification with noble metals, and surface complexation with bidentate benzene derivatives to improve the transfer of the charge carriers [1–3]. Simultaneously, the challenges of surface trapping and recombination of the generated charge carriers were also highlighted.

However, concerning these early studies, the exact geometry of the photocatalyst surface was not considered at this point, and ultrafine particles were studied without well-defined geometry. In the last years, significant progress has been made both in the preparation procedures of the TiO_2 particles and in the application of computational methods to simulate the geometry and properties of such interfaces at the atomic scale. As a result, stabilization of a specific interface structure can be achieved during the photocatalyst preparation, leading to the formation of faceted particles terminated with specific, well-defined crystal planes. At present, the application of such single-crystalline particles can be considered a state-of-the-art approach for investigating the details of photocatalytic reactions. Moreover, when systematically studied, it allows for the optimization of the final structure of the photocatalyst and increases its activity in a specific reaction [4]. This is primarily a result of a preferred electronic structure of the exposed facet. Therefore, photogenerated electrons and holes may accumulate at different crystal facets leading to improved charge carriers' separation and more selective photocatalytic reactions. In this regard, this chapter concisely highlights recent state-of-the-art progress in (1) the synthesis of crystal-facet exposed anatase, rutile, and brookite, (2) crystal facet-dependent properties of TiO_2, and (3) the correlation of surface properties with photocatalytic activity in photodegradation of emerging pollutants present in water, H_2 generation, and CO_2 reduction into valuable chemicals. Furthermore, by reviewing the research progress on crystal facet engineering of TiO_2, we hope to provide directions for future selective semiconductor modification with electron-donor or electron-acceptor to improve the overall efficiency in photocatalytic reaction kinetics and mechanism.

2. Computed geometries and energies of the TiO_2 surfaces in vacuum

Recently, detailed atomic structures of the possible crystal facet terminations and their corresponding surface energies are preferably studied using a computational approach. This allows us to describe actual surface geometry and investigate how it will change in a different environment. Therefore, energetic stabilization of the specific facet might be predicted to help obtain it experimentally. However, the analyzed models are often studied in a vacuum, which enables researchers to compare the relative stability of different structures and provides a useful starting point for further investigations. Concerning TiO_2 crystal planes, corresponding surface models were analyzed in detail for both anatase and rutile polymorphs, including (0 0 1), (1 0 0), (1 0 1), (1 0 3), (1 0 5), (1 1 0), and (1 1 2) structures for anatase [5–11], as well as (0 0 1), (0 1 1), (1 0 0), (1 1 0), and (1 1 1) for rutile [10–14]. On the other hand, brookite surfaces are rarely investigated. Nevertheless, a detailed study on different brookite surfaces was reported by Gong and Selloni [15], as well as additional information can be found about the (1 0 1), (1 2 1), and (2 0 1) structures [16]. Moreover, in the case of some models, different atomic geometries or possible reconstructions have been suggested. This specifically includes (1 x 4) reconstruction of the anatase

(0 0 1) surface, reconstruction of the rutile (0 0 1), (0 1 1), and (1 0 0) surfaces, as well as two possible geometries of anatase (1 0 3), including smooth $(1\ 0\ 3)_s$ and faceted $(1\ 0\ 3)_f$ terminations. The following description is a summation of these works. Finally, a different number of atoms exposed to the environment are discussed for different surfaces. For example, atoms from the second (or deeper) atomic layer are not always considered as "exposed" by the authors, although they are needed to complete surface structure. Here, if the surface structure has visible steps that give possible access to such atoms, we have always considered them as exposed ones. Nevertheless, it should be minded that such atoms are more or less covered with the ones located closer to the top of the surface and might be less accessible in practice.

2.1 Unreconstructed anatase surfaces

The atomic structures of the perfect anatase terminations with different crystal planes are presented in **Figure 1**. Starting from the top-left side, the (0 0 1) surface is

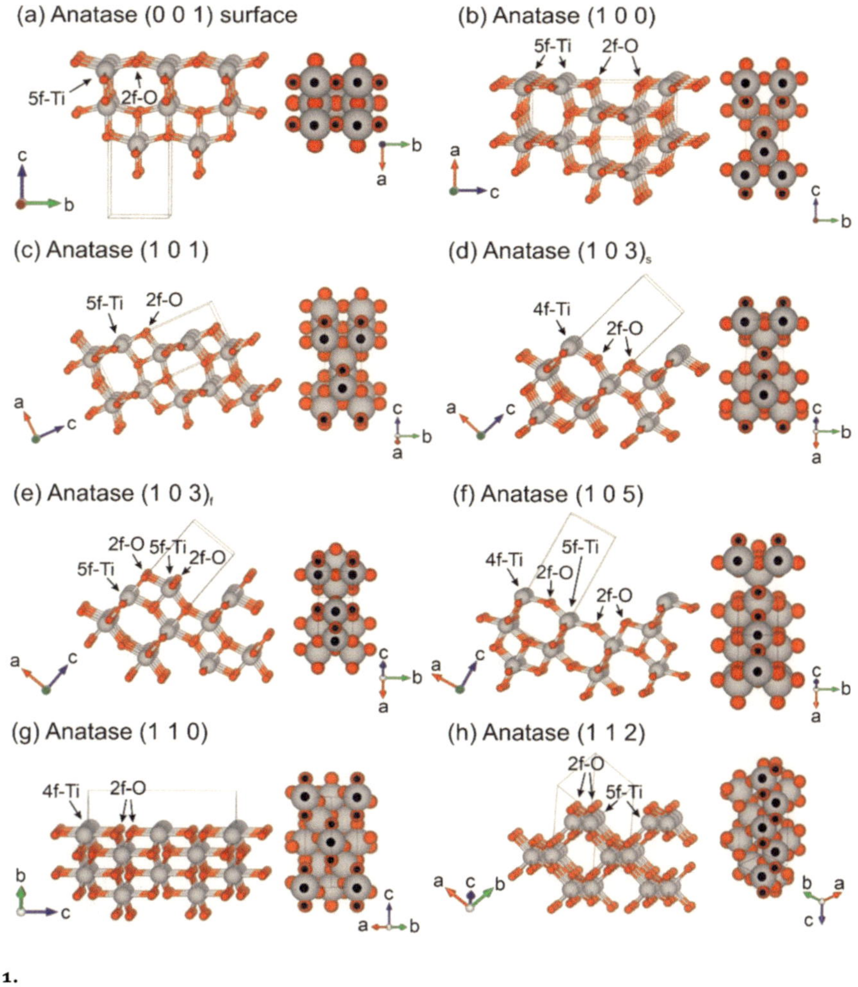

Figure 1.
Perfect, bulk-cut atomic geometries of different anatase surfaces in their isomeric (ball-and-stick models) and top (space-filling models) views. Ti atoms are large, gray spheres and O atoms are small red ones. In the top view, the undercoordinated atoms are marked with dots. The bulk crystal unit cell and the unit cell vectors are presented to help navigate between different structures. Visualizations were performed with the VESTA program.

flat, exposing all of the Ti atoms as 5-fold coordinated sites (5f-Ti) and ½ of the O as 2-fold coordinated ones (2f-O). The high density of the undercoordinated species leads to a high surface energy of about 0.90–1.08 $J{\cdot}m^{-2}$. Noticeable relaxation of the atomic structure, compared to the bulk (0 0 1) crystal plane, includes breaking of the bond symmetry between 5f-Ti and two neighboring 2f-O, resulting in one Ti-O bond being visibly longer than the other. For the (1 0 0) surface, exposed atoms consist of both undercoordinated 5f-Ti and 2f-O, as well as fully coordinated 6f-Ti and 3f-O. Here, all Ti atoms exposed directly at the first atomic layer are 5f-coordinated, while only ½ of oxygens are 2f-coordinated, as shown in **Figure 1b**. Moreover, visible steps inward of the crystal structure give access to the second-layer atoms, which are always fully coordinated. Including these sites as surface, atoms give a final fraction of 5f-Ti is ½ and 2f-O is ⅓. The relaxed structure of the (1 0 0) facet shows outward relaxation of fully coordinated O atoms and inward relaxation of 5f-Ti, creating an oxygen-rich interface, while 6f-Ti relaxes outward. Corresponding surface energy is usually reported in the range of 0.53 to 0.79 $J{\cdot}m^{-2}$.

Furthermore, the (1 0 1) surface of anatase is energetically the most stable, with the reported surface energy in a vacuum being 0.44–0.65 $J{\cdot}m^{-2}$, and it is commonly observed in nature. The corresponding atomic structure consists of a sawtooth profile with ½ of Ti atoms being 5f-coordinated and ⅓ of O atoms being 2f-coordinated, as shown in **Figure 1c**. Here, the most exposed atoms are undercoordinated oxygens, and specifically, further relaxation leads to the outward rise of the 3f-O above the undercoordinated Ti. Similarly to the (1 0 0) surface, the (1 0 1) can be considered oxygen-rich. Next, the (1 0 3) surface is often discussed in two possible terminations, either smooth one $(1\ 0\ 3)_s$ or faceted $(1\ 0\ 3)_f$, both presented in **Figure 1d-e**, respectively. The smooth (1 0 3) termination is one of the few facets that exposes 4-fold coordinated Ti atoms (4f-Ti) at the top of the surface, being ⅓ of the total surface Ti. Other Ti sites located below the 4f-Ti "tooth" are fully coordinated. Here, the 2f-O represents ⅖ of all exposed O and two 2f-O are always bonded to the single 6f-Ti. The outermost O atoms are 3f-coordinated and, similar to other facets, undergo significant relaxation outward of the crystal structure. On the contrary, on the faceted (1 0 3) termination relatively large fraction of Ti atoms is 5f-coordinated, being ⅔ of all Ti. Two different 5f-Ti sites exist on this surface, one bonded to a single 2f-O and the other bonded to three 2f-O atoms. The total fraction of 2f-O is the same as for the smooth (1 0 3), that is, ⅖. Similar to the (0 0 1) surface, significant symmetry breaking is observed for the 2f-O bridging two equivalent 5f-Ti sites (the one bonded to three 2f-O), resulting in one bond being longer than the other. Analogically to other facets, 3f-coordinated O relaxes outward; however, the change is smaller than for a $(1\ 0\ 3)_s$ structure. Energetically, Lazzeri et al. reported that the $(1\ 0\ 3)_f$ is slightly more stable than $(1\ 0\ 3)_s$, with surface energies being 0.83 and 0.93 $J{\cdot}m^{-2}$, respectively [5]. However, different results were presented by Zhao et al. with analogical energies of 1.14 and 1.05 $J{\cdot}m^{-2}$ [6].

The (1 0 5) surface structure is rarely discussed; however, Jiang et al. have proposed possible geometry of such facet with a corresponding surface energy of 0.84 $J{\cdot}m^{-2}$. In their model, the atomic structure of the (1 0 5) facet composes of both 4f-Ti (¼) and 5f-Ti (¼) exposed, together with 2f-O ($^3/_7$), as shown schematically in **Figure 1f** [17]. Although performed, relaxation of the geometry was not described. Furthermore, the (1 1 0) surface has a characteristic layered structure with deep cavities running along the [0 0 1] direction. Such structure exposes up to four atomic Ti layers and three O layers at different depths. As shown in **Figure 1g**, the first layer composes only of 4f-Ti and 2f-O atoms, while further layers are fully coordinated. As described by Zhao et al., this surface also shows a "layered" relaxation pattern, where O atoms relax outward for the odd-numbered letters and inward for the even-numbered ones [6]. Simultaneously, Ti relaxes in the opposite pattern, ultimately leading to the breaking of the perfect 2D symmetry of the surface structure (i.e., the surface stops being perfectly flat, in contrast to the bulk crystal plane). Following this description and maximum exposition of the first four Ti and first three O layers, the fraction of undercoordinated species is ¼ for 4f-Ti and ⅓ for 2f-O. Corresponding surface energy is commonly reported between 01.02 and 1.33 $J{\cdot}m^{-2}$.

Finally, the (1 1 2) surface is also rarely considered; however, Mino et al. analyzed its surface model with surface energy between 0.95 and 0.98 $J{\cdot}m^{-2}$, depending on the possible relaxation of the cell parameters (exact value was approximated by the authors from the graph) [9]. Similar to the (1 1 0) surface, the (1 1 2) geometry exposes the first layer of undercoordinated species (5f-Ti and 2f-O, both being ½ of all Ti/O atoms) and the second layer of fully coordinated atoms, as shown in **Figure 1h**. A detailed relaxation pattern was not described.

2.2 Unreconstructed rutile surfaces

Concerning the rutile crystal structure of TiO_2, its possible terminations with different crystal planes are shown in **Figure 2**, analogically to the anatase. Starting from the (0 0 1) crystal plane, the corresponding termination shows structure similar to anatase (1 1 0), with the first atomic layer composed fully of undercoordinated 4f-Ti and 2f-O species and the second layer being fully coordinated. Assuming that the second layer is also partially exposed to the environment, the final fraction of undercoordinated sites is again ½ for both Ti and O. The relaxation of this surface is also similar to that of anatase (1 1 0), where the 4f-Ti atoms on the surface relax inward toward the bulk structure, while the 2f-O and 6f-Ti atoms relax outward. The high density of the broken bonds results in high surface energy, with reported values between 1.21 and 1.58 $J{\cdot}m^{-2}$, making it one of the least stable rutile facets. Furthermore, **Figure 1b** shows (0 1 1) surface structure of the rutile phase with a little corrugated profile. Here, the uppermost part composes fully of 5f-Ti and 2f-O, while subsurface 3f-O atoms are also partially exposed, ultimately being ½ of all O atoms. Surface relaxation described by Barnard et al. showed that all 5f-Ti, 2f-O, and 3f-O atoms relax outward on this surface; however, a bit different pattern was also reported by Ramamoorthy et al. [14]. Along the (0 0 1) surface, the (0 1 1) is reported to be one of the least stable rutile surfaces, with surface energies reported in the range of 0.95 to 1.11 $J{\cdot}m^{-2}$.

Furthermore, the (1 0 0) surface shows a layered structure with rows of 5f-Ti and 2f-O atoms at the top of the exposed "tooths," as well as fully coordinated species in the cavities. The corresponding fraction of the undercoordinated atoms is ½ for Ti and ⅓ for O. Relaxation pattern described by Ramamoorthy et al. includes the moving of the surface Ti and O atoms in opposite directions along the [0 1 0] axis [14]. Simultaneously, surface Ti relaxes outward the crystal structure. Corresponding surface energy was reported in the range between 0.60 and 0.83 $J{\cdot}m^{-2}$. Next, the (1 1 0) structure is energetically the most stable surface of the rutile polymorph, with the reported surface energy between 0.34 and 0.59 $J{\cdot}m^{-2}$. Its surface structure exposes both 5f-Ti atoms (½) and 2f-O (⅓) in the subsequent rows, as presented schematically in **Figure 2d**. Relaxation of the (1 1 0) structure includes an inward shift of all the undercoordinated species and outward of the 6f-Ti and 3f-O, therefore making the final structure more puckered.

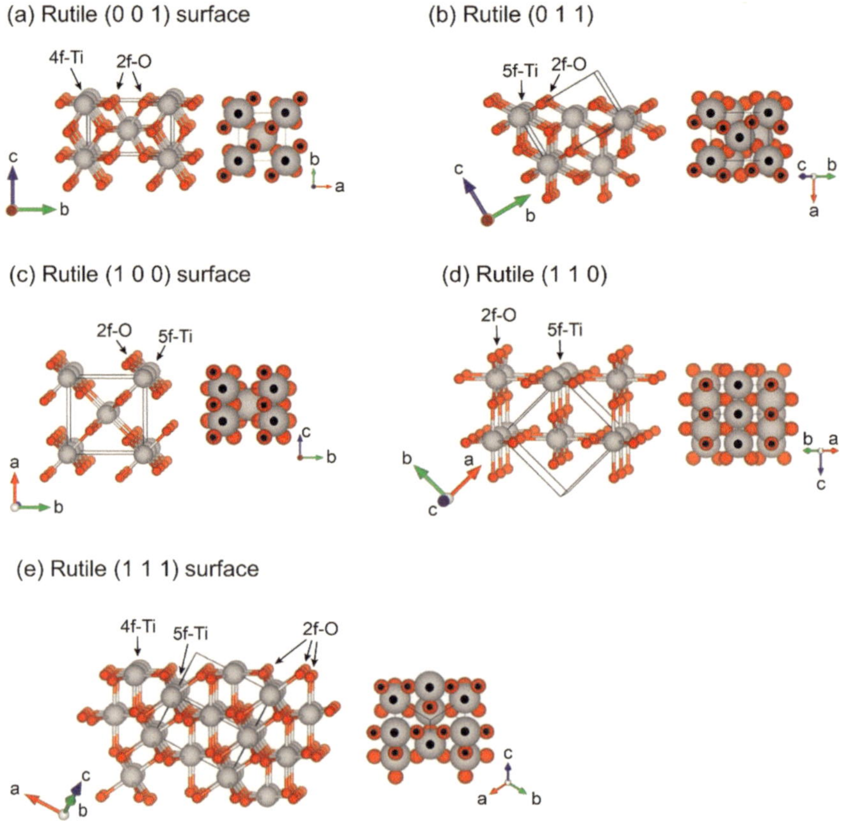

Figure 2.
Perfect, bulk-cut atomic geometries of different rutile surfaces. Figure features are analogical to the anatase one.

Finally, the rutile (1 1 1) surface is rarely discussed; however, a detailed analysis of its possible structure was presented by Wang et al. [18]. For the unreconstructed, bulk-cut, stoichiometric surface, the corresponding surface energy was found to be in the range of 1.34 $J{\cdot}m^{-2}$, exposing ⅓ of Ti atoms as 4f-Ti in the first layer, ⅓ of Ti as 5f-Ti in the second one, and hypothetically ⅓ of 6f-Ti in the third one. Here, all O atoms in the first layers are 2f-coordinated and 3f-O can be found only below the 5f-Ti sites, as shown in **Figure 2e**. According to Jiang et al., the 4f-Ti atoms show significant inward relaxation for such a structure [13]. However, very high surface energy and density of undercoordinated atoms generally lead to the instability of such "perfect" (1 1 1) surface, and further stabilization by hydroxylation was discussed in the following parts.

2.3 Unreconstructed brookite surfaces

Finally, **Figure 3** shows unreconstructed brookite surfaces, which are analogical to the anatase and rutile. Compared to the other polymorphs, brookite surfaces form rather complex structures, often exposing 4f-Ti. Due to such complexity, the calculated fraction of undercoordinated species for brookite surfaces is presented only for Ti atoms. Moreover, in the case of most surfaces, multiple terminations can be considered. Here, we have highlighted only those, which were found to be most stable, or whose reconstructed geometries showed significant stabilization.

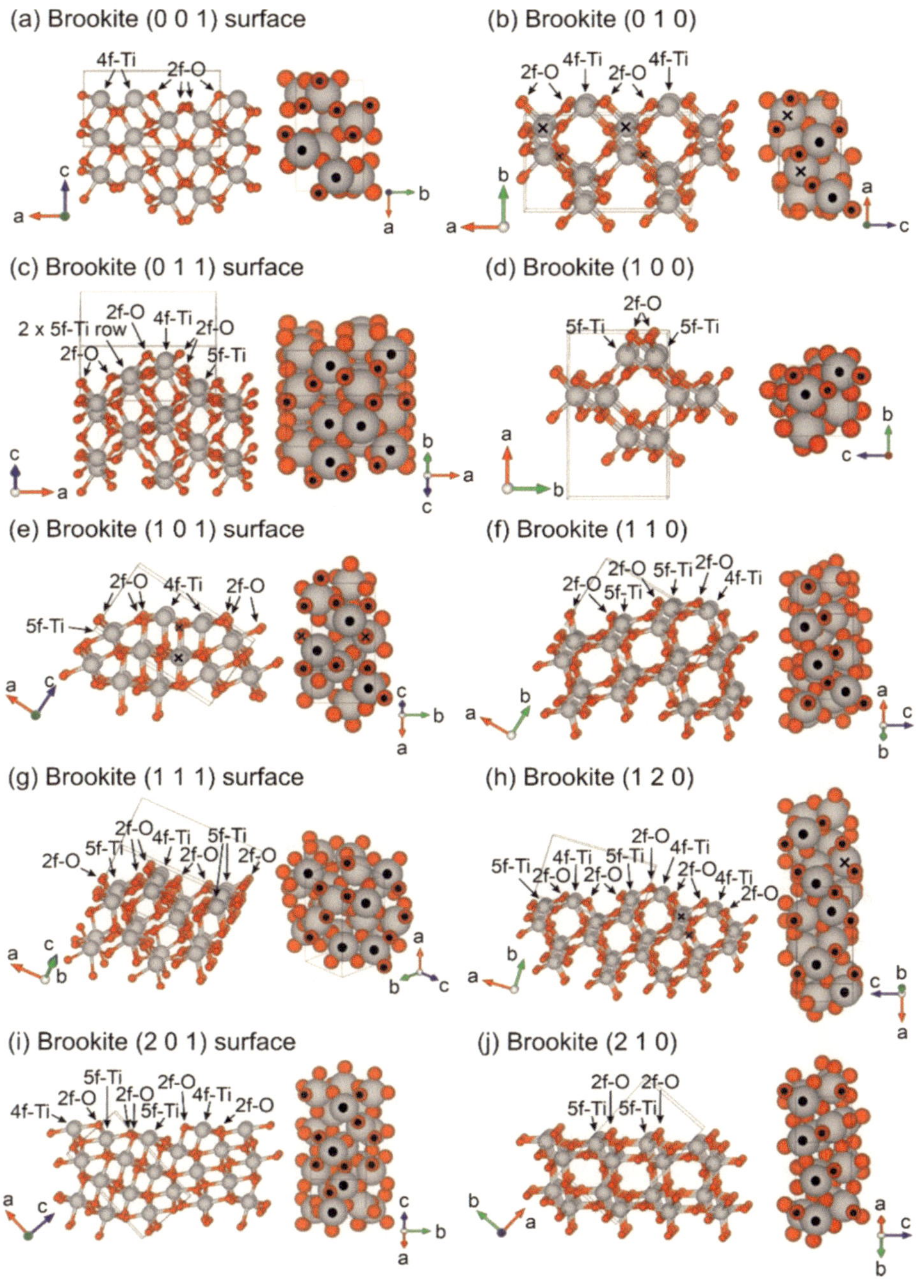

Figure 3.
Perfect, bulk-cut atomic geometries of different brookite surfaces. Figure features are analogical to the anatase one. The "X" mark indicates possible bond breaking after relaxation.

The (0 0 1) brookite surface directly exposes 4f-Ti and 6f-Ti; however, some additional 6f-Ti, that are partially exposed can be found in the deeper parts of the surface, as presented in **Figure 3a** (after including deeper-layer Ti fraction of 4f-Ti is ¼). The top Ti atoms are bridged with a network of 2f-O and saturated O can be found only in the deeper parts of the surface. The calculated energy of such geometry after straightforward optimization was found to be 1.18 J·m^{-2}; however, possible reconstruction is known to stabilize alternate geometry, which will be discussed later.

Furthermore, the (0 1 0) brookite surface shows structure similar to the anatase (1 1 0) and rutile (0 0 1), with a top layer of (2f-O)-(4f-Ti)-(2f-O) bonding to the 6f-Ti in cavities. Assuming exposition up to the first three Ti layers, the fraction of 4f-Ti is ¼ for such geometry. Moreover, reported relaxation has shown that the uppermost 6f-Ti significantly rises to the top, breaking one bond with a deeper layer 3f-O, as highlighted with "X" in **Figure 3b**. Therefore, additional 5f-Ti is expected to be exposed on the relaxed surface. The final surface energy of such a structure was found to be 0.77 J·m^{-2}. Focusing on the next structure, the (0 1 1) brookite surface is shown in **Figure 3c**, exposing a combination of both 4f-Ti (⅛), 5f-Ti (⅜), and 6f-Ti (½), which are connected by a quite complex network of both undercoordinated and fully coordinated O atoms. Relaxation of the structure leads to outward displacement of 6f-Ti, while undercoordinated titanium relaxes inward the crystal structure, giving final surface energy of 0.74 J·m^{-2}. The (1 0 0) brookite surface is similar to the anatase (1 1 2), with a sawtooth-like profile, as shown in **Figure 3d**. The top of the surface directly exposes a row of 5f-Ti (½) atoms connected by the 2f-O, while deeper layers are fully coordinated. It can be noted that this structure can possess two very similar, but not strictly identical, configurations, where the top of the surface is composed of either the first or the second Ti layer, as shown in **Figure 3d**. Gong and Selloni [15] reported that these two configurations differ slightly in surface energy (either 0.88 or 0.93 J·m^{-2}). Nevertheless, their overall geometry is very similar, and both configurations differ mostly due to slight changes in the exact bond lengths and angles at the very top. Next, **Figure 3e** shows the structure of the perfect (1 0 1) brookite surface, again exposing a combination of 4f-Ti (¼), 5f-Ti (⅛), and 6f-Ti (⅝), connected with both saturated and 2f-coordinated O atoms. Significant changes in the perfect structure during the relaxation include bond breaking between 3f-O that bridges two surface 4f-Ti atoms and the subsurface 6f-Ti, highlighted in **Figure 3e** with "X." In this case, the resulting 5f-Ti is located in the subsurface region; however, additional 2f-O appears exposed directly on the surface top. The resulting surface energy for such geometry is 0.87 J·m^{-2}. The surface structure of brookite (1 1 0) presented in **Figure 3f** is characterized by terraces that end with a 4f-Ti edge and further step-down. Again a variety of Ti species are exposed at this surface ($^1/_7$ of 4f-Ti and $^2/_7$ of 5f-Ti). Undercoordinated atoms undergo large relaxation at the edges, with O atoms relaxing outward the crystal structure and Ti atoms relaxing inward. The corresponding surface energy was found to be 0.85 J·m^{-2}. The (1 1 1) surface of brookite can be considered in a variety of different, complex terminations, which were studied in detail by Gong and Selloni. Following their results, **Figure 3g** shows the most probable (1 1 1) structure, exposing ⅛ of Ti sites as 4-fold coordinated and ⅜ as 5f-Ti. No significant relaxation changes were reported for such a structure and the corresponding surface energy was 0.72 J·m^{-2}. Next, the (1 2 0) brookite surface is shown in **Figure 3h**, exposing 4f-Ti ($^3/_{13}$) and 5f-Ti ($^2/_{13}$). This is another brookite surface that shows possible bond breaking between saturated Ti and O after the relaxation, as indicated in the surface model with "X," which results in the appearance of the additional 5f-Ti site at the surface's top. The corresponding surface energy was 0.82 J·m^{-2}. Furthermore, the (2 0 1) brookite surface was not investigated in detail through a computational approach; however, it has gained significant experimental attention due to reported relatively high photocatalytic activity. The atomic model of the (2 0 1) surface, proposed by Lin et al., and Zhou et al., is shown in **Figure 3i** [16, 19]. This surface composes of little up-and-down terraces, exposing either 4f-Ti

($^2/_7$) or 5f-Ti ($^2/_7$), respectively. On the "flat" parts, the two 5f-Ti are bridged with two 2f-O, while two 4f-Ti are bridged with one 2f-O and one 3f-O. Additional 2f-O are present at the edge of each step. Finally, **Figure 3j** shows the energetically most stable brookite surface (2 1 0), with reported surface energy of 0.70 $J \cdot m^{-2}$. The atomic geometry of this surface shows similarities with anatase (1 0 1), as both expose characteristic steps of undercoordinated 5f-Ti bonded to the 2f-O at the edge of the step. Lower parts of the surface, below the (5f-Ti)-(2f-O) steps, are fully coordinated. The final fraction of 5f-Ti for this surface is ½. Similar to other TiO_2 surfaces, the 6f-Ti and topmost 3f-O show visible relaxation outward of the crystal structure, while 5f-Ti relax inward.

2.4 Surface energies and reconstructions

The above description shows possible terminations of the TiO_2 crystals by different crystal planes, resulting in different surface energies calculated for relaxed models. Concerning analyses in a vacuum, these energies roughly correspond to the density of theoretical bonds that needs to be broken to form particular termination. However, as presented in **Figure 4**, considering different values reported in the literature, the relationship is not always strict and should be considered more as a guidance than an actual rule. It should also be noted that different computational details will lead to different computed energy values, which should be especially minded.

As presented in **Figure 4**, some of the reported energies can achieve quite large values, which is commonly the reason why such hypothetical structures are not necessarily observed experimentally. This may include the fact that analyzed crystals/nanoparticles do not terminate in such orientation or that the final structure does not correspond to such particular atomic geometry. The second option is generally known as surface reconstruction, where interface atoms adopt geometry different from the corresponding crystal plane to minimize final surface energy.

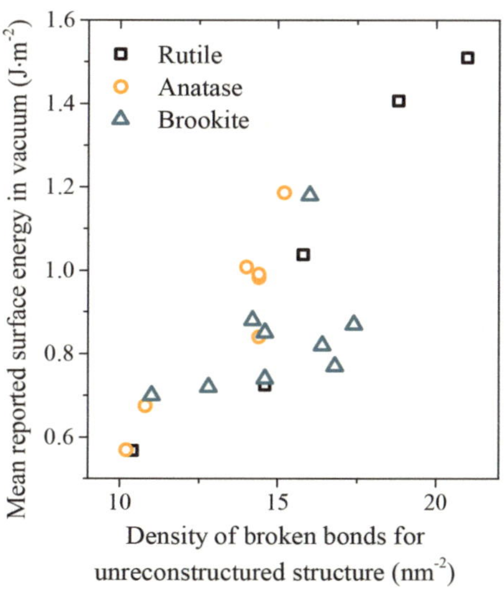

Figure 4.
Relation between reported surface energies of different, unreconstructed TiO2 surfaces in vacuum and a density of broken bonds, needed to form the surface from bulk crystal. To give better comparison between values obtained in different studies (no single study reports all considered surfaces), final values are presented as a mean.

Surface reconstruction was reported as an important process for several possible TiO_2 terminations. Probably, the most notable is the (1 x 4) reconstruction of the anatase (0 0 1) surface, which was described in the early 2000s and should be especially considered under ultra-high vacuum conditions [20–22]. The probable structure of such termination was proposed by Lazzeri and Selloni with the "ad-molecule" model, which was found to be energetically more stable than the unreconstructed surface (0.51 $J{\cdot}m^{-2}$ for the reconstructed geometry, vs. 0.90 $J{\cdot}m^{-2}$ for the unreconstructed one) [8]. In their model, they propose that every fourth of the (5f-Ti)-(2f-O)-(5f-Ti) periodic units, shown in **Figure 1a**, is replaced by the row of TiO_3 bridging species that rise above the perfect surface. The stability of such a configuration was explained due to the stress relief induced by the change in the bond length between the 5f-Ti and 2f-O atoms "left" in the unreconstructed part of the surface. Specifically, the (1 x 4) periodicity of such reconstruction resulted in the bond length being the closest to the "natural" length and, in consequence, the lowest surface energy. Nevertheless, despite the energetical stability of the proposed model, such a structure was not found to be completely in agreement with the experiment. This was mostly due to the relatively low activity of such structures, despite the expected exposition of the 4f-Ti atoms at the formed bridges, which should act as good adsorption and dissociation centers. This has led to further refinement of the proposed geometry by Wang et al. where they suggested that exposed 4f-Ti became oxidized to the fully-coordinated 6f-Ti [23]. The atomic geometry of their model is shown in **Figure 5**, where 5a and 5b correspond to the non-defected surface, while other images show different defect sites observed during the scanning tunneling microscopy (STM) analysis. In conclusion, they have shown that oxidized reconstruction is chemically inert and only reduced 4f-Ti sites show considerable activity.

However, despite a lot of attention being given to such (1 x 4) reconstruction, the actual geometry of the anatase (0 0 1) surface during growth and in aqueous suspensions is still probable to be unreconstructed. This fact is firstly justified by the fluorine stabilization of the (0 0 1) surface in its unreconstructed form, which is commonly used during the preparation procedure of such nanocrystals (details of this stabilization are described in the following section). After such preparation, fluorine has to be removed from the surface, as well an energy barrier needs to be overcome to induce the reconstruction. Both of these processes are known to occur in temperatures above 500–600°C, and below this temperature, reconstruction is not obvious [24–26]. Moreover, the possible stability of the unreconstructed geometry was experimentally confirmed by DeBenedetti et al. in the aqueous carboxylic acid solution [27]. Therefore, although (1 x 4) reconstruction of the anatase (0 0 1) surface should be minded, the actual geometry should be carefully considered, depending on the experimental details and available techniques.

Furthermore, different reconstructions were also proposed for other TiO_2 surfaces, and their general summation is presented in **Table 1**. Noticeably, a variety of rutile surfaces are expected to reconstruct, including (0 0 1), (0 1 1), and (1 0 0) terminations. Interestingly, these surfaces are also expected to be present in the equilibrium shape of the rutile crystal, as shown later. The (0 0 1) rutile surface is generally known to reconstruct or facet, especially in higher temperatures [28–30]. Commonly, the {0 1 1} and {1 1 4} facets are reported to form at the (0 0 1) surface, from which the {0 1 1} facets are also expected to adopt the geometry of the (2 x 1) reconstruction of the (0 1 1) surface itself. Detailed studies of this (2 x 1) reconstruction were presented by several authors, generally proposing a "brookite (0 0 1)-like" atomic structure [31–33]. This structure shows a characteristic topmost zigzag, composed of

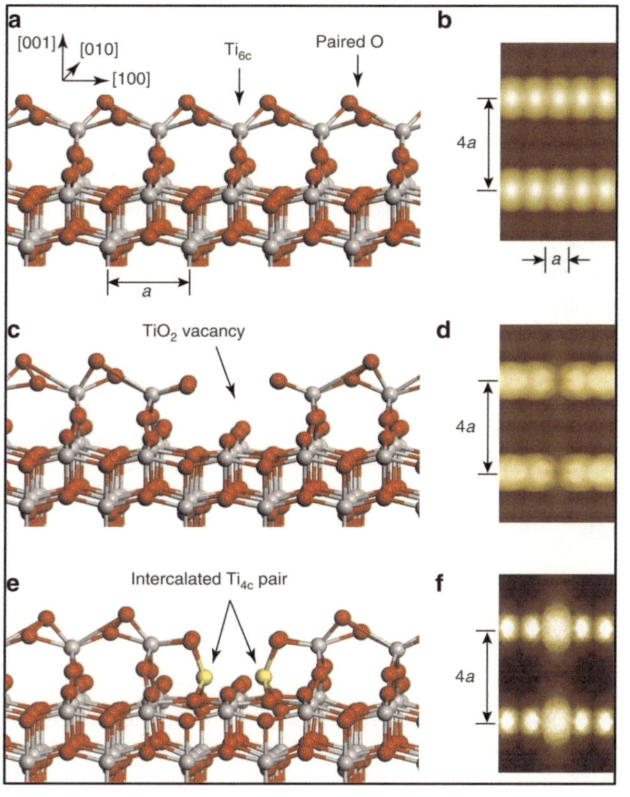

Figure 5.
Atomic models of the anatase (0 0 1) surface after the oxidized (1 x 4) reconstruction proposed by Wang et al. [23]. Non-defected (a) and defected (c, e) structures. Panels (b, d, and f) show corresponding simulated scanning tunneling microscope images. Please note that models (a, c, and e) are shown in parallel to the (1 x 4) periodicity of the reconstruction. Reprinted from [23] under a creative commons attribution 3.0 Unported license.

Polymorph and surface	Periodicity	Geometry	Specific formation conditions
Anatase (0 0 1)	(1 × 4)	Row of 6f-coordinated Ti above the bulk-like surface.	Ultra-high vacuum and annealing in 500–600°C.
Rutile (0 0 1)	Not discussed	{0 1 1} and {1 1 4} microfacets.	Annealing in 750–780°C.
Rutile (0 1 1)	(2 × 1)	Brookite (0 0 1)-like.	Not discussed.
Rutile (1 0 0)	(1 × 3), (1 × 5), (1 x 7)	{1 1 0} microfacets, partially reduced.	Annealing in the ultra-high vacuum.
Brookite (0 0 1)	(1 × 1)	Several bonds breaking between surface and subsurface species, topmost exposition of the 2f-O.	Not discussed.

Table 1.
Summation of the most important reported reconstructions of the TiO2 surfaces.

the 5f-Ti and 2f-O, observed in the STM images. As reported by Gong et al., surface energy of such a configuration was found to be $0.42\ J{\cdot}m^{-2}$, while the unreconstructed surface of about $0.89\ J{\cdot}m^{-2}$ is under the same computational parameters [33]. Furthermore, the (1 0 0) rutile surface can especially reconstruct after annealing in the ultra-high vacuum conditions, showing the formation of the {1 1 0} microfacets with several reconstruction patterns, such as a (1 x 3), (1 x 5), and (1 x 7). Commonly, the (1 x 3) reconstruction is considered, which was recently studied by Balzaretti et al. concerning its surface energy and interactions with water [34]. Interestingly, they observed that reconstructed geometry is slightly less stable (0.04 $J{\cdot}m^{-2}$ difference) than unreconstructed one. It is also noteworthy that photoemission experiments have shown that such annealed, reconstructed surface is partially reduced, which might influence its stability [29]. Finally, reports about brookite surfaces are relatively rare. Therefore, possible reconstructions of its geometries are also not discussed. Nevertheless, Gong and Selloni reported an important reconstruction of the (0 0 1) surface, which results in a reduction of its surface energy from 1.18 to 0.62 $J{\cdot}m^{-2}$, making it one of the most exposed surfaces in the equilibrium shape [15]. This structure is very similar to the proposed (2 x 1) reconstruction of the rutile (0 1 1) surface. In this reconstruction, the 3f-O atoms, that were originally bridging the 4f-Ti atoms (see **Figure 2a**), break their bonds with subsurface 6f-Ti and rise above the 4f-Ti. Simultaneously, two subsurface 3f-O break their bonds with surface 6f-Ti and move toward each other, ultimately locating below the 4f-Ti pair.

3. Facets stabilization, crystal shapes, and synthesis strategies

According to general thermodynamics, the total surface energy of the crystal should be minimized for the whole system to achieve minimum energy. Therefore, the exposition of some crystal planes is not expected. However, to consider such a lowest-energy shape, it is first necessary to consider the formation of surfaces equivalent to the specific model. For example, for the anatase polymorph, the (1 0 0), ($\bar{1}$ 0 0), (0 1 0), and (0 $\bar{1}$ 0) have the same atomic arrangement and are equivalent. Therefore, if the (1 0 0) surface became stable at the considered conditions, it is expected that four analogical facets will form in the crystal, with their orientation being the same as the orientation between the crystal planes. The family of such equivalent crystal planes is denoted using brackets and their Miller indices, which are also used to index observed crystal facets. The most important TiO_2 facets and their corresponding equivalent crystal planes are listed in **Table 2**.

Finally, after considering possible surface terminations, their energies, and equivalent planes, the 3-dimensional minimum-energy crystal shape can be obtained, according to the Gibbs-Curie-Wulff theorem [35]. This shape is also called the Wulff construction. Concerning the anatase, rutile, and brookite, their corresponding Wulff constructions are shown in **Figure 6**, according to the existing studies [5, 14, 15].

Constructions shown in **Figure 6** can be seen as a perfect case, and therefore, it is not unusual that experimentally obtained micro- or nanostructures can exhibit a variety of very different shapes. This results from two important aspects of the preparation procedure. Firstly, the adsorption of a different species can drastically change the energy of a final surface. Moreover, and more importantly, the relative energy of different surfaces might also change, leading to a situation where minimum-energy construction will expose a completely new set of different facets. Such energetic stabilization is often achieved by the addition of specific capping agents, pH control, or growing on a substrate. The second aspect is the kinetics of such growth. Particularly, very fast nucleation of the substrate can lead to the situation where final nanoparticles will not form a well-defined geometry, despite a thermodynamic preference to grow in some specific direction. In the case of TiO_2, this can typically be

Polymorph	Facet {*h k l*}	No. of equivalent crystal planes	Crystal planes
Anatase	{0 0 1}	2	(0 0 1), $(0\ 0\ \bar{1})$
	{1 0 0}	4	(1 0 0), $(\bar{1}\ 0\ 0)$, (0 1 0), $(0\ \bar{1}\ 0)$
	{1 0 *l*}	8	$(1\ 0\ l)$, $(0\ 1\ l)$, $(\bar{1}\ 0\ l)$, $(0\ \bar{1}\ l)$, $(1\ 0\ \bar{l})$, $(0\ 1\ \bar{l})$, $(\bar{1}\ 0\ \bar{l})$, $(0\ \bar{1}\ \bar{l})$
	{1 1 0}	4	(1 1 0), $(\bar{1}\ 1\ 0)$, $(1\ \bar{1}\ 0)$, $(\overline{1 1}\ 0)$
Rutile	{0 0 1}	2	(0 0 1), $(0\ 0\ \bar{1})$
	{0 1 1}	8	(0 1 1), (1 0 1), $(0\ \bar{1}\ 1)$, $(\bar{1}\ 0\ 1)$, $(0\ 1\ \bar{1})$, $(1\ 0\ \bar{1})$, $(0\ \bar{1}\ \bar{1})$, $(\bar{1}\ 0\ \bar{1})$
	{1 0 0}	4	(1 0 0), $(\bar{1}\ 0\ 0)$, (0 1 0), $(0\ \bar{1}\ 0)$
	{1 1 0}	4	(1 1 0), $(\bar{1}\ 1\ 0)$, $(1\ \bar{1}\ 0)$, $(\overline{1 1}\ 0)$
	{1 1 1}	8	(1 1 1), $(1\ \bar{1}\ 1)$, $(\bar{1}\ 1\ 1)$, $(\bar{1}\ \bar{1}\ 1)$, $(1\ 1\ \bar{1})$, $(1\ \bar{1}\ \bar{1})$, $(\bar{1}\ 1\ \bar{1})$, $(\bar{1}\ \bar{1}\ \bar{1})$
Brookite	{*h* 0 0}, {0 *k* 0}, {0 0 *l*}	2	Combination of all possible $\bar{h}$, $\bar{k}$, and $\bar{l}$
	{*h k* 0}, {*h* 0 *l*}, {0 *k l*}	4	
	{*h k l*}	8	

Table 2.
Summation of the most important TiO2 facets and their corresponding crystal planes.

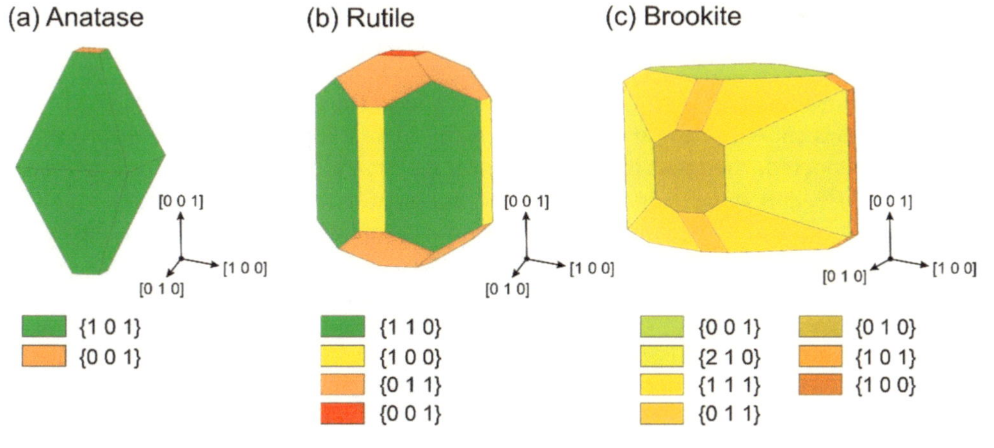

Figure 6.
Wulff constructions of the (a) anatase, (b) rutile, and (c) brookite TiO2 crystals, based on the existing studies [5, 14, 15]. Adapted colors generally follow the reported surface energy.

addressed using "dissolution-recrystallization" processes, where nucleated seeds are dynamically dissolved and then recrystallized in the rebuilt, stable crystal shape [36]. Alternatively, Ti-precursors, which will nucleate slower, can be used to achieve more stable growth. For example, when using Ti-alkoxides as precursors, it is known that a longer carbon chain should result in slower nucleation [37]. Ultimately, both kinetic and thermodynamic aspects should be considered when designing a synthesis route

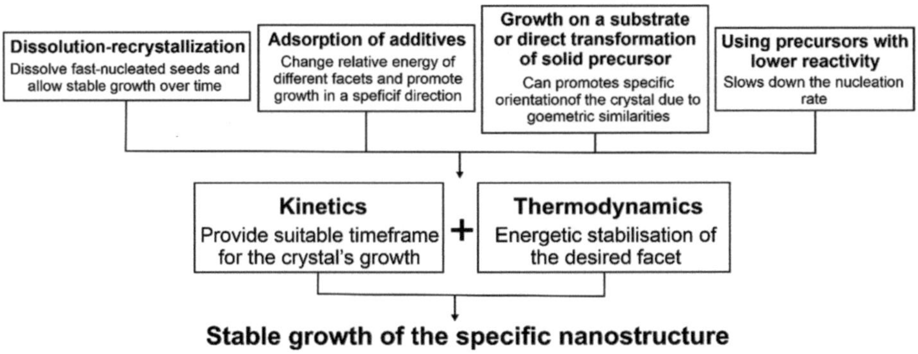

Figure 7.
Summation of the most important aspects and some strategies used for the preparation of faceted TiO2 nanostructures.

for a specific facet exposition. It is also noteworthy that multiple effects can be ascribed to the same additive. For example, during the HF-assisted growth of anatase nanostructures, hydrofluoric acid enables both dissolution of the TiO_2 nuclei and thermodynamically stabilized {0 0 1} facets [26]. A summation of general synthesis strategies is presented in **Figure 7**.

The hydrothermal process is usually used under specific synthesis conditions to obtain desired crystal shapes, considering the appropriate precursor, capping agent, and solvent. Various reaction substrates or ions can play a role as a capping agent in the synthesis. Therefore, in the next subsections, the most important synthesis strategies are described, together with highlighting the key factors.

3.1 Anatase crystal facets and shapes

The evolution of the anatase crystal shapes due to the exposition of different crystal facets is shown in **Figure 8**. Although the formation of a range of different facets was reported, the most investigated anatase crystal facets in the literature are {1 0 1}, {1 0 0}, and {0 0 1}, from which the first two facets are low energetic. Due to the symmetry of anatase crystal structure, nanocrystals with exposed {1 0 1} facets are octahedral, whereas the {0 0 1} facets form anatase nanosheets. Moreover, the combination of {1 0 1} and {0 0 1} facets (decahedral anatase nanostructures) and {0 0 1} and {1 0 0} (anatase cuboids) are also extensively studied. Decahedral nanocrystals may also undergo further flattening to nanosheets, resulting in dominant {0 0 1} facets. As reported by Barnard and Curtiss, the relative stability of these three facets depends heavily on the hydrogenation/oxygenation of the surface and, therefore, is strongly affected by the pH of the solution [11]. One of the consequences is that {1 0 1} and {1 0 0} exposing structures can be prepared in similar conditions, but {1 0 0} requires a higher pH. Alternatively, due to our best knowledge, the formation of the {0 0 1} exposing structures in the basic pH was not yet reported.

Anatase with exposed {1 0 1} facets is characterized by the lowest surface energy. Alone they form octahedral nanostructures, which synthesis methods are presented in **Table 3**. The most described procedure is a two-step synthesis, in which the first step is the fabrication of potassium titanate nanowires from a hydrothermal reaction of TiO_2 P25 (or other commercial TiO_2) in KOH solution. The second step is not always the same. For example, Amano et al. proposed direct hydrothermal treatment of titanate

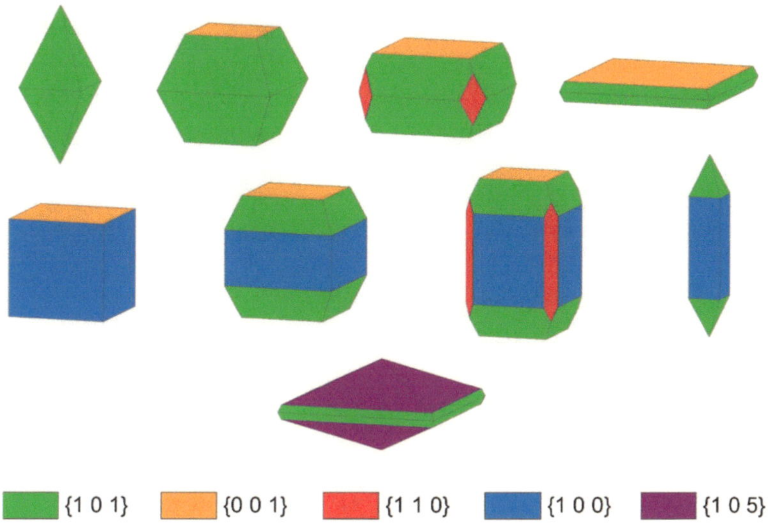

Figure 8.
The possible anatase crystal shapes resulting from the exposition of different crystal facets.

Precursor	Synthesis route and conditions	Comments	Ref.
Potassium titanate nanowires	Hydrothermal; 170°C, 24 hours	The proportion of regular octahedral bipyramids was ca. 70%	[38]
Potassium titanate nanowires, NH_4NO_3, HMTA	Hydrothermal; 200°C, 24 hours	Relative high yield of regular octahedral bipyramids >80%	[39]
$TiCl_4$, oleylamine	Synthesis in Schlenkline; quickly heating at 290°C	—	[40]
$TiOF_2$, hydrazine	Hydrothermal; 210°C, 24 hours	—	[41]
$Ti(SO_4)_2$, hydrazine	Hydrothermal; 200°C, 12 hours	The productivity of anatase octahedra was over 80%	[42]

Table 3.
Selected synthesis of octahedral anatase nanocrystals with exposed {1 0 1} facets.

nanowires. This synthesis resulted in the formation of mesoparticles with exposed {1 0 1} facets. The proportion of regular octahedral bipyramids in the nanostructures was about 70% [38]. A different route is the production of ammonium-exchanged titanate nanowires from $K_2Ti_6O_{13}$ precipitates obtained during the first step [39]. The above description considers mainly octahedral anatase nanostructures with the exposed {1 0 1} facets. However, according to Wang et al., these bipyramidal facets were obtained under hydrothermal conditions using potassium titanate nanowires as a precursor, hydrogen peroxide, and hydrofluoric acid as capping agents [43]. Alternatively, hydrazine-assisted formation of the {1 0 1} octahedrons was reported when starting from the precursors like $TiOF_2$ and $Ti(SO_4)_2$ [41, 42].

Another low-energetic anatase crystal facet is {1 0 0}, which synthesis methods are presented in **Table 4**. This crystal facet usually co-exists with other crystal facets and forms cuboids or rectangular prisms with truncated prisms [40, 41]. However, some

Precursor	Synthesis route and conditions	{1 0 0} facets exposition (%)	Comments	Ref.
$TiCl_4$, HCl, $NaBF_4$	Hydrothermal; 130°C, 12 hours	45	55% {0 0 1}	[44]
TiF_4, ethanol, oleic acid	Solvothermal; 200°C, 40 minutes	95	Anatase Nanosheets	[45]
Na-titanate	Hydrothermal; 120°C, 24 hours	Not mentioned	Nanorods	[46]
$TiOF_2$, NH_3	Hydrothermal; 210°C, 24 hours	Not mentioned	Co-exist with {1 0 1} and {0 0 1}	[41]

Table 4.
Selected synthesis of anatase nanocrystals with exposed {1 0 0} facets.

syntheses with a high percentage of {1 0 0} crystal were also successfully performed. For example, Xu et al. reported the preparation of anatase nanosheets with exposed {1 0 0} facets [44]. Previously, nanosheets and nanocrystals were combined with {0 0 1} facets exposition. The second possible shape of nanostructures is nanorod, which synthesis was described by Li and Xu. The precursor was sodium titanate obtained from a facile hydrothermal route during the reaction of P25 in sodium hydroxide solution. Anatase nanorods with {1 0 0} facets were transformed from Na-titanate *via* exchanging alkali-ions with protons under alkaline conditions to form the H-titanates [45].

The starting point for the investigation on the high-energetic anatase crystal facets was the research of Yang et al. [26], who proved that {0 0 1} facets could be energetically preferable to {1 0 1}, although the surface energy of {0 0 1} facets is, in general, higher than {1 0 1}. The main requirement was the addition of fluorine to the reaction environment. In these theoretical studies, among the surface termination using 12 elements (H, B, C, N, O, F, Si, P, S, Cl, Br, or I), only fluorine-terminated surface allowed to stabilize {0 0 1} facets rather than {1 0 1}. These calculations were completed by experiments in which anatase nanostructures with exposed {0 0 1} facets were successfully synthesized using the hydrothermal approach with hydrofluoric acid as a capping agent. However, in the above experiments, {0 0 1} facets accounted for only ~47% of all exposed crystal facets. In the meantime, Wen et al. [46] showed the synthesis of anatase nanocrystals with exposed {0 0 1} facets using 1-butanol as a solvent. This procedure allowed obtaining of large-sized well-defined anatase nanosheets wholly dominated with {0 0 1} and {1 0 0} facets, which had a percentage of 98.7% and 1.3%, respectively. The results can be explained by the alcohol stabilization effect associated with fluorine adsorption over the (0 0 1) surface. The role of particular alcohols, especially aliphatic with different chain lengths, was systematically studied recently [37, 47, 48].

The comparison of the anatase nanosheets with exposed {0 0 1} facets is presented in **Table 5**. In most studies concerning TiO_2 with exposed {0 0 1} facets, hydrofluoric acid was used in the experimental procedure. However, other fluoride-based reagents were also investigated. For example, ionic liquids (IL) were applied for stabilization of these high-energetic facets, e.g., 1-butyl-3-methylimidazolium hydrogen sulfate [Bmim]HSO_4 and 3-methyl-1-(3-sulfonyl propyl) imidazolium trifluoro methane sulfonate [$HO_3S(CH_2)_3$MIM][CF_3SO_3] [18]. Moreover, the fluorine atoms can be delivered by using an appropriate Ti source. An example of the compound in the Ti-O-F

Precursor and F source	Synthesis route and conditions	{0 0 1} facets exposition (%)	Comments	Ref.
TiF_4, HF, 1-butanol	Solvothermal; 210°C, 24 hours	98.7	Large size in length (ca. 4.14 μm).	[47]
TBT, HF	Hydrothermal; 250°C, 24 hours	75	—	[49]
TBT, HF, isobutyl alcohol	Solvothermal; 180–200°C, 20 hours	97	Flower-like structure with ca. 2.0 mm	[50]
Degussa P25, HF, H_2O_2	Hydrothermal; 180°C, 10 hours	65	Truncated bipyramidal anatase	[51]
$TiOF_2$	Calcination at 300–900°C, 2 hours	>83	—	[52]
TiF_4, diethylene glycol, acetic acid	Solvothermal; 180°C, 8 hours	>90	—	[53]
$HFTiO_3$, HF	Vapor-phase hydrothermal; 230°C, 3 hours	98.2	—	[54]
TiF_4, $[bmin]^+ [BF4]^-$	Microwave-assisted; 210°C, 1.5 hours	80	—	[55]
TiF_4, 1-Methylimidazolium tetrafluoroborate	Microwave-assisted; 210°C, 1.5 hours	50	—	[56]
$TiCl_4$, ethylene glycol	Solvothermal; 240°C, 48 hours	55	Hexagonal nanoplatelets	[57]

Table 5.
Selected synthesis of anatase nanocrystals with exposed {0 0 1} facets.

system is titanium oxyfluoride ($TiOF_2$), which can transform to TiO_2 *via* a solvothermal process [36] or simple calcination at a temperature above 600°C [58].

Moreover, next to {0 0 1} and {1 0 1} facets, a small rhombus originating from {1 1 0} facet exposition with a surface energy of about 1.09 $J \cdot m^{-2}$ can be formed. Liu et al. reported that the hydrothermal treatment of metallic Ti powder together with the combination of hydrofluoric acid and hydrogen peroxide allowed the formation of these highly energetic facets. It was described that HF was responsible for Ti dissolution, whereas H_2O_2 reacted with Ti^{4+} to obtain peroxotitanium acid. This complex slows down the hydrolysis rate, which is necessary to pack the Ti–O–Ti chains and finally form {1 1 0} facets [59]. Similar results were obtained by Li et al., who used $TiCl_3$ instead of Ti powder for the fabrication of {1 1 0} facets of anatase [49].

Excluding {1 1 0} facets, other synthesis procedures of anatase crystals with exposed high-index facets have been reported in the literature. For example, Xu et al. reported the synthesis of anatase single crystals with the exposed {1 1 1} facets. According to the density functional theory calculations, their surface energy was 1.61 $J{\cdot}m^{-2}$, which was explained by a high percentage of undercoordinated Ti and O atoms on the (1 1 1) surface [50]. Finally, Jiang et al. performed a gas-phase oxidation process using $TiCl_4$ as a precursor, which led to obtaining anatase single crystals with exposed high-index {1 0 5} facets. The regulation of the Ti/O ratio in the reaction system enabled inhibition of the growth of other crystal facets like {1 0 1} or {1 0 3}

[17]. Therefore, the research direction to obtain high-index crystal facets with relatively high surface energy is still under further investigation. However, there is a probability that non-equilibrium conditions and sophisticated synthesis will be necessary to preserve this surface.

3.2 Rutile and brookite crystal facets and shapes

Compared to anatase, preparation procedures of both rutile and brookite polymorphs are far less investigated. Moreover, observed nanocrystals are commonly not consistent with a theoretical Wulff construction. In the case of rutile, above all, this is due to the commonly observed appearance of the {1 1 1} facets after the hydrothermal processes. As discussed previously, the (1 1 1) surface in its bulk-terminated form is very energetic, and its stabilization can be achieved only through hydroxylation, as reported by Wang et al. [18]. Interestingly, their calculations suggested that after the hydroxylation, the surface energy of the {1 1 1} facets can be similar or even lower than the most stable {1 1 0}. Ultimately, commonly prepared nanocrystals expose a combination of these two facets [51, 52]. However, some authors report even 100% of the {1 1 1} exposition. For example, Wu et al. reported that wholly exposition of the {1 1 1} can be achieved with a suitably high addition of NaF to the reaction solution [53]. Truong et al. synthesized the rutile nanocrystals with unusual {3 3 1} facets [54]. Their preparation route was based on the solvothermal treatment of titanium-glycolate complex in the presence of picolinic acid as an additive. The resulting product possessed a specific aggregated flower-like structure with facets exposed along the (3 3 1) plane. Based on the detailed experimental investigation, it was further proposed that {3 3 1} facets are composed with the periodically repeating {1 1 0} and {1 1 1} microfacets. Other rutile crystals with a less common shape were synthesized by Chen and Lou, who have reported stabilization of the {0 0 1} rutile facets during the hydrothermal growth in the presence of amorphous MoO_3 [55]. The detailed procedure involved the hydrothermal treatment of mixed TiF_4 + HCl and $(NH_4)_6Mo_7O_{24}{\cdot}4H_2O$ + HNO_3 solutions for 5 hours at 180°C. The final product comprised nanosized rutile platelets with {0 0 1} facets exposed aggregated to approximately 500-nm diameter spheres. The summation of the observed rutile crystal shapes is presented in **Figure 9**.

The research on brookite with exposed facets is the most overlooked issue in facet engineering of TiO_2. Three main challenges can be distinguished: firstly, difficulties in obtaining pure brookite phase. Secondly, the pristine brookite is supposed to be

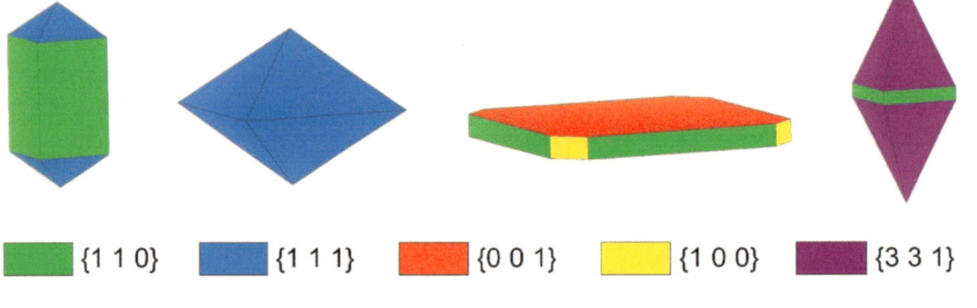

Figure 9.
Scheme of some of the experimentally observed rutile crystal shapes.

photocatalytic inactive. Finally, this TiO_2 polymorph is metastable and undergoes a thermal transition to rutile at high temperatures [56].

However, the rising investigations about brookite in recent years have led to the recognition of this metastable phase as an active photocatalyst. One of the first studies by Lin et al. reported single-crystalline nanosheets surrounded with four {2 1 0}, two {1 0 1}, and two {2 0 1} facets. These nanostructures exhibited higher photocatalytic performance toward methylene orange removal and hydroxyl radical production than commercial TiO_2 P25 [16]. Furthermore, {2 1 0} facets were also predominant in nanocrystals described by Xu et al., who demonstrated a tunable synthesis of brookite nanomaterials with the following shapes: quasi-octahedral, ellipsoid-tipped, and wedge-tipped nanorods [57]. The above results can be explained by the lowest surface energy of (2 1 0) surface among brookite crystals. However, Zhao et al. synthesized and investigated brookite nanostructures with exposed {1 2 1} and {2 1 1} facets. Particularly, TiO_2 with a majority of {1 2 1} facets exposition, which had many undercoordinated atoms on the surface and a lower VB potential, exhibited enhanced photocatalytic activity toward Rhodamine B degradation under simulated solar light. Therefore, the presented examples from the literature proved that crystal facets engineering is a promising approach to obtaining photocatalytic active material from the inactive phase [60].

4. Application in environmental photocatalysis

Following the ongoing demand for sustainable technologies, faceted TiO_2 nanocrystals are primarily studied as possible photocatalysts in various environmental remediation processes. Due to the relatively higher photocatalytic activity, the majority of these studies focus on the anatase polymorph; however, some interesting findings are also reported for rutile and brookite.

Concerning fundamental aspects of reactivity of different facets, it is often desired to compare pristine photocatalysts with the majority of one specific facet exposed. Comparison of their relative activities can give the so-called activity order of the investigated surfaces [61–63]. Initially, it was generally noticed that an increase of the {0 0 1} content on the anatase nanoparticles increases its photocatalytic activity both for water splitting and for degradation of organic pollutants [64–66]. This was straightforwardly connected with the high surface energy of the (0 0 1) surface, which was expected to provide a high density of potentially active sites for the photocatalytic reactions. However, further studies have presented opposite results, leading to a significant reexamination of the problem. For example, studies by Gordon et al. [67], Pan et al. [61], Mino et al. [68], and Mao et al. [63] have shown relatively low photocatalytic activity of the {0 0 1} facets in different reactions. An interesting study was also reported by Günnemann et al., who studied a variety of different TiO_2 surfaces cut from single crystal samples [69]. Their conclusions support observations of relatively lower photocatalytic activity of the anatase {0 0 1} facets, while {1 0 0} were the most active for ·OH generation (using terephthalic acid as a probe) and {1 0 1} showed the highest activity for methanol oxidation. Ultimately, these results showed that, at present, the photocatalytic activity of different crystal facets is hardly connected *a priori* with its surface energy or high density of undercoordinated species, as initially assumed. Instead, possible adsorption, detailed electronic interactions as well as density of charge trapping and transfer are further considered as crucial for the activity of a specific facet. This makes an overall problem very case-specific and due to our best knowledge, general conclusions are not possible to draw at this moment. Nevertheless, some of the recent details, key factors, and suggested mechanisms can be discussed for specific applications.

4.1 Water treatment from organic pollutants

Concerning photocatalytic degradation of organic pollutants, it is first noteworthy that these studies can be sub-categorized into 3 categories: degradation of dyes, degradation of non-color compounds, and generation of reactive oxygen species (ROS). Particularly, it should be minded that due to possible sensitization, dye degradation can be initiated by a different mechanism than other pollutants, therefore producing possibly different results. In this regard, it is not recommended to use dyes as a model pollutant, when assessing photocatalytic activity toward the degradation of organic compounds in general [70]. Here, we will focus on the reports and mechanisms discussing the degradation of photochemically inactive compounds and the generation of ROS, which is the main issue for current advanced oxidation technologies.

Concerning degradation of persistent pollutants and ROS generation, water, oxygen, and pollutant itself are the main substrates that can react at the photocatalyst surface. Usually, it is assumed that the process is initiated by the photogenerated holes (h^+) that can either oxidize the pollutant, inducing its further transformation, or produce $\cdot OH$ radicals from H_2O [71, 72]. Simultaneously, excited electrons are often expected to reduce oxygen to the $\cdot O_2^-$, which can also contribute to the final degradation rate; however, their reactivity is much lower than h^+ or $\cdot OH$ [73]. Based on this description, it could be expected that the photocatalyst with the highest photooxidation ability should achieve the highest degradation rates. Focusing on the anatase, this is in accordance with some of the reported studies showing that the {1 0 0} facets are highly active, especially concerning $\cdot OH$ generation [45, 69]. This is also in accordance with the simulations performed by Ma et al., who have shown that h^+ localization is the most favored on this facet, compared to the {0 0 1} and {1 0 1} [74]. However, many studies have also reported {1 0 1} facets to be the most photocatalytic active in the degradation process, which cannot be connected to higher h^+ reactivity on this surface. Moreover, our recent studies have shown that {1 0 1} facets revealed higher mineralization efficiency measured as a total organic carbon (TOC) removal during the phenol degradation process, independently of the degradation rate [41]. Interestingly, both of these facts can be attributed to the increased reduction ability of the {1 0 1} facets. First of all, while the reactivity of the $\cdot O_2^-$ is lower than h^+ or $\cdot OH$, they are good ring-opening agents, which might promote the efficient conversion of the aromatic compounds to CO_2 [75]. Moreover, possible multi-electron oxygen reduction can also promote the formation of the $\cdot OH$, as well as proton transfer from organic compounds to the adsorbed -OH groups, which might initiate the degradation. This problem was specifically investigated in detail for the anatase {1 0 1} facets, which have shown that a combination of O_2 and H_2O on the (1 0 1) surface results in the formation of surface -OH groups [76]. The process was especially favorable in the presence of two excess electrons in the reaction model, therefore connecting it with a possible 2-electron reduction. Importantly, this shows that on the reduced (1 0 1), H_2O can dissociate, forming the final -OH, which is not occurring spontaneously on the perfect surface. These findings have a fundamental meaning for the reactivity of the {1 0 1} facets, especially for water treatment processes, as the -OH groups are a preferable source of the $\cdot OH$ formation compared to H_2O itself [77]. This is in good agreement with a recent study by Hwang et al., who confirmed that a significant amount of free ·OH is formed through the oxygen reduction process, based on the $^{18}O_2$ incorporation into the product [78]. Although utilized samples were not strictly faceted during this study, the exposition of the {1 0 1} structures might be expected due to their energetic stability. Furthermore, a recent study by Dudziak et al. showed that a very good correlation could be observed between the activity of the {1 0 1} enclosed anatase samples for degradation of aromatic compounds and higher probability of both h^+ and e^- trapping on these facets, compared to the {0 0 1} and {1 0 0} ones [79]. The combination of these studies suggests a possible mechanism of the {1 0 1} reactivity as the result of e^- induced H_2O dissociation and further generation of ·OH with photogenerated holes. However, more detailed studies might still be needed to clarify it. Finally, recent reports have also shown that the application of nanostructures exposing {1 0 1} facets might result in lower toxicity of the final solution, during the naproxen degradation process, than {0 0 1} ones [80]. Therefore, at this moment, a combination of high reactivity, high TOC removal, and low toxicity makes {1 0 1} a preferable choice for the degradation of organic compounds, especially micropollutants with aromatic structure and high photostability.

Compared to the anatase facets, {1 0 1} in particular, other TiO_2 structures are not studied in detail and generally show markedly lower photoactivity. Nevertheless, a few important findings are worth noticing. First of all, the same reductive pathway of ·OH generation, reported for anatase by Hwang et al., was not observed for rutile, suggesting that especially for the {1 1 0} rutile facets, ·OH generation occurs strictly through H_2O oxidation [78]. Furthermore, the study by Günnemann et al. showed different activities of rutile {0 0 1}, {0 1 1}, and {1 1 1} facets in the methanol oxidation and ·OH generation. Specifically, they reported that {0 1 1} facets exhibited lower methanol oxidation ability, while {1 1 1} generated lower amounts of hydroxyl radicals. Besides, rutile activity in both reactions was fairly similar and generally worse than anatase [69]. Nevertheless, their study did not consider rutile {1 1 0} facets, which on the other hand, were studied by Kobayashi et al. for the oxidation of oxalic acid [52]. In this study, {1 1 0} revealed higher activity than {0 0 1}, which is also in some agreement with the oxidative ·OH generation by this facet. Finally, concerning brookite facets, it is especially worth mentioning that structures co-exposing {2 1 0}, {2 0 1}, and {1 0 1} facets result in significant activity increase for ·OH generation and methyl orange degradation, otherwise not observed for the control brookite samples [16].

4.2 Solar fuel production

In recent years, semiconductor-based materials have been extensively studied for energy applications that can contribute to reducing greenhouse gases. Storing solar energy into the chemical bonds of fuel seems to be a promising way to replace the traditional combustion of fossil fuels with environmentally friendly technology. Specifically, much attention has been paid to photocatalytic H_2 production from water and CO_2 reduction to valuable chemical compounds.

Regarding H_2 generation from water, it should be noted that pristine structures show low activity, and therefore, surface modification with co-catalysts needs to be applied. In the case of the faceted particles, the overall problem deals with specific interactions between the surface and co-catalyst, as well as possible charge separation between different co-exposed facets. Focusing on the single-facet, important findings were reported by Gordon et al., who noticed the higher activity of the anatase modified with Pt for octahedrons exposing {1 0 1} than nanosheets with exposed {0 0 1}

[67]. Similar results were obtained by Wang et al. for anatase nanoparticles modified with Mo_XC as a co-catalyst [81]. This fact is attributed to the increased reduction ability of the (1 0 1) surface, which results in the synergy of TiO_2 and co-catalyst. Recently, specifically, the problem of interactions between (1 0 1) and different possible metal co-catalyst was investigated in detail by Wang and Gong in their computational study [82]. Based on the obtained results, they have proposed alloyed Cu/Pt and Rh/Pt co-catalysts as promising candidates for hydrogen evolution. This concept was based on optimizing the electron transfer between (1 0 1) anatase surface and Cu or Rh as the electron-acceptor and further exposition of Pt as the active part of the co-catalyst. Considered models and their electronic structures are shown in **Figure 10**.

Furthermore, a combination of {1 0 1} with other co-exposed facets can increase the activity of the TiO_2 materials for H_2 generation. For example, Wei et al. presented a detailed comparison between octahedral {1 0 1} and decahedral {1 0 1}/{0 0 1} anatase particles modified with Cu, Ag, and Au nanoparticles. Particularly, a combination of both {1 0 1} and {0 0 1} facets resulted in a slightly higher activity when modified with Ag or Au, as well as a significantly higher activity when modified with Cu [83]. Furthermore, Meng et al. reported increased H_2 production using the decahedral {1 0 1}/{0 0 1} anatase samples, when both facets were selectively modified by Pt and Co_3O_4, respectively [84]. It is especially noteworthy that such a combination of the facet co-exposition and selective modification with optimized co-catalysts was recently proposed to achieve almost 100% of quantum efficiency during water splitting reaction over $SrTiO_3$ photocatalyst [4]. Therefore, it confirms the

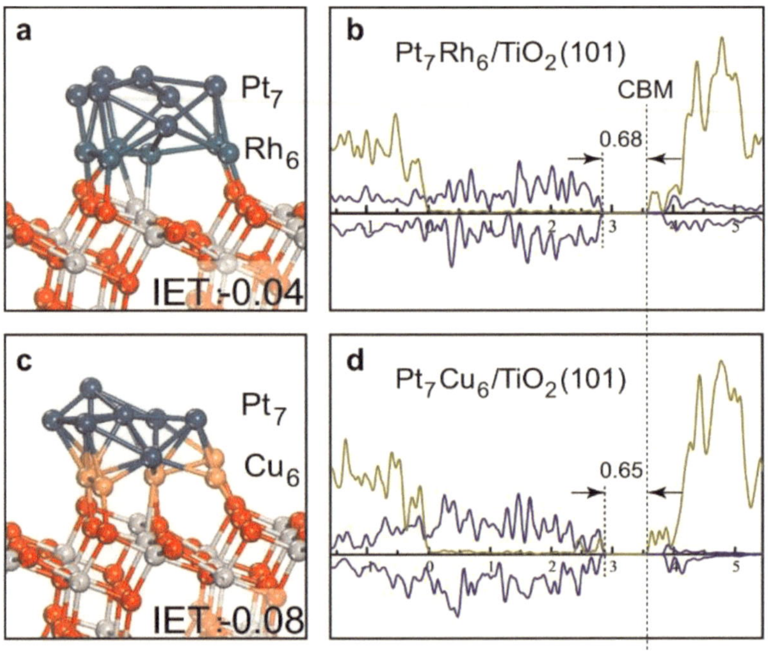

Figure 10.
Optimal models (a, c) for electron transfer between TiO2 (1 0 1) anatase surface and alloyed metal cocatalysts, proposed by Wang and Gong and their corresponding density of states distribution (b, d). In images (b, d), blue line shows the states of the metal cluster, while yellow-green is TiO2. Reprinted from the [92] under a creative commons attribution 4.0 international license. IET refers to the energy of intrinsic electron transfer in eV.

importance of optimizing facet-facet exposition and further facet-co-catalyst interactions to optimize the final performance.

Furthermore, recent studies also focus on the photocatalytic reduction of carbon (IV) oxide to valuable chemical compounds. This reaction begins with the adsorption of CO_2 and H_2O molecules, which was investigated theoretically by Mishra and Nanda. Using DFT calculations, they examined the chemical restructuring of CO_2 and H_2O molecules during the process of adsorption, co-adsorption, and conversion on (0 0 1), (1 0 0), and (1 0 1) surfaces [85]. They observed that the energy barrier of bicarbonate complex formation, which resulted from the co-adsorption of carbon dioxide and water, was the lowest for the (0 0 1) surface. Therefore, {0 0 1} facets are supposed to be the most reactive anatase facets for CO_2 photocatalytic reduction. However, if this surface undergoes reconstruction, the number of active sites is reduced. Therefore, experimental conditions like temperature and high vacuum will be crucial for the photocatalytic performance of anatase nanocrystals.

Although the photocatalytic reaction depends on the adsorption of reactants, the investigations provided by Ma et al., in the application of CO_2 reduction to formic acid, showed different behavior of anatase crystal facets compared with previous studies [86]. The surface electron transfer for (0 0 1) and (1 0 1) surfaces was characterized by similar barrier levels. However, the reductive ability of electrons generated on the (1 0 1) plane is higher than that on the (0 0 1) plane; therefore, electrons may be transferred more easily to reactants for low-energetic facets. Moreover, HCOOH on the (0 0 1) surface can replace water and, in consequence, occupy the active sites, hindering the reaction. On the contrary, formic acid seemed to remain undissociated on (1 0 1) surfaces, so more suitable product adsorption properties led to a higher photocatalytic performance.

The above-reported studies were theoretical, so the experimental results may not be consistent with DFT calculations. Therefore, Liu et al. demonstrated the blue anatase nanocrystals with exposed {0 0 1}, {1 0 1}, and a combination of {1 0 1} and {0 0 1} facets [87]. They reported that oxygen-deficient TiO_2 nanostructures with co-exposed {1 0 1} and {0 0 1} facets exhibited relatively high quantum yield for CO_2 reduction to CO (0.31% under UV–vis light and 0.134% under visible light). Moreover, this photocatalyst demonstrated more than four times higher visible light activity in comparison with {0 0 1} or {1 0 1}. This high photocatalytic activity was a result of two effects. Firstly, co-exposed {0 0 1} and {1 0 1} facets had increased the capacity of reversible CO_2 adsorption. Secondly, the created surface junction between facets enhanced the charge separation and hindered the recombination processes. Similar results were obtained by Yu et al., who investigated the mist-efficient content ratio of {0 0 1} and {1 0 1} facets [88]. The decahedral-shaped sample with 58% content of {0 0 1} facets exhibited the highest methane production from CO_2. {1 0 1} facets acted as reduction sites, whereas {0 0 1} facets were the oxidation sites on the photocatalyst surface. However, a too high amount of the {0 0 1} facets on the anatase surface may have caused an electron overflow effect toward {1 0 1} facets, so the migration of electrons to {1 0 1} facets is more difficult than in the previous case.

Carbon dioxide may be further converted to methane, which generally gives rise to operational risks and environmental problems [89, 90]. Therefore, selective oxidation to CH_3OH is a promising way to CH_4 storage. Feng et al. reported the facet-dependent selectivity of $CH_4 \rightarrow CH_3OH$ conversion over anatase nanocrystals. They showed that silver-decorated TiO_2 with predominant {0 0 1} facets exhibited a selectivity of approximately 80%, which was significantly better than the sample with dominated {1 0 1} facets. This high selectivity resulted from oxygen vacancy generation by

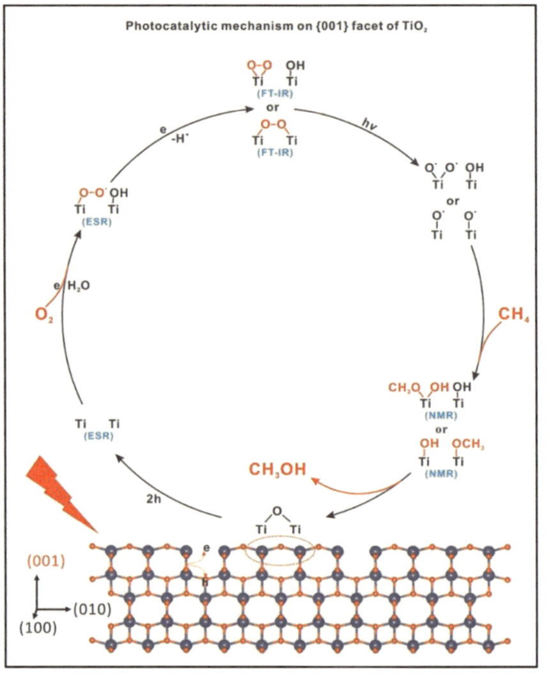

Figure 11.
The proposed mechanism of methane photocatalytic conversion over {0 0 1} anatase crystal facet. Reprinted from [101] under a creative commons attribution 4.0 international license.

photoinduced holes, which played a crucial role in avoiding the formation of $\cdot CH_3$ and $\cdot OH$ radicals. Therefore, the undesired overoxidation to CO was limited, in opposite to TiO_2 exposing {1 0 1} facets [91]. The proposed mechanism of CH_4 oxidation on the {0 0 1} facets is presented in **Figure 11**.

5. Conclusions

The surface structure of TiO_2 is constantly being recognized as a crucial factor regarding its photocatalytic activity. Control over this interface can be achieved by exposing the specific crystal facets, which allows detailed insight into the reaction kinetics and mechanism. Experimentally, such high-quality structures can be prepared through energetic stabilization of the specific facet, with simultaneous control of nucleation and growth rates. However, contrary to the preparation procedures, surface energy is now recognized as a less important factor for the final photocatalytic activity. Here, more attention is given to the electronic properties, detailed interactions with substrates, and local defects. However, as the presented chapter aimed to provide a general introduction to the problem of TiO_2 facets, only some of the recent findings were stressed here, while these problems are still under constant intensive research. Particularly, it should be highlighted that further modifications and charge separation between different co-exposed facets are now used to achieve remarkable final photocatalytic efficiency of different materials. In this regard, we believe that the presented findings will be useful for further studies in this direction, providing a survey of different aspects of crystal facet engineering of TiO_2, a photocatalyst that is still the most studied in this field.

Acknowledgements

The research was financially supported by the Polish National Science Centre grant no. NCN 2021/43/B/ST5/02983.

Author details

Szymon Dudziak, Marta Kowalkińska and Anna Zielińska-Jurek*
Department of Process Engineering and Chemical Technology, Gdansk University of Technology, Gdansk, Poland

*Address all correspondence to: annjurek@pg.edu.pl

References

[1] Gerischer H, Heller A. The role of oxygen in photooxidation of organic molecules on semiconductor particles. The Journal of Physical Chemistry. 1991; **95**(13):5261-5267

[2] Wang CM, Heller A, Gerischer H. Palladium catalysis of O_2 reduction by electrons accumulated on TiO2 particles during Photoassisted oxidation of organic compounds. Journal of the American Chemical Society. 1992; **114**(13):5230-5234

[3] Moser J, Punchihewa S, Infelta PP, Grätzel M. Surface complexation of colloidal semiconductors strongly enhances interfacial electron-transfer rates. Langmuir. 1991;**7**(12): 3012-3018

[4] Takata T, Jiang J, Sakata Y, Nakabayashi M, Shibata N, Nandal V, et al. Photocatalytic water splitting with a quantum efficiency of almost unity. Nature. 2020;**581**(7809): 411-414

[5] Lazzeri M, Vittadini A, Selloni A. Structure and energetics of stoichiometric TiO_2 anatase surfaces. Physical Review B - Condensed Matter and Materials Physics. 2001;**63**(15): 1554091-1554099

[6] Zhao Z, Li Z, Zou Z. Surface properties and electronic structure of low-index stoichiometric anatase TiO_2 surfaces. Journal of Physics. Condensed Matter. 2010;**22**(17):175008

[7] Arrouvel C, Digne M, Breysse M, Toulhoat H, Raybaud P. Effects of morphology on surface hydroxyl concentration: A DFT comparison of anatase-TiO_2 and γ-alumina catalytic supports. Journal of Catalysis. 2004; **222**(1):152-166

[8] Lazzeri M, Selloni A. Stress-driven reconstruction of an oxide surface: The anatase TiO_2 (001)-(1×4) surface. Physical Review Letters. 2001;**87**(26): 266105-1-266105-4

[9] Mino L, Ferrari AM, Lacivita V, Spoto G, Bordiga S, Zecchina A. CO adsorption on anatase nanocrystals: A combined experimental and periodic DFT study. Journal of Physical Chemistry C. 2011;**115**(15): 7694-7700

[10] Labat F, Baranek P, Adamo C. Structural and electronic properties of selected rutile and anatase TiO_2 surfaces: An ab initio investigation. Journal of Chemical Theory and Computation. 2008;**4**(2):341-352

[11] Barnard AS, Curtiss LA. Prediction of TiO_2 nanoparticle phase and shape transitions controlled by surface chemistry. Nano Letters. 2005;**5**(7): 1261-1266

[12] Perron H, Domain C, Roques J, Drot R, Simoni E, Catalette H. Optimisation of accurate rutile TiO_2 (110), (100), (101) and (001) surface models from periodic DFT calculations. Theoretical Chemistry Accounts. 2007; **117**(4):565-574

[13] Jiang F, Yang L, Zhou D, He G, Zhou J, Wang F, et al. First-principles atomistic Wulff constructions for an equilibrium rutile TiO_2 shape modeling. Applied Surface Science. 2018;**436**: 989-994

[14] Ramamoorthy M, Vanderbilt D, King-Smith RD. First-principles calculations of the energetics of stoichiometric TiO_2 surfaces. Physical Review B. 1994;**49**(23): 16721-16727

[15] Gong XQ, Selloni A. First-principles study of the structures and energetics of stoichiometric brookite TiO_2 surfaces. Physical Review B. Condensed Matter and Materials Physics. 2007;**76**(23):1-11

[16] Lin H, Li L, Zhao M, Huang X, Chen X, Li G, et al. Synthesis of high-quality brookite TiO_2 single-crystalline nanosheets with specific facets exposed: Tuning catalysts from inert to highly reactive. Journal of the American Chemical Society. 2012;**134**(20): 8328-8331

[17] Jiang HB, Cuan Q, Wen CZ, Xing J, Wu D, Gong XQ, et al. Anatase TiO_2 crystals with exposed high-index facets. Angewandte Chemie - International Edition. 2011;**50**(16):3764-3768

[18] Wang Y, Sun T, Liu X, Zhang H, Liu P, Yang H, et al. Geometric structure of rutile titanium dioxide (111) surfaces. Physical Review B. Condensed Matter and Materials Physics. 2014;**90**(4):1-6

[19] Zhou G, Jiang L, Dong Y, Li R, He D. Engineering the exposed facets and open-coordinated sites of brookite TiO_2 to boost the loaded Ru nanoparticle efficiency in benzene selective hydrogenation. Applied Surface Science. 2019;**486**:187-197

[20] Hengerer R, Bolliger B, Erbudak M, Grätzel M. Structure and stability of the anatase TiO_2 (101) and (001) surfaces. Surface Science. 2000;**460**(1–3):162-169

[21] Liang Y, Gan S, Chambers SA, Altman EI. Surface structure of anatase (001) reconstruction, atomic steps, and domains. Physical Review B. Condensed Matter and Materials Physics. 2001; **63**(23):1-7

[22] Herman GS, Sievers MR, Gao Y. Structure determination of the two-Domain (1 x 4) Anatase TiO_2 (001) surface. Physical Review Letters. 2000; **84**:3354

[23] Wang Y, Sun H, Tan S, Feng H, Cheng Z, Zhao J, et al. Role of point defects on the reactivity of reconstructed anatase titanium dioxide (001) surface. Nature Communications. 2013;**4**:1-8

[24] Selçuk S, Selloni A. Surface structure and reactivity of anatase TiO_2 crystals with dominant {001} facets. Journal of Physical Chemistry C. 2013;**117**(12): 6358-6362

[25] Yuan W, Wang Y, Li H, Wu H, Zhang Z, Selloni A, et al. Real-time observation of reconstruction dynamics on TiO_2 (001) surface under oxygen via an environmental transmission electron microscope. Nano Letters. 2016;**16**(1): 132-137

[26] Yang HG, Sun CH, Qiao SZ, Zou J, Liu G, Smith SC, et al. Anatase TiO_2 single crystals with a large percentage of reactive facets. Nature. 2008;**453**(7195): 638-641

[27] DeBenedetti WJI, Skibinski ES, Jing D, Song A, Hines MA. Atomic-scale understanding of catalyst activation: Carboxylic acid solutions, but not the acid itself, increase the reactivity of Anatase (001) faceted Nanocatalysts. Journal of Physical Chemistry C. 2018; **122**(8):4307-4314

[28] Wang Y, Lee S, Vilmercati P, Lee HN, Weitering HH, Snijders PC. Atomically flat reconstructed rutile TiO_2 (001) surfaces for oxide film growth. Applied Physics Letters. 2016;**108**(9): 091604

[29] Tait RH, Kasowski RV. Ultraviolet photoemission and low-energy-electron diffraction studies of TiO_2 (rutile) (001) and (110) surfaces. Physical Review B. 1979;**20**(12):5178-5191

[30] Firment LE. Thermal faceting of the rutile TiO_2 (001) surface. Surface Science. 1982;**116**(2):205-216

[31] Zhou R, Li D, Qu B, Sun X, Zhang B, Zeng XC. Rutile TiO_2 (011)-2 × 1 reconstructed surfaces with optical absorption over the visible light Spectrum. ACS Applied Materials & Interfaces. 2016;**8**(40):27403-27410

[32] Wu L, Wang Z, Xiong F, Sun G, Chai P, Zhang Z, et al. Surface chemistry and photochemistry of small molecules on rutile TiO_2 (001) and TiO_2 (011)-(2 × 1) surfaces: The crucial roles of defects. The Journal of Chemical Physics. 2020;**152**:044702

[33] Gong XQ, Khorshidi N, Stierle A, Vonk V, Ellinger C, Dosch H, et al. The 2 × 1 reconstruction of the rutile TiO2(0 1 1) surface: A combined density functional theory, X-ray diffraction, and scanning tunneling microscopy study. Surface Science. 2009;**603**:138-144

[34] Balzaretti F, Gupta V, Ciacchi LC, Aradi B, Frauenheim T, Köppen S. Water reactions on reconstructed rutile TiO_2: A density functional theory/density functional tight binding approach. Journal of Physical Chemistry C. 2021;**125**(24):13234-13246

[35] Wulff G. Zur Frage der Geschwindigkeit des Wachsthums und der Auflösung der Krystallflächen. Zeitschrift für Krystallographie und Mineralogie. 1901;**34**:449-530

[36] Huang Z, Wang Z, Lv K, Zheng Y, Deng K. Transformation of $TiOF_2$ cube to a hollow nanobox assembly from anatase TiO_2 nanosheets with exposed {001} facets via solvothermal strategy. ACS Applied Materials & Interfaces. 2013;**5**(17):8663-8669

[37] Dudziak S, Kowalkińska M, Karczewski J, Pisarek M, Siuzdak K, Kubiak A, et al. Solvothermal growth of {0 0 1} exposed anatase nanosheets and their ability to mineralize organic pollutants. The effect of alcohol type and content on the nucleation and growth of TiO_2 nanostructures. Applied Surface Science. 2021;**563**:150360

[38] Amano F, Yasumoto T, Prieto-Mahaney OO, Uchida S, Shibayama T, Ohtani B. Photocatalytic activity of octahedral single-crystalline mesoparticles of anatase titanium(IV) oxide. Chemical Communications. 2009;**17**:2311-2313

[39] Li J, Yu Y, Chen Q, Li J, Xu D. Controllable synthesis of TiO_2 single crystals with tunable shapes using ammonium-exchanged titanate nanowires as precursors. Crystal Growth & Design. 2010;**10**(5):2111-2115

[40] Lai Z, Peng F, Wang Y, Wang H, Yu H, Liu P, et al. Low temperature solvothermal synthesis of anatase TiO_2 single crystals with wholly {100} and {001} faceted surfaces. Journal of Materials Chemistry. 2012;**22**(45):23906-23912

[41] Kowalkińska M, Dudziak S, Karczewski J, Ryl J, Trykowski G, Zielińska-Jurek A. Facet effect of TiO_2 nanostructures from $TiOF_2$ and their photocatalytic activity. Chemical Engineering Journal. 2021;**404**:126493

[42] Gai L, Mei Q, Qin X, Li W, Jiang H, Duan X. Controlled synthesis of anatase TiO_2 octahedra with enhanced photocatalytic activity. Materials Research Bulletin. 2013;**48**:4469-4475

[43] Wang F, Sun L, Li Y, Zhan W, Wang X, Han X. Hollow Anatase TiO_2 Octahedrons with Exposed High-Index

{102} Facets for Improved Dye-Sensitized Photoredox Catalysis Activity. Inorganic Chemistry. 2018;**57**(8): 4550-4555

[44] Xu H, Ouyang S, Li P, Kako T, Ye J. High-active anatase TiO_2 nanosheets exposed with 95% {100} facets toward efficient H_2 evolution and CO_2 photoreduction. ACS Applied Materials & Interfaces. 2013;**5**(4):1348-1354

[45] Li J, Xu D. Tetragonal faceted-nanorods of anatase TiO_2 single crystals with a large percentage of active {100} facets. Chemical Communications. 2010; **46**(13):2301-2303

[46] Wen CZ, Zhou JZ, Jiang HB, Hu QH, Qiao SZ, Yang HG. Synthesis of micro-sized titanium dioxide nanosheets wholly exposed with high-energy {001} and {100} facets. Chemical Communications. 2011;**47**(15): 4400-4402

[47] Yang HG, Liu G, Qiao SZ, Sun CH, Jin YG, Smith SC, et al. Solvothermal synthesis and photoreactivity of anatase TiO2 nanosheets with dominant {001} facets. Journal of the American Chemical Society. 2009;**131**(11):4078-4083

[48] Zheng Y, Wang J, Yang P. Anatase TiO_2 nanosheets exposed {001} facet: Solvent effects on the photocatalytic performance. Journal of Nanoscience and Nanotechnology. 2017;**17**(2): 1204-1209

[49] Li Q, Li T, Chang S, Tao Q, Tian B, Zhang J. Enlarging {110} exposed facets of anatase TiO_2 by the synergistic action of capping agents. Cryst Eng Comm. 2016;**18**(27):5074-5078

[50] Xu H, Reunchan P, Ouyang S, Tong H, Umezawa N, Kako T, et al. Anatase TiO_2 single crystals exposed with high-reactive {111} facets toward efficient H_2 evolution. Chemistry of Materials. 2013;**25**(3):405-411

[51] Lai Z, Peng F, Wang H, Yu H, Zhang S, Zhao H. A new insight into regulating high energy facets of rutile TiO_2. Journal of Materials Chemistry A. 2013;**1**(13):4182-4185

[52] Kobayashi M, Petrykin V, Kakihana M, Tomita K. Hydrothermal synthesis and photocatalytic activity of whisker-like rutile-type titanium dioxide. Journal of the American Ceramic Society. 2009;**92**:S21-S26

[53] Wu T, Kang X, Kadi MW, Ismail I, Liu G, Cheng HM. Enhanced photocatalytic hydrogen generation of mesoporous rutile TiO_2 single crystal with wholly exposed {111} facets. Chinese Journal of Catalysis. 2015;**36**: 2103-2108

[54] Truong QD, Hoa HT, Le TS. Rutile TiO_2 nanocrystals with exposed {3 3 1} facets for enhanced photocatalytic CO_2 reduction activity. Journal of Colloid and Interface Science. 2017;**504**:223-229

[55] Chen JS, Lou XW. Unusual rutile TiO_2 nanosheets with exposed (001) facets. Chemical Science. 2011;**2**(11): 2219-2223

[56] Perego C, Wang YH, Durupthy O, Cassaignon S, Revel R, Jolivet JP. Nanocrystalline brookite with enhanced stability and photocatalytic activity: Influence of lanthanum (III) doping. ACS Applied Materials & Interfaces. 2012;**4**(2):752-760

[57] Xu Y, Lin H, Li L, Huang X, Li G. Precursor-directed synthesis of well-faceted brookite TiO_2 single crystals for efficient photocatalytic performances. Journal of Materials Chemistry A. 2015; **3**(44):22361-22368

[58] Shi T, Duan Y, Lv K, Hu Z, Li Q, Li M, et al. Photocatalytic oxidation of acetone over high thermally stable TiO_2 nanosheets with exposed (001) facets. Frontiers in Chemistry. 2018;**6**(MAY): 1-10

[59] Liu M, Piao L, Zhao L, Ju S, Yan Z, He T, et al. Anatase TiO_2 single crystals with exposed {001} and {110} facets: Facile synthesis and enhanced photocatalysis. Chemical Communications. 2010;**46**(10): 1664-1666

[60] Zhao M, Xu H, Chen H, Ouyang S, Umezawa N, Wang D, et al. Photocatalytic reactivity of {121} and {211} facets of brookite TiO_2 crystals. Journal of Materials Chemistry A. 2015; **3**(5):2331-2337

[61] Pan J, Liu G, Lu GQ, Cheng HM. On the true photoreactivity order of {001}, {010}, and {101} facets of anatase TiO_2 crystals. Angewandte Chemie - International Edition. 2011;**50**(9): 2133-2137

[62] Ye L, Mao J, Peng T, Zan L, Zhang Y. Opposite photocatalytic activity orders of low-index facets of anatase TiO_2 for liquid phase dye degradation and gaseous phase CO_2 photoreduction. Physical Chemistry Chemical Physics. 2014;**16**(29):15675-15680

[63] Mao J, Ye L, Li K, Zhang X, Liu J, Peng T, et al. Pt-loading reverses the photocatalytic activity order of anatase TiO_2 {001} and {010} facets for photoreduction of CO_2 to CH_4. Applied Catalysis B: Environmental. 2014;**144**: 855-862

[64] Li M, Chen Y, Li W, Li X, Tian H, Wei X, et al. Ultrathin anatase TiO_2 nanosheets for high-performance photocatalytic hydrogen production. Small. 2017;**13**(16):1604115

[65] Xiang Q, Lv K, Yu J. Pivotal role of fluorine in enhanced photocatalytic activity of anatase TiO_2 nanosheets with dominant (0 0 1) facets for the photocatalytic degradation of acetone in air. Applied Catalysis B: Environmental. 2010;**96**(3–4):557-564

[66] Wu Q, Liu M, Wu Z, Li Y, Piao L. Is photooxidation activity of {001} facets truly lower than that of {101} facets for anatase TiO_2 crystals? Journal of Physical Chemistry C. 2012;**116**(51):26800-26804

[67] Gordon TR, Cargnello M, Paik T, Mangolini F, Weber RT, Fornasiero P, et al. Nonaqueous synthesis of TiO_2 nanocrystals using TiF_4 to engineer morphology, oxygen vacancy concentration, and photocatalytic activity. Journal of the American Chemical Society. 2012;**134**(15): 6751-6761

[68] Mino L, Pellegrino F, Rades S, Radnik J, Hodoroaba VD, Spoto G, et al. Beyond Shape Engineering of TiO_2 Nanoparticles: Post-Synthesis Treatment Dependence of Surface Hydration, Hydroxylation, Lewis Acidity and Photocatalytic Activity of TiO2 Anatase Nanoparticles with Dominant {001} or {101} Facets. ACS Applied Nano Materials. 2018;**1**(9):5355-5365

[69] Günnemann C, Haisch C, Fleisch M, Schneider J, Emeline AV, Bahnemann DW. Insights into different photocatalytic oxidation activities of Anatase, Brookite, and rutile single-crystal facets. ACS Catalysis. 2019;**9**(2): 1001-1012

[70] Amalia FR, Takashima M, Ohtani B. Are you still using organic dyes? Colorimetric formaldehyde analysis for true photocatalytic-activity evaluation. Chemical Communications. 2022;**58**: 11721-11724

[71] Agrios AG, Pichat P. Recombination rate of photogenerated charges versus surface area: Opposing effects of TiO_2 sintering temperature on photocatalytic removal of phenol, anisole, and pyridine in water. Journal of Photochemistry and Photobiology A: Chemistry. 2006;**180**(1–2):130-135

[72] Lv K, Guo X, Wu X, Li Q, Ho W, Li M, et al. Photocatalytic selective oxidation of phenol to produce dihydroxybenzenes in a TiO_2/UV system: Hydroxyl radical versus hole. Applied Catalysis B: Environmental. 2016;**199**:405, 405-411, 411. DOI: 10.1016/j.apcatb.2016.06.049

[73] Bahnemann DW, Hilgendorff M, Memming R. Charge carrier dynamics at TiO_2 particles: Reactivity of free and trapped holes. The Journal of Physical Chemistry. A. 1997;**101**(21):4265-4275

[74] Ma X, Dai Y, Guo M, Huang B. Relative photooxidation and photoreduction activities of the {100}, {101}, and {001} Surfaces of Anatase TiO_2. Langmuir. 2013;**29**(44): 13647-13654

[75] Antunes CSA, Bietti M, Salamone M, Scione N. Early stages in the TiO_2-photocatalyzed degradation of simple phenolic and non-phenolic lignin model compounds. Journal of Photochemistry and Photobiology A: Chemistry. 2004; **163**(3):453-462

[76] Setvin M, Aschauer U, Hulva J, Simschitz T, Daniel B, Schmid M, et al. Following the reduction of oxygen on TiO_2 Anatase (101) step by step. Journal of the American Chemical Society. 2016; **138**(30):9565-9571

[77] Chen J, Li YF, Sit P, Selloni A. Chemical dynamics of the first proton-coupled electron transfer of water oxidation on TiO_2 anatase. Journal of the American Chemical Society. 2013; **135**(50):18774-18777

[78] Hwang JY, Hee MG, Kim B, Tachikawa T, Majima T, Hong S, et al. Crystal phase-dependent generation of mobile OH radicals on TiO_2: Revisiting the photocatalytic oxidation mechanism of anatase and rutile. Applied Catalysis B: Environmental. 2021;**November 2020**(286):119905

[79] Dudziak S, Kowalkińska M, Karczewski J, Pisarek M, Gouveia JD, Gomes JRB, et al. Surface and trapping energies as predictors for the photocatalytic degradation of aromatic organic pollutants. Journal of Physical Chemistry C. 2022;**126**: 14859-14877

[80] Kowalkińska M, Sikora K, Łapiński M, Karczewski J, Zielińska-Jurek A. Non-toxic fluorine-doped TiO_2 nanocrystals from $TiOF_2$ for facet-dependent naproxen degradation. Catalysis Today. 2022;**413–415**:113959

[81] Wang Y, Mino L, Pellegrino F, Homs N, Ramírez de la Piscina P. Engineered MoxC/TiO_2 interfaces for efficient noble metal-free photocatalytic hydrogen production. Applied Catalysis B: Environmental. 2022;**318**:121783

[82] Wang D, Gong XQ. Function-oriented design of robust metal cocatalyst for photocatalytic hydrogen evolution on metal/titania composites. Nature Communications. 2021;**12**:1-6

[83] Wei Z, Janczarek M, Endo M, Wang K, Balčytis A, Nitta A, et al. Noble metal-modified faceted anatase titania photocatalysts: Octahedron versus decahedron. Applied Catalysis B: Environmental. 2018;**237**:574-587

[84] Meng A, Zhang J, Xu D, Cheng B, Yu J. Enhanced photocatalytic

H_2-production activity of anatase TiO_2 nanosheet by selectively depositing dual-cocatalysts on (101) and (001) facets. Applied Catalysis B: Environmental. 2016;**198**:286-294

[85] Mishra SB, Nanda BRK. Facet dependent catalytic activities of anatase TiO_2 for CO2 adsorption and conversion. Applied Surface Science. 2020;**531**:147330

[86] Ma S, Song W, Liu B, Zhong W, Deng J, Zheng H, et al. Facet-dependent photocatalytic performance of TiO_2: A DFT study. Applied Catalysis B: Environmental. 2016;**198**:1-8

[87] Liu L, Jiang Y, Zhao H, Chen J, Cheng J, Yang K, et al. Engineering Coexposed {001} and {101} Facets in Oxygen-Deficient TiO_2 Nanocrystals for Enhanced CO2 Photoreduction under Visible Light. ACS Catalysis. 2016;**6**(2): 1097-1108

[88] Yu J, Low J, Xiao W, Zhou P, Jaroniec M. Enhanced Photocatalytic CO_2-Reduction Activity of Anatase TiO2 by Co-exposed {001} and {101} Facets. Journal of the American Chemical Society. 2014;**136**:8839

[89] Keller N, Ivanez J, Highfield J, Ruppert AM. Photo−/thermal synergies in heterogeneous catalysis: Towards low-temperature (solar-driven) processing for sustainable energy and chemicals. Applied Catalysis B: Environmental. 2021;**296**:120320

[90] Fang F, Liu Y, Sun X, Fu C, Prakash Bhoi Y, Xiong W, et al. TiO2 facet-dependent reconstruction and photocatalysis of CuOx/TiO_2 photocatalysts in CO_2 photoreduction. Applied Surface Science. 2021;**564**:150407

[91] Feng N, Lin H, Song H, Yang L, Tang D, Deng F, et al. Efficient and selective photocatalytic CH_4 conversion to CH_3OH with O_2 by controlling overoxidation on TiO_2. Nature Communications. 2021;**12**(1):4652

Chapter 4

Nanostructured Titanium Dioxide ($NS-TiO_2$)

Bochra Bejaoui, Imen Bouchmila, Khaoula Nefzi, Imen Belhadj Slimen, Sidrine Koumbad, Patrick Martin, Nicolas Joly and Naceur M'Hamdi

Abstract

During the past decade, research in the area of synthesis and applications of nanostructured titanium dioxide (NS TiO_2) has become tremendous. NS TiO_2 materials have shown great potential and a wide range of applications. The decrease in the particle size and the increase of the surface/volume ratio lead to the increase of the specific surface and the modification of the physicochemical properties and the appearance of new interesting properties (photocatalytic, optical, magnetic, electronic...). Their new morphology even allows the appearance of new biological properties. NS TiO_2 can thus be used for the same applications as those known for their precursors before transformation and their nanostructures are accompanied by new properties allowing applications. This chapter briefly describes the synthesis process of the different NS TiO_2, their chemical and surface modifications, and their application. The preparation of NS TiO_2, including nanoparticles, nanorods, nanowires, nanosheets, nanofibers, and nanotubes is described. This chapter discusses the effects of precursor properties and synthesis conditions on the structure, crystallinity, surface specificity, and morphology of titanium dioxide nanoparticles. Recent advances in NS TiO_2 in nano-biosensing, medical implants, drug delivery, and antibacterial fields, pharmaceutical applications, as well as their toxicity and biocompatibility, were presented.

Keywords: titanium dioxide nanoparticle, syntheses process, chemical methods, physical methods, biosynthesis, environmental applications, biomedical applications, biocompatibility

1. Introduction

Nanotechnology encompasses biology, chemistry, materials science, medicine and physics. Today, With the advent of nanoscience and titanium dioxide nanostructured materials nanotechnology, nanostructured materials are an important research area due to their various unique properties. Among all transition metal oxides, TiO_2 nanostructures are the most attractive materials in modern science and technology [1]. TiO_2 is used commercially in donuts, cosmetics, pigments [2], catalysts, sunscreens [3, 4], solar cells [5], water splitting, and more. TiO_2 is used in plastics, paints, varnishes, paper, pharmaceuticals, inks, pharmaceuticals, toothpaste, food, and industry [6, 7]. Nanostructured titanium dioxide (NS-TiO_2) is a non-toxic, environmentally friendly, inexpensive, and efficient functional material with a broad range of applications [8–11]. In the past decade, nanostructured TiO_2, which can have either a stoichiometric or nonstoichiometric composition, has attracted increasing attention from researchers around the world as a promising highly efficient photocatalyst for the synthesis of organic compounds that meets the principles of green chemistry [12–17]. Today, nano-structured materials are an important area of research due to their several unique characteristic features. Among all the transition metal oxides, TiO_2 nanostructures are the best-looking materials in modern science and technology [1]. Nano-TiO_2 nanostructures include titanium dioxide nanoparticles (TiO_2-NPs) and titanium dioxide nanotubes (TNTs) [18]. With the advent of nanotechnology, NS–TiO_2 has found many applications. Nanoscale titanium dioxide (nano-TiO_2) has been widely used in environmental protection, cosmetics, antibacterial agents, self–cleaning coatings and cancer treatment, solar cells, photocatalysis, and composite nanofillers [19–21]; due to the fact of its unique size and high specific surface area, nano-TiO_2 has more stable physical and chemical properties compared to titanium dioxide. In addition, nano-TiO_2 has great application potential in biomedical fields [22, 23] due to the fact of its good antibacterial activity, favorable biocompatibility, and unique photocatalytic activity [24]. Research has shown that nanostructured TiO_2 elicits a favorable molecular response and osseointegration, with better bone formation than non-nanostructured materials [25–27]. The unique physicochemical properties of all these forms of NS–TiO_2, render this material a promising future in many applications. Several reviews and reports on different aspects of titanium dioxide, including its properties, preparation, modification, and application, have been published. However, despite advances in the development of nanostructured TiO_2 systems for bone repair, review articles addressing this topic are still scanty [28].

The purpose of this chapter is to introduce and discuss the properties [29], fabrication, modification, and applications of nanostructured titanium dioxide (NS-TiO_2). With the advent of nanotechnology, NS-TiO_2 has found many applications.

2. Synthesis process of NS TiO_2

Various synthesis methods such as sol-gel, hydrothermal and solvothermal methods, vapor deposition, electrochemical deposition, oxidation, and sonochemical and micro-waves methods are used to obtain high-quality TiO_2 nanostructures [12, 15]. In this section, we will analyze the most used methods for the preparation of TiO_2 nanostructures.

2.1 Chemical and physical methods

2.1.1 Sol: gel process

Sol-gel is a versatile method used for the synthesis of TiO_2 nanostructures of different morphologies such as sheets, tubes, particles, wires, rods, mesoporous, and aerogels [30–32]. Mehrotra and Singh [33] suggested different steps and conditions that can control the morphology of the final products in the sol-gel process (**Figure 1**).

The sol-gel method can use two ways of synthesis: the inorganic or colloidal route in which the precursors used are metal salts such as chlorides, nitrates, and oxychlorides in an aqueous solution. The Metallo-organic or polymeric route: obtained from metal

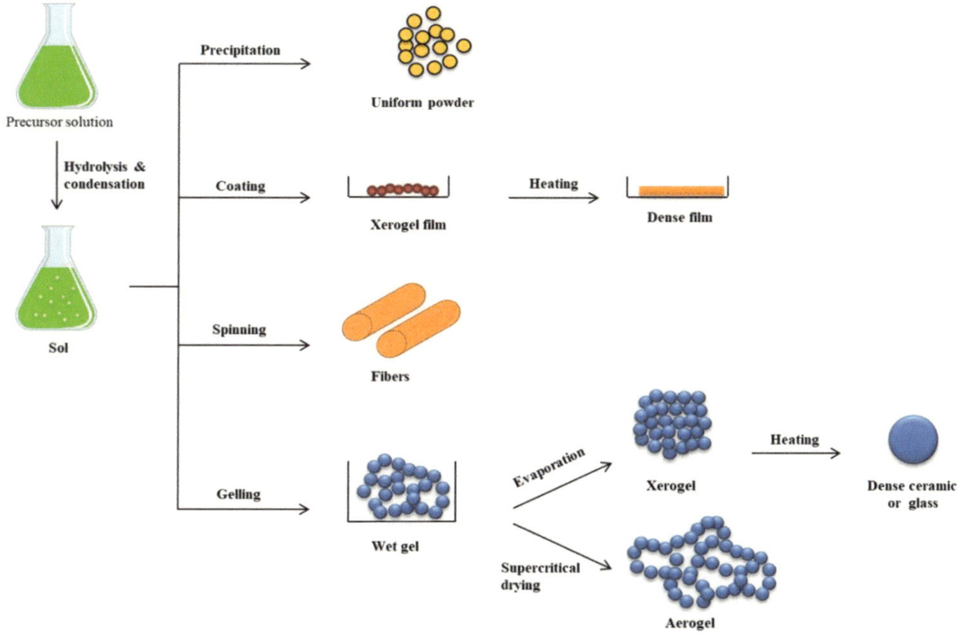

Figure 1.
Sol-gel process steps for the synthesis of TiO$_2$ nanostructures [30, 32, 33].

alkoxides in organic solutions. The polymerization reaction to obtain titanium dioxide takes place in two steps, namely hydrolysis, and condensation.

2.1.2 Hydrothermal and solvothermal processing

These two methods of synthesis are quite similar. The hydrothermal method is considered one of the most promising techniques for obtaining nanostructured TiO_2 at stable temperatures and pressures. It has the advantage of following simple steps and being inexpensive. The hydrothermal technique allows the production of high-quality 1D nanostructures, especially nanorods. By adapting the synthesis parameters, it is possible to control the morphology of the structures. However, the disadvantages of this method include the high capital requirement for instrumentation, the inability to monitor crystal growth, and the method can only be performed under supercritical solvent conditions [32, 34, 35]. Solvothermal methods use non-aqueous solvents with very high boiling points. When synthesizing with the solvents, better control of the properties of the titanium dioxide particles is achieved. The physicochemical characteristics (viscosity, polarity, boiling point, thermal conductivity, dielectric constant) of the solvent have a great influence on the nanostructures of the product [36]. Kathirvel et al. [37] prepared TiO_2 nanocrystals by the solvothermal method using six alcohols of different classes (primary, secondary, and tertiary). The synthesis was carried out using titanium isopropoxide as a precursor at a temperature of 150°C for 8 h. The crystallinity and morphology of TiO_2 nanocrystals varied depending on the chain length and the class of alcohol [37]. Li et al. [38] on the other hand, used the solvothermal method to obtain TiO_2 microspheres with suitable size without surfactant in a single step. The synthesis was performed using titanocene dichloride and acetone, heated at 180°C for 12 h [39]. It has been shown that the addition of surfactants to the synthesis effectively controls the growth of the particles [40–42].

2.1.3 Vapor deposition

Deposition methods form high-quality solid materials by condensing materials in a vaporous state. The deposition process is usually performed at low pressure in a vacuum chamber. If a chemical reaction occurs, it is called chemical vapor deposition (CVD) and if no reaction occurs, it is called physical vapor deposition (PVD). In this process, a precursor (solid or liquid) is heated to form an active gaseous reactant that is transferred to the reaction chamber. When the substrate is exposed to the volatile precursor, a reaction occurs on the surface of the substrate and the deposition process begins to produce the desired product. The precursors used in this method are highly volatile, non-toxic, and pyrophoric. The by-products formed during this process are degraded through the reaction chamber by the gas flow. This technique proved to be suitable to prepare TiO_2 nanostructures with tailored morphologies [43, 44].

2.1.4 Oxidation method

The principle of this method is to oxidize metallic titanium into titanium oxide by anodization or by the use of oxidants. Anodization or anodic oxidation consists in performing a surface treatment to form a titanium structure of pores/nanotubes on TiO_2. Oxidation of titanium can be achieved by using oxygen sources such as hydrogen peroxide, pure oxygen, acetone, and a mixture of argon and oxygen [30]. Mohan et al. [45] used this technique to synthesize self-organized titanium oxide nanotube layers from titanium alloys in electrolyte mixtures. The length and diameter of the nanotubes were controlled by playing on different anodization parameters such as temperature and time. Significant results were observed at 25°C. Indeed, at this temperature compared to others, smooth and circular nanotube arrays, with no apparent defects in their morphology were obtained [45]. From a previously treated titanium plate dissolved in 30% hydrogen peroxide, titanium dioxide nanorods were obtained by a dissolution precipitation mechanism. The addition of inorganic sodium salts can lead to the formation of anatase (NaF and Na_2SO_4) or rutile (NaCl addition) titanium dioxide nanorods [46].

2.1.5 Electrochemical anodization/electrodeposition process

Electrochemical anodization is an electrochemical process used to manufacture nanoparticles such as titanium nanotubes and nanopores. This method consists in growing the oxide layer on the metal surface. This process is performed in a standard two-electrode system immersed in a first, second, or third-generation electrolyte solution. The titanium forms the anode electrode and the platinum the cathode.

2.1.6 Sonochemical synthesis

Sonochemical synthesis has proven to be an efficient method to obtain nanoparticles with interesting properties in a short time [47]. The chemical effects observed during this technique are attributed to acoustic cavitation phenomena. Indeed, during cavitation in a liquid medium, there is formation, growth, and collapse of bubbles in the liquid. The violent implosion of the bubbles in less than a microsecond generates short-lived hot spots with a temperature of about 5000 K, pressures close to 1000 atm, and cooling rates higher than 109 K/s. Under these conditions, metal ions are reduced to metal or metal oxide nanoparticles [48]. The main advantage of this method is that the reaction times are reduced and the manipulations are performed under ambient conditions. In addition, it is a simple technique to implement and energy efficient. The nanostructures obtained are ultrafine particles. Studies have shown that ultrasonic synthesis of TiO_2 nanostructures can improve their properties. This technique is more efficient than other methods including microwaves [49].

2.1.7 Microwave method

The microwave-assisted synthesis method also uses electromagnetic waves such as sonication. Titanium dioxide can be synthesized by this technique at frequencies ranging from 0.3 to 300 GHz and wavelengths from 0.001 to 1 m. Two different mechanisms can be involved in microwave chaffing: dipolar polarization and ionic conduction [50]. Any material or substance containing mobile electric charges such as polar molecules or conducting ions can be heated using microwaves. In the dipolar polarization mechanism, microwave energy allows molecules to try to orient themselves with the electric field oscillating billions of times per second. The constant rotary motion of the molecule trying to align itself with the field causes friction and collisions.

3. Physical and chemical properties of NS TiO_2

Titanium dioxide is one of the most studied and well-researched compounds in materials science, due to its outstanding and exceptional properties which include stability of its chemical structure, biocompatibility, physical, optical, and electrical properties, nontoxicity, corrosion resistance, and low cost [51–53]. Generally, the morphology and physical/chemical properties of TiO_2 nanostructures depend on the synthesis process, precursor type, and concentration, use of capping agents, synthesis temperature, pressure, and time [31]. Titanium dioxide, CI 77891, also known as Titanium (IV) oxide or Titania, CAS No: 13463-67-7 is a naturally occurring oxide with the chemical formula TiO_2 and a molecular weight of 79.87 g mol^{-1}. It belongs to the family of transition metal oxides [54]. The most important titanium minerals are rutile (TiO_2), ilmenite ($FeTiO_3$), and titanite ($CaTiSiO_5$) [54]. In nature, titanium dioxide occurs mainly in three crystalline forms: rutile, anatase, and brookite. In addition, other polymorphs have also been reported (**Figure 2**) [32]. In addition, there are at least 3 reported non-crystalline TiO_2 phases: a low-density amorphous TiO_2 and two high-density amorphous TiO_2 types. TiO_2 (II) and TiO_2 (H) are high-pressure forms that have been synthesized from the rutile phase [31, 54–56].

In various technologically relevant applications, nano-size-scaled materials have shown beneficial properties related not only to their chemical composition but also to the small dimensions and the large surface-to-volume ratio. Generally, a material is defined as a nanomaterial when it has a specific surface area by volume greater than 60 m^{2}cm^{-3}, excluding materials consisting of particles with a size lower than 1 nm [57].

The high surface area brought about by small particle size is a crucial parameter for the high performance of many TiO_2-based devices. It provides more active sites and a large interface for any type of reaction/interaction between the device and the interacting media. Thus, the performance of TiO_2-based devices is largely influenced by the size of TiO_2 building units. For example, high surface area TiO_2 nanomaterials can guarantee good accessibility and contact with the electrolyte in lithium-ion batteries. Small primary crystals offer short diffusion paths for lithium and are beneficial for short charging–discharging times in batteries. Anatase, which has a

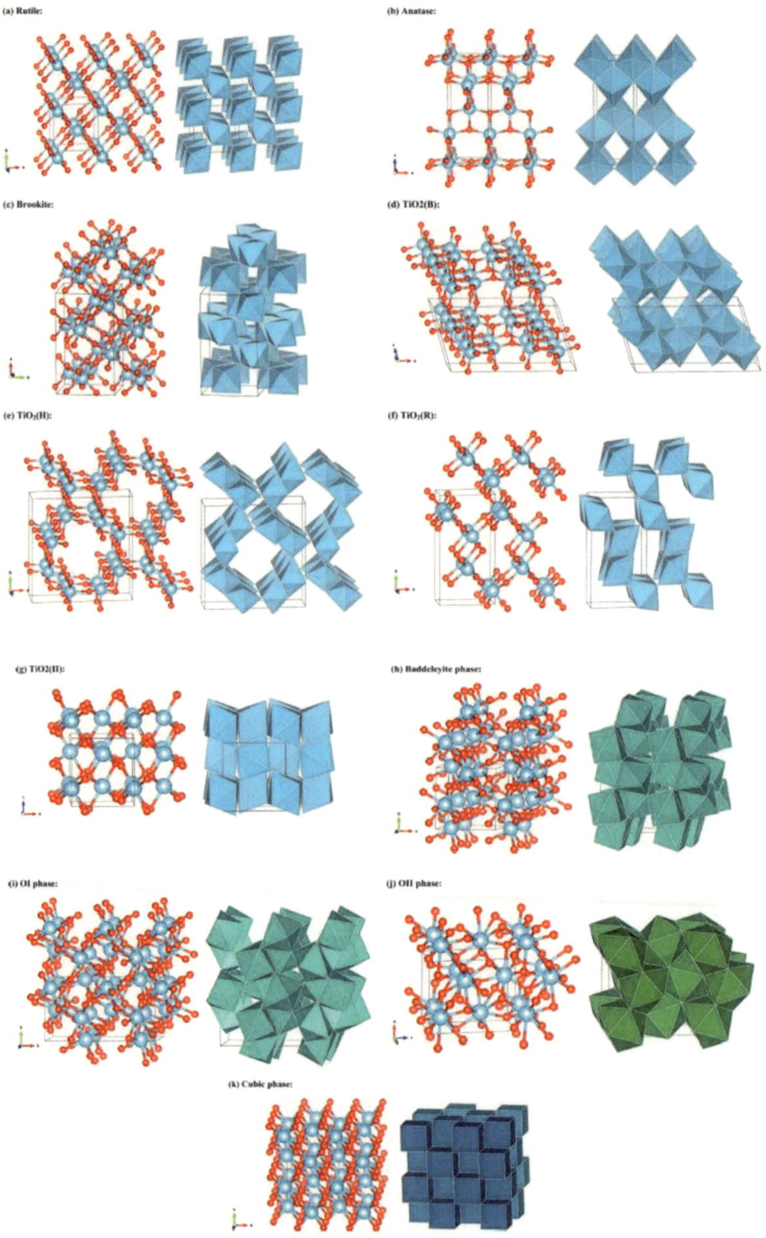

Figure 2.
Structures of TiO_2 phases: (a) rutile, (b) anatase, (c) brookite, (d) TiO_2 (B), (e) TiO_2 (II), (f) TiO_2(R), (g) TiO_2 (II), (h) baddeleyite TiO_2, (i) TiO_2-OI, (j) TiO_2-OII and (f) cubic TiO_2 [32].

greater surface area than its counterparts, is widely used as a photocatalyst in photon–electron transfer, whereas rutile is used for light scattering [57]. Surface charge is an important property of nanoparticle dispersions. When nanoparticles are dispersed in an aqueous solution, surface ionization and adsorption of cations or anions generate a surface charge, creating an electric potential between the particle surface and the bulk of the dispersion medium [58]. Depending on the measurement technique, the surface charge can be expressed either as surface charge density (potentiometric titration) or zeta potential (electrokinetic method). The point where the surface charge density is zero is defined as the point of zero charges (ZPC), and the point where the zeta potential is zero is defined as the isoelectric point (IEP) [58].

The surface of TiO_2 nanoparticles dispersed in aqueous media or humid atmosphere can react immediately with water molecules, and reasonable amounts of hydroxyl groups are formed as shown in Eq. 1 [30, 58].

$$Ti^{IV} + H_2O \rightarrow Ti^{IV} - OH + H^{+} \quad (1)$$

When the surface of TiO_2 is fully hydroxylated, the oxide ions in the oxide and water absorbed on the surface would distribute electrons and form equal quantities of two types of hydroxyl groups [30].

The surface charge of titania is a function of solution pH, which is affected by the reactions that occur on the particle surface as shown in Eqs. 2 and 3.

$$Ti^{IV} - OH + H^{+} \rightarrow Ti^{IV} - OH_2^{\ +} \quad (2)$$

$$Ti^{IV} - OH \rightarrow Ti^{IV} - O^{-} + H^{+} \quad (3)$$

A variety of nanostructured TiO_2 materials with fascinating morphologies have been reported. The synthesis methods used for the fabrication of these nanostructures have a significant effect on their dimensions. In general, nanostructure forms of TiO_2 have been classified into 0D (nanospheres, quantum dots), 1D (nanowires, rods, and tubes), 2D (layers and sheets), and 3D (nanoparticles, nanoflowers, etc.) architectures, which are summarized in **Figure 3** [61, 62].

Dissolution is defined as the dynamic process during which constituent molecules of the dissolving solid migrate from the surface to the bulk solution through a diffusion layer. The thermodynamic parameter that controls this process is described as solubility and along with the concentration gradient between the particle surface and the bulk, the solution acts as the driving force of particle dissolution [36]. Both solubility and rate of dissolution are dependent on a particle's chemical and surface properties such as surface area, surface morphology, and surface energy, as well as size. Crystallinity and crystal structure also need to be considered. They depend also on the possible adsorbed species, and the state of aggregation of the nanoparticles

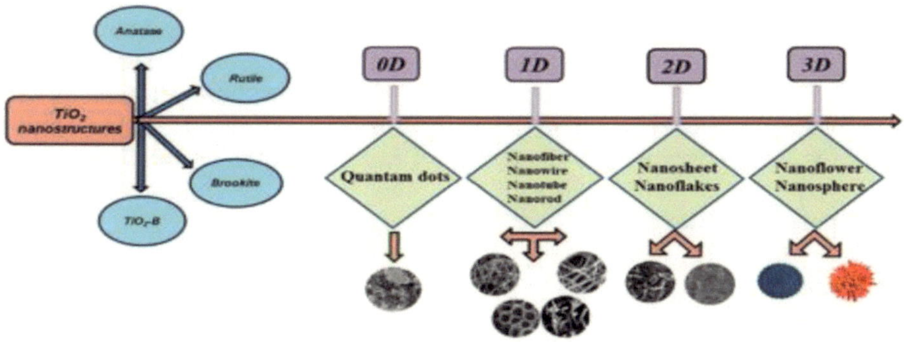

Figure 3.
Categorization of hierarchical TiO_2 nanostructure form [59, 60].

and are further impacted by the surrounding media (properties of the diffusion layer and the possible solute concentration) [36, 63].

Studies have shown that TiO_2 nanoparticles tend to aggregate and their aggregation has a strong influence on nanoparticle behavior due to the nature and size of the aggregates (i.e., the packing density of the nanoparticles), and aggregation can potentially impact their reactivity, nanoparticle-cellular interactions, and toxicity [43]. There are two types of aggregations: homo-aggregation and hetero-aggregation. Homoaggregation refers to the aggregation of two particles of identical characteristics (i.e., NP–NP attachment). Heteroaggregation refers to the aggregation of particles with different physical or chemical characteristics (e.g., NP–clay particle attachment). In the natural environment of aquatic systems, the state of aggregation of the nanoparticles is greatly influenced by diverse conditions such as ionic strength (IS), ionic composition, co-existing colloids, natural organic matter (NOM) (e.g., humic y fulvic and humic substances), pH, and other physicochemical factors [64].

4. Potential and applications of NS TiO_2

Metal oxide nanoparticles (NPs) have found a variety of applications in numerous industrial, medical, and environmental fields s, attributable to recent advances in the nanotechnology field.

4.1 Photocatalytic applications

Photocatalysis is the decomposition and degradation of pollutants under the action of light rays on the surface of a catalyst, usually titanium dioxide (TiO_2). It allows the destruction of volatile organic compounds, inorganic pollutants, and microorganisms. The finalized process produces essential water and carbon dioxide [65]. All current applications of photocatalysis use TiO_2 as a semiconductor for several reasons [66]. Titanium dioxide, in its current commercial forms, is not toxic (apart from recent reservations about the use of reservations regarding the use of nanoparticles) and, due to its photostability in air and water, does not release toxic elements [67]. As titanium is a relatively abundant element, the cost of TiO2 is not too high, at least for some applications. The most widely used crystallographic form is the anatase form because TiO_2 with a rutile structure (although having a lower band gap value allowing it to absorb light in the early visible spectrum) is significantly less active. The most effective commercial composition at present is TiO_2 Degussa P25 (80% anatase, 20% rutile) [68]. For practical industrial applications of semiconductor photocatalysts, Sharma et al. [52] proved the development of research of new semiconductor materials in visible-light active TiO_2/SnX (X = S and Se) and their application as photocatalysts since it is a new area of scientific interest. Indeed, they focused on the addition of TiO_2 composites with SnX (X = S, Se) as potential candidates for environmental purification.

4.2 Photovoltaic applications

In the current global scenario, the rise in technological demands of the world's population has caused a rapid increase in energy consumption, which in turn has led to an exponential increase in environmental pollution, which we have witnessed seriously in the last decades. To surmount this situation, the efficient use of green energy has become a hot topic worldwide. On the other hand, intelligent materials are also of great value in the current market due to their multipurpose for a variety of applications. Among the green energy alternatives available today, solar energy provides more promising perspectives as the sun can deliver the ultimate solution to the prevailing sustainable energy supply challenge. Among the different solar cell technologies currently available, dye-sensitized solar cells have drawn a lot of attention due to their promising prospects. On the Other Side, photocatalysis has also made a strong case for itself due to its promising opportunities for clean, green, and sustainable development in environmental technology applications [69, 70].

4.3 Sensing applications

In recent years, gas sensors have become extremely important for environmental and industrial atmosphere monitoring [71]. Gas detection techniques are based on resistance sensing, electrochemical and optical methods, gas and liquid chromatography, and acoustic waves. Nevertheless, certain sensors have various drawbacks: they consume energy and time, they are wide in size, they are expensive, and they display slow response and low selectivity [72, 73]. Consequently, special attention has been given to chemoresistive sensors, which are formed by metal oxides, carbon-based materials, and conducting polymers. Among these materials, semiconducting metal oxides have been extensively investigated and explored due to the potential for different valences, morphologies, and physicochemical characteristics [74]. They are becoming more complex than pure metals, with bonding going from ionic to highly covalent to metallic. For this reason, metal oxide nanoparticles are attracting considerable attention from industry for use in diverse applications such as catalytic processes, magnetic storage media, electronics, sensors, and solar energy conversion.

4.4 Hydrogen production and storage

Hydrogen (H_2) generation has become viral in the last few decades due to hydrogen as a future energy source and its capacity to replace expensive and polluting fossil fuels [75]. In addition, hydrogen also contributes to the development of a green world due to its zero emissions and minimizes dependence on non-renewable resources. In general, hydrogen production processes can be divided into two categories based on the usage of renewable and non-renewable resources. The methods for utilizing renewable energy resources are photoelectrolysis, thermal and photocatalytic water splitting, and steam reforming and gasification. Steam reforming and gasification methods are processes that depend on non-renewable resources [76]. Among carbon materials, activated carbon (AC) can be produced easily from agricultural residues such as hardwoods, coconut shells, fruit pits, walnut shells, and lignite. Which makes CA abundantly available and less expensive. CA also has characteristics such as a high surface area and a porous structure [77]. Such as high surface area and porous structure [77]. Due to these characteristics, AC-TiO_2 nanocomposites have been extensively investigated for the photocatalytic decomposition of dyes [78]. As an example, Mahadwad et al. [79] decomposed the reactive black dye 5 under mercury vapor light with AC-TiO_2 nanocomposites. Recently, Xing et al. [80] reported the H_2 generation activity with different types of simulated seawater with Rh/Cr_2O_3GaN nanowire photocatalyst [81]. Reddy et al. [82] have developed a low-cost nanocomposite such as AC-TiO_2 by a one-step hydrothermal method, which is a potential catalyst for H_2 generation under sunlight. In the photocatalytic H_2 generation process, sacrificial agents have a crucial role in consuming the valence band (VB) holes.

4.5 Environmental applications

TiO_2 is an environmental-friendly material that has been widely used in the photodegradation of a large number of pollutants. Nanostructured TiO_2 was used in pollution abatement, energy conversion (i.e. hydrogen production and solar cells), and energy storage (i.e. lithium batteries and supercapacitors). Its practical interest was also described in water purification, self-cleaning, self-sterilization of surfaces, as well as light-assisted H2 production [83]. In the textile field, Gaminian and Montazer [84] assessed the self-cleaning effects of Cu_2O/TiO_2 on polyester fabric and concluded that both washed and unwashed samples showed significant photodegradation properties of methylene blue. Production of the reducing agent ethylene glycol as a product of the alkaline hydrolysis for the synthesis of Cu nanoparticles was reported indeed. In another trial, Harifi and Montazer [85] developed Fe3 + –doped Ag/TiO_2 nanostructures for photocatalytic uses under the UV-vis light spectrum. The photodegradation activity assessed using methylene blue was confirmed under both UV and visible light regions. Zhou et al. [86] explored the degradation of acetone in the air using iron-doped mesoporous TiO_2 nanoparticles. Their findings showed a high degradation rate of this organic pollutant. In the same way, El-Roz et al. [87] reported an enhanced photocatalytic activity of luffa/TiO_2 nanocomposites against methanol. Píšťková et al. [88] investigated the photodecomposition of acebutolol, propranolol, atenolol, nadolol, and metoprolol, which are β-blockers, using immobilized TiO_2 in an aqueous media. Their results showed a complete photodegradation in 2 h of all tested β-blockers. Coronado et al. [89] described some TiO_2 applications in water purification. This application is argued by the excellent optical and catalytic properties of nanostructured TiO_2, allowing oxidation and reduction catalysis of both organic and inorganic contaminants. The photo-generated free radicals and e−/h + pairs are highly implicated in degrading organic substances, water pollutants, and harmful microorganisms [90]. In this trend, nanocomposite TiO_2 thins films ($P/Ag/Ag_2O/Ag_3PO_4$) were able to decompose up to 90% of rhodamine B under solar light exposure [91, 92].

4.6 Biomedical applications

Nanomedicine is defined as "the development of nanoscale (1–100 nm) or nanostructured objects/nano-robots/skin patches and their use in medicine for diagnostic and therapeutic purposes based on the use of their structure, which has unique medical effects" [93]. It relies on the use of nanodevices and nanostructures operating at the cellular level, providing therefore comprehensive monitoring, control, repair, and enhancement of biological systems at the molecular level. The use of nanoparticles is deep-rooted in the history of medicine. The application of nanosilver to overcome bacterial infections and the use of nanosized agents to modulate immune response are some examples. TiO_2 nanostructures are one of the most plentiful nanomaterials having a broad spectrum of applications in nanomedicine. TiO_2 is not only a cost-effective and highly biocompatible nanoparticle [94], but it is also a non-toxic substance [95], which use in food and drugs has been approved by the American Food and Drug Administration (FDA) to be [96].

5. Future challenges and perspectives

In this chapter, the use of nanostructured titanium dioxide is an effective and attractive alternative for fabricating flexible devices for multiple applications, which can be explored based on TiO_2 properties, fabrication, and modification. A further challenge is to enhance the spectral sensitivity of these structures to the visible and near-infrared regions and the biocompatibility of TiO_2 nanostructures. Therefore, future studies focused on long-term, constant photoactivity are greatly needed. These can be achieved by changing the synthesis route. Nonmetal-doped TiO_2 nanostructures exhibit low photocatalytic activity under visible UV light. Some materials, such as polymers, glasses, ceramics, and metals, therefore serve as magical identities for economical and environmentally friendly applications in this field. Future research requires the development of new synthetic methods and nanostructures with higher surface states. This can be serviced by techniques compatible with non-lithographic complementary metal oxide semiconductors. This technique has potential applications in new dopant materials, incorporation of dopants into TiO_2 nanostructures, and environmental and alternative energy applications. Therefore, there is a great need to improve the structure and properties of these materials. Basic knowledge of chemistry, physics, and computer modeling will help you accomplish your task.

6. Conclusion

Many reviews and reports have been published on various aspects such as the properties, production, modification, and application of titanium dioxide. This chapter provided a detailed overview of the synthesis, properties, and applications of nanostructured titanium dioxide (NS-TiO_2). Moreover, Titanium dioxide nanoparticles have gained a lot of attention because of their numerous applications. The formation of TiO_2 from various biological sources (plants, microorganisms, and related bioproducts) has been discussed. Furthermore, the mechanism of their uptake, translocation, and accumulation in plants is explored. The potential impact of TiO_2 has also been reported. Titanium dioxide nanoparticles have found a variety of applications in numerous industrial, medical, and environmental fields, attributable to recent advances in the nanotechnology field.

Author details

Bochra Bejaoui[1,2*], Imen Bouchmila[1], Khaoula Nefzi[3], Imen Belhadj Slimen[4], Sidrine Koumbad[5], Patrick Martin[5], Nicolas Joly[5] and Naceur M'Hamdi[6]

1 National Institute of Research and Pysico-chemical Analysis (INRAP), Laboratory of Useful Materials, Technopark of Sidi Thabet, Ariana, Tunisia

2 Faculty of Sciences of Bizerte, University of Carthage, Department of Chemistry, Zarzouna, Bizerte, Tunisia

3 Carthage University, National Research Institute of Rural Engineering, Water and Forests (INRGREF), LR11INRGREF0 Laboratory of Management and Valorization of Forest Resources, Ariana, Tunisia

4 Department of Animal Sciences, National Agronomic Institute of Tunisia, Carthage University, Tunis, Tunisia

5 Unité Transformations and Agroressources, Université d'Artois-Uni LaSalle, Bethune, France

6 Research Laboratory of Ecosystems and Aquatic Resources, National Agronomic Institute of Tunisia, Carthage University, Tunis, Tunisia

*Address all correspondence to: bochrabej@yahoo.fr

References

[1] Ali I, Suhail M, Alothman ZA, Alwarthan A. Recent advances in syntheses, properties, and applications of TiO2 nanostructures. RSC Advances. 24 Aug 2018;**8**(53):30125-30147. DOI: 10.1039/c8ra06517a

[2] Pfaff G, Reynders P. Angle-dependent optical effects deriving from submicron structures of films and pigments. Chemical Reviews. 1999;**99**(7):1963-1981

[3] Newman MD, Stotland M, Ellis JI. The safety of nanosized particles in titanium dioxide–and zinc oxide–based sunscreens. Journal of the American Academy of Dermatology. 2009;**61**(4):685-692

[4] Salvador A, Pascual-Martı MC, Adell JR, Requeni A, March JG. Analytical methodologies for atomic spectrometric determination of metallic oxides in UV sunscreen creams. Journal of Pharmaceutical and Biomedical Analysis. 2000;**22**(2):301-306

[5] Fujishima A, Honda K. Electrochemical photolysis of water at a semiconductor electrode. Nature. 1972;**238**(5358):37-38

[6] Patri A, Umbreit T, Zheng J, NagashimaK, GoeringP, Francke-CarrollS, et al. Energy dispersive X-ray analysis of titanium dioxide nanoparticle distribution after intravenous and subcutaneous injection in mice. Journal of Applied Toxicology: An International Journal. 2009;**29**(8):662-672

[7] Sul Y-T. Electrochemical growth behavior, surface properties, and enhanced in vivo bone response of TiO_2 nanotubes on microstructured surfaces of blasted, screw-shaped titanium implants. International Journal of Nanomedicine. 2010:87-100

[8] Auguglia V, Caronna T, Paola A, Marcì G, Pagliaro M, Palmisano G, et al. Environmentally Benign Photocatalysts. New York: Springer; 2010. pp. 623-645

[9] Rossi S, Savi M, Mazzola M, Pinelli S, Alinovi R, Gennaccaro L, et al. Particle and Fibre Toxicology. 2019;**16**:1

[10] Zada A, Qu Y, Ali S, Sun N, Lu H, Yan R, et al. Improved visible-light activities for degrading pollutants on TiO2/g-C3N4 nanocomposites by decorating SPR Au nanoparticles and 2,4-dichlorophenol decomposition path. Journal of Hazardous Materials. 2018;**342**:715-723

[11] Zhao L, Zhu Y, Chen Z, Xu H, Zhou J, Tang S, et al. Nanotoxicology. 2018;**12**:169

[12] Ali W, Ullah H, Zada A, Muhammad W, Ali S, Shaheen S, et al. Synthesis of TiO_2 modified self-assembled honeycomb ZnO/SnO_2 nanocomposites for exceptional photocatalytic degradation of 2,4-dichlorophenol and bisphenol A. Science of the Total Environment. 2020;**746**:141291

[13] Ali S, Li Z, Chen S, Zada A, Khan I, Khan I, et al. Synthesis of activated carbon-supported TiO2-based nano-photocatalysts with well recycling for efficiently degrading high-concentration pollutants. Catalysis Today. 2019;**335**:557-564

[14] Kozlova EA, Rempel AA, Valeeva AA, Gorbunova TI, Kozhevnikova NS, Cherepanova SV, et al. Photoactivity of TiO 2/CdS and SiO 2/CdS hybrid nanostructured systems in the partial oxidation of ethanol under irradiation with visible light. Kinetics and Catalysis. 2015;**56**:515-522

[15] Liu C, Raziq F, Li Z, Qu Y, Zada A, Jing L. Synthesis of TiO2/g-C3N4 nanocomposites with phosphate-oxygen functional bridges for improved photocatalytic activity. Chinese Journal of Catalysis. 2017;**38**(6):1072-1078

[16] Utepova IA, Trestsova MA, Chupakhin ON, Charushin VN, Rempel AA. Green Chemistry. 2015;**17**:4401

[17] Utepova IA, Chupakhin ON, Trestsova MA, Khina AAM, Kucheryavaya DA, Charushin VN, et al. Russian Chemical Bulletin. 2016;**65**:445

[18] Liu S, Chen X, Yu M, Li J, Liu J, Xie Z, et al. Applications of titanium dioxide nanostructure in stomatology. Molecules. 2022;**27**:3881. DOI: 10.3390/molecules27123881

[19] Cuddy MF, Poda AR, Moser RD, Weiss CA, Cairns C, Steevens JA. A weight-of-evidence approach to identify nanomaterials in consumer products: A case study of nanoparticles in commercial sunscreens. Journal of Exposure Science & Environmental Epidemiology. 2016;**26**:26-34

[20] Shah R, Ali S, Raziq F, Ali S, Ismail PM, Shah S, et al. Exploration of metal organic frameworks and covalent organic frameworks for energy-related applications. Coordination Chemistry Reviews. 2023;**477**:214968

[21] Weir A, Westerhoff P, Fabricius L. Titanium dioxide nanoparticles in food and personal care products. Environmental Science & Technology ES&T. 2012;**46**:2242-2250

[22] Ai JW, Liu B, Liu WD. Folic acid-tagged titanium dioxide nanoparticles for enhanced anticancer effect in osteosarcoma cells. Materials Science & Engineering. C, Materials for Biological Applications. 2017;**76**:1181-1187

[23] Ding L, Li J, Huang R, Liu Z, Li C, Yao S, et al. Salvianolic acid B protects against myocardial damage caused by nanocarrier TiO_2; and synergistic anti-breast carcinoma effect with curcumin via codelivery system of folic acid-targeted and polyethylene glycol-modified TiO2 nanoparticles. International Journal of Nanomedicine. 2016;**11**:5709-5727

[24] Jia L, Qiu J, Du L, Li Z, Liu H, Ge S. TiO_2 nanorod arrays as a photocatalytic coating enhanced antifungal and antibacterial efficiency of Ti substrates. Nanomedicine. 2017;**12**:761-776

[25] Das K, Bose S, Bandyopadhyay A. TiO_2 nanotubes on Ti: Influence of nanoscale morphology on bone cell–materials interaction. Journal of Biomedical Materials Research Part A: An Official Journal of The Society for Biomaterials, The Japanese Society for Biomaterials, and The Australian Society for Biomaterials and the Korean Society for Biomaterials. 2009;**90**(1):225-237

[26] Garcia-Lobato M, Mtz-Enriquez A, Garcia C, Velazquez-Manzanares M, Avalos-Belmontes F, Ramos-Gonzalez R, et al. Applied Surface Science. 2019;**484**:975

[27] Wang N, Li H, Lü W, Li J, Wang J, Zhang Z, et al. Biomaterials. 2011;**32**:690

[28] Bezerra IA, Neta MF, Mota HL, Lira GA, Neves RRM. Nanostructured titanium dioxide for use in bone implants: A short review. Cerâmica. 2020;**66**(380)

[29] Zada A, Khan M, Khan MA, Khan Q, Yangjeh AH, Dang A, et al. Review on the hazardous applications and photodegradation mechanisms of chlorophenols over different photocatalysts. Environmental Research. 2021;**195**:110742

[30] Arun J, Nachiappan S, Rangarajan G, Alagappan RP, Gopinath KP, Lichtfouse E.

Synthesis and application of titanium dioxide photocatalysis for energy, decontamination, and viral disinfection: A review. Environmental Chemistry Letters. 2022, 2022. DOI: 10.1007/s10311-022-01503-z

[31] Kumar A, Pandey G. Different methods used for the synthesis of TiO_2 based nanomaterials: A review. American Journal of Nano Research and Applications. 2018;**6**(1):1-10

[32] Nyamukamba P, Okoh O, Heroe H, Taziwa R, Zinya S. Synthetic Methods for Titanium Dioxide Nanoparticles: A Review. Titanium Dioxide-material for a Sustainable Environment. 2018

[33] Mehrotra RC, Singh A. Recent trends in metal alkoxide chemistry. In: Progress in Inorganic Chemistry. John Wiley & Sons, Ltd; 1997. pp. 239-454. DOI: 10.1002/9780470166475.ch4

[34] Kaur N, Singh M, Moumen A, Duina G, Comini E. 1D titanium dioxide: Achievements in chemical sensing. Materials (Basel). 2020;**13**:2974

[35] Liu N, Chen X, Zhang J, Schwank JW. A review on TiO_2-based nanotubes synthesized via hydrothermal method: Formation mechanism, structure modification, and photocatalytic applications. Catalysis Today. 2014;**225**:34-51

[36] Guo R, Bao Y, Kang Q, Liu C, Zhang W, Zhu Q. Solvent-controlled synthesis and photocatalytic activity of hollow TiO2 microspheres prepared by the solvothermal method. Colloids and Surfaces A: Physicochemical and Engineering Aspects. 2022;**633**:127931

[37] Kathirvel S, Pedaballi S, Su C, Chen B-R, Li W-R. Morphological control of TiO_2 nanocrystals by solvothermal synthesis for dye-sensitized solar cell applications. Applied Surface Science. 2020;**519**:146082

[38] Li Z-Q, Chen W-C, Guo F-L, Mo L-E, Hu L-H, Dai S-Y. Mesoporous TiO_2 yolk-shell microspheres for dye-sensitized solar cells with a high efficiency exceeding 11%. Scientific Reports. 2015;**5**(14178):1-8

[39] Li G, Song J, Meng C, Wang D, Du Z, Sun W, et al. Single source, surfactant-free, and one-step solvothermal route synthesized TiO2 microspheres for highly efficient mesoscopic perovskite solar cells. Solar RRL. 2020;**4**:2000519

[40] Melo ACCA, De Jeus RA, Olivera de M, Claudia A, et al. Effect of non-ionic surfactant in the solvothermal synthesis of anatase TiO_2 nanoplates with a high percentage of exposed {001} facets and its role in the photocatalytic degradation of methylene blue dye. Environmental Research. 2022;**214**:114094

[41] Paschalidou P, Theocharis CR. Tuning the porosity and surface characteristics of nanoporous titania using non-ionic surfactant reverse micelles. RSC Advances. 2018;**8**(52):29890-29898

[42] Qiao L, Swihart MT. Solution-phase synthesis of transition metal oxide nanocrystals: Morphologies, formulae, and mechanisms. Advances in Colloid and Interface Science. 2017;**244**:199-266

[43] Lepcha A, Maccato C, Mettenbörger A, Andreu T, Mayrhofer L, Walter M, et al. Electrospun black titania nanofibers: Influence of hydrogen plasma-induced disorder on the electronic structure and photoelectrochemical performance. The Journal of Physical Chemistry C. American Chemical Society. 2015;**119**:18835-18842

[44] Perraudeau A, Dublanche-Tixier C, Tristant P, Chazelas C. Dynamic mode optimization for the deposition

of homogeneous TiO2 thin film by atmospheric pressure PECVD using a microwave plasma torch. Applied Surface Science. 2019;**493**:703-709

[45] Mohan L, Dennis C, Padmapriya N, Anandan C, Rajendran N. Effect of electrolyte temperature and anodization time on formation of TiO_2 nanotubes for biomedical applications. Materials Today Communications. 2020;**23**:101103

[46] Wu J-M. Low-temperature preparation of titania nanorods through direct oxidation of titanium with hydrogen peroxide. Journal of Crystal Growth. 2004;**269**:347-355

[47] Kamali M, Dewil R, Appels L, Aminabhavi TM. Nanostructured materials via green sonochemical routes - Sustainability aspects. Chemosphere. 2021;**276**:130146

[48] Savun B. A review on sonochemistry and its environmental applications. Acoustics. 2020;**2**:766-775

[49] Moreira AJ, Campos LO, Maldi CP, Dias JA, Paris EC, Giraldi TR, et al. Photocatalytic degradation of Prozac® mediated by TiO2 nanoparticles obtained via three synthesis methods: Sonochemical, microwave hydrothermal, and polymeric precursor. Environmental Science and Pollution Research. 2020;**27**:27032-27047

[50] Zhu Y-J, Chen F. Microwave-assisted preparation of inorganic nanostructures in liquid phase. Chemical Review. American Chemical Society. 2014;**114**:6462-6555

[51] Sadek O, Touhtouh S, Rkhis M, Anoua R, El Jouad M, Belhora F, et al. Synthesis by sol-gel method and characterization of nano-TiO2 powders. Materials Today: Proceedings. 2022;**66**:456-458

[52] Sharma R, Sarkar A, Jha R, Kumar Sharma A, Sharma D. Sol-gel–mediated synthesis of TiO2 nanocrystals: Structural, optical, and electrochemical properties. International Journal of Applied Ceramic Technology. 2020;**17**:1400-1409

[53] Nateq MH, Ceccato R. Sol-gel synthesis of TiO_2 nanocrystalline particles with enhanced surface area through the reverse micelle approach. Advances in Materials Science and Engineering. Hindawi. 2019;**2019**:e1567824

[54] Lukong VT, Ukoba KO, Jen TC. Heat-assisted sol-gel synthesis of TiO_2 nanoparticles structural, morphological, and optical analysis for self-cleaning application. Journal of King Saud University - Science. 2022;**34**:101746

[55] Keshari AK, Choudhary P, Shukla VK. Precursor-induced evolution in single anatase phase synthesis of TiO_2 nanoparticles for water treatment and dye-sensitized solar cell. Physica B: Condensed Matter. 2022;**631**:413716

[56] Lu H, Zhong J, Ji C, Zhao J, Li D, Zhao R, et al. Fabricating an optimal rutile TiO_2 electron transport layer by delicately tuning TiCl4 precursor solution for high-performance perovskite solar cells. Nano Energy. 2020;**68**:104336

[57] Kaur A, Bajaj B, Kaushik A, Saini A, Sud D. A review on template-assisted synthesis of multi-functional metal oxide nanostructures: Status and prospects. Materials Science and Engineering: B. 2022;**286**:116005

[58] Allende P, Orera A, Laguna-Bercero MÁ, Valenzuela ML, Díaz C, Barrientos L. Insights of the formation mechanism of nanostructured titanium oxide polymorphs from different macromolecular metal-complex precursors. Heliyon. 2021;**7**:e07684

[59] Dharma HNC, Jaafar J, Widiastuti N, Matsuyama H, Rajabsadeh S, Othman MHD, et al. A review of titanium dioxide (TiO2)-based photocatalyst for oilfield-produced water treatment. Membranes. 2022;**12**(3):345

[60] Reghunath S, Pinheiro D, KR SD. A review of hierarchical nanostructures of TiO2 : Advances and applications. Applied Surface Science Advances. 2021;**3**:100063

[61] Quintero Y, Mosquera E, Diosa J, García A. Ultrasonic-assisted sol-gel synthesis of TiO2 nanostructures: Influence of synthesis parameters on morphology, crystallinity, and photocatalytic performance. Journal of Sol-Gel Science and Technology. 2020;**94**:477-485

[62] Razak KA, et al. Factors of controlling the formation of titanium dioxide (TiO_2) synthesized using sol-gel method – A short review. Journal of Physics: Conference Series. IOP Publishing. 2022;**2169**(1)

[63] Cao R, Gao J, Tao H. One-step hydrothermal synthesis of multiphase nano TiO2. Materials Letters. 2022;**317**:132121

[64] Khizir HA, Abbas TA-H. Hydrothermal synthesis of TiO_2 nanorods as sensing membrane for extended-gate field-effect transistor (EGFET) pH sensing applications. Sensors and Actuators A: Physical. 2022;**333**:113231

[65] Basavarajappa PS, Patil SB, Ganganagappa N, Reddy KR, Raghu AV, Reddy CV. Recent progress in metal-doped TiO_2, non-metal-doped/codoped TiO_2, and TiO_2 nanostructured hybrids for enhanced photocatalysis. International Journal of Hydrogen Energy. 2020;**45**(13):7764-7778

[66] Zhao Y, Linghu X, Shu Y, Zhang J, Chen Z, Wu Y, et al. Classification and catalytic mechanisms of heterojunction photocatalysts and the application of titanium dioxide (TiO2)-based heterojunctions in environmental remediation. Journal of Environmental Chemical Engineering. 2022;**108077**

[67] Crețescu I, Lutic D. Advanced removal of crystal violet dye from aqueous solutions by photocatalysis using commercial products containing titanium dioxide. Comptes Rendus Chimie. 2022;**25**(S3):1-12

[68] Qin S, Denisov N, Sarma BB, Hwang I, Doronkin DE, Tomanec O, et al. Pt single atoms on TiO2 polymorphs—Minimum loading with a maximized photocatalytic efficiency. Advanced Materials Interfaces. 2022;**9**(22):2200808

[69] Calogero G, Bartolotta A, Di Marco G, Di Carlo A, Bonaccorso F. Vegetable-based dye-sensitized solar cells. Chemical Society Reviews. 2015;**44**(10):3244-3294. DOI: 10.1039/c4cs00309h

[70] Ukoba KO, Eloka-Eboka AC, Inambao FL. Review of solar energy inclusion in Africa: A case study of Nigeria. 2017

[71] Ren P, Qi L, You K, Shi Q. Hydrothermal synthesis of hierarchical SnO2 nanostructures for improved formaldehyde gas sensing. Nanomaterials. 2022;**12**:228

[72] Biswas P, Zhang C, Chen Y, Liu Z, Vaziri S, Zhou W, et al. A portable micro-gas chromatography with integrated photonic crystal slab sensors on chip. Biosensors. 2021;**11**:236

[73] Zampolli S, Elmi I, Cardinali GC, Masini L, Bonafè F, Zardi F. Compact-GC platform: A flexible system integration strategy for a completely microsystems-based gas-chromatograph. Sensors and Actuators B: Chemical. 2020;**305**:127444

[74] Zhai T, Fang X, Liao M, Xu X, Zeng H, Yoshio B, et al. A comprehensive review of one-dimensional metal-oxide nanostructure photodetectors. Sensors. 2009;**9**:6504-6529

[75] Sadanandam G, Lalitha K, Kumari VD, Shankar MV, Subrahmanyam M. Cobalt-doped TiO_2: A stable and efficient photocatalyst for continuous hydrogen production from glycerol: Water mixtures under solar light irradiation. International Journal of Hydrogen Energy. 2013;**38**:9655-9664. DOI: 10.1016/j.ijhydene.2013.05.116

[76] Ramesh RN, Bhargav U, Mamatha Kumari M, Cheralathan KK, Sakar M. Review on the interface engineering in the carbonaceous titania for the improved photocatalytic hydrogen production. International Journal of Hydrogen Energy. 2020;**45**:7584-7615. DOI: 10.1016/j.ijhydene.2019.09.041

[77] Soliman AM, Elwy HM, Thiemann T, Majedi Y, Labata FT, AlRawashdeh NAF. Removal of Pb(II) ions from aqueous solutions by sulphuric acid-treated palm tree leaves. Journal of the Taiwan Institute of Chemical Engineers. 2016;**58**:264-273. DOI: 10.1016/j.jtice.2015.05.035

[78] Singh J, Sahu K, Pandey A, Kumar M, Ghosh T, Satpati B, et al. Atom beam sputtered Ag-TiO2 plasmonic nanocomposite thin films for photocatalytic applications. Applied Surface Science. 2017;**411**:347-354

[79] Mahadwad OK, Parikh PA, Jasra RV, Patil C. Photocatalytic degradation of reactive black-5 dye using TiO2 impregnated ZSM-5. Bulletin of Materials Science. 2011;**34**:551-556

[80] Xing X, Zhu H, Zhang M, Xiao L, Li Q, Yang J. Effect of heterojunctions and phase-junctions on visible-light photocatalytic hydrogen evolution in BCN-TiO2 photocatalysts. Chemical Physics Letters. 2019;**727**:11-18

[81] Guan X, Chowdhury FA, Pant N, Guo L, Vayssieres L, Mi Z. Efficient unassisted overall photocatalytic seawater splitting on GaN-based nanowire arrays. Journal of Physical Chemistry C. 2018;**122**:13797-13802. DOI: 10.1021/acs.jpcc.8b00875

[82] Reddy NR, Bharagav U, Kumari MM, Cheralathan KK, Ojha PK, Shankar MV, et al. Inclusion of low-cost activated carbon for improving hydrogen production performance of TiO2 nanoparticles under natural solar light irradiation. Ceramics International. 2021;**47**(7):10216-10225

[83] Fujishima A, Zhang X, Tryk DA. TiO_2 photocatalysis and related surface phenomena. Surface Science Reports. 2008;**63**:515-582

[84] Gaminian H, Montazer M. Enhanced self-cleaning properties on polyester fabric under visible light through a single-step synthesis of cuprous oxide doped nano-TiO2. Photochemistry and Photobiology. 2015;**91**:1078-1087

[85] Harifi T, Montazer M. Fe3+: Ag/TiO2 nanocomposite: Synthesis, characterization, and photocatalytic activity under UV and visible light irradiation. Applied Catalysis A. 2014;**473**:104-115

[86] Zhou M, Yu J, Cheng B, Yu H. Preparation and photocatalytic activity of Fe-doped mesoporous titanium dioxide nanocrystalline photocatalysts. Materials Chemistry and Physics. 2005;**93**:159-163

[87] El-Roz M, Haidar Z, Lakiss L, Toufaily J, Thibault-Starzyk F. Immobilization of TiO_2 nanoparticles on natural Luffa cylindrica fibers for photocatalytic applications. RSC Advances. 2013;**3**:3438-3445

[88] Píšťková V, Tasbihi M, Vávrová M, Štangar UL. Photocatalytic degradation of β-blockers by using immobilized titania/silica on glass slides. Journal of Photochemistry and Photobiology A: Chemistry. 2015;**305**:19-28

[89] Coronado JM, Sanchez B, Portela R, Suarez S. Influence of catalyst properties and reactor configuration on the photocatalytic degradation of trichloroethylene under sunlight irradiation. Journal of Solar Energy Engineering, Transactions of the ASME. 2008;**130**:041012

[90] Selmi W, Hosni N, Naceur JB, Maghraoui-Meherzi H, Chtourou R. Titanium Dioxide Thin Films for Environmental Applications. In: Titanium Dioxide - Advances and Applications. London, United Kingdom: IntechOpen; 2022

[91] Yan S, Li Y, Xie F, Wu J, Jia X, Yang J, et al. Environmentally safe and porous MS@TiO2@PPy monoliths with superior visible-light photocatalytic properties for rapid oil–water separation and water purification. ACS Sustainable Chemistry & Engineering. 2020;**8**:5347-5359

[92] Zhu Q, Hu X, Stanislaus MS, Zhang N, Xiao R, Liu N, et al. A novel P/Ag/Ag2O/Ag3PO4/TiO2 composite film for water purification and antibacterial application under solar light irradiation. Science of the Total Environment. 2017;**577**:236-244

[93] Guin D, Thakran S, Singh P, Ramachandran S, Hasija Y, Kukreti R. Chapter 1 - Translational biotechnology: A transition from basic biology to evidence-based research. In: Translational Biotechnology. Hasija Y, editor. Elsevier, Academic Press; 2021:3-24

[94] Molaeirad A, Janfaza S, Karimi-Fard A, Mahyad B. Photocurrent generation by adsorption of two main pigments of halobacterium salinarum on TiO_2 nanostructured electrode. Biotechnology and Applied Biochemistry. 2015;**62**:121-125

[95] McCullagh C, Robertson JM, Bahnemann DW, Robertson PK. The application of TiO_2 photocatalysis for disinfection of water contaminated with pathogenic micro-organisms: A review. Research on Chemical Intermediates. 2007;**33**:359-375

[96] Bonetta S, Bonetta S, Motta F, Strini A, Carraro E. Photocatalytic bacterial inactivation by TiO2-coated surfaces. AMB Express. 2013;**3**:59

Chapter 5

Synthesis and Properties of Titanium Dioxide Nanoparticles

Mohsen Mhadhbi, Houyem Abderazzak and Barış Avar

Abstract

Natural titanium dioxide (TiO_2) occurs in three distinct polymorphs (rutile, anatase, and brookite). Currently, TiO_2 gained the attention of several researchers around the world. TiO_2 is used in several applications because of its excellent properties (structural, optical, electrical, chemical, non toxic, etc.). Thus, the applications are influenced by its surface, size, morphology, and crystal phase. TiO_2 as photocatalyst is widely used in energy and eco-friendly applications involving water purification, hydrogen production, phenol degradation, etc. The novelty of the present chapter lies in explaining the recently reported methods that are used to synthesize TiO_2 nanoparticles, such as sol-gel, hydrothermal, precipitation, etc. The different properties of TiO_2 are also provided in this chapter.

Keywords: titanium dioxide nanoparticles, properties, synthesis, structure, natural titanium dioxide (TiO_2)

1. Introduction

Titanium dioxide, with the chemical formula TiO_2, is one of the most valuable raw material and has been used in several applications including photocatalysis, medicine, sensors, paints, environment, solar energy, and others. TiO_2 has excellent corrosion resistance, good thermal and chemical stability, and low cost [1].

With the development of nanotechnology, TiO_2 nanoparticles (NPs), with attractive properties, have been widely fabricated and developed. In the past decades, the demand of titanium dioxide NPs observed remarkable growth because of its specific properties. Moreover, titania is accepted as a pharmaceutical and food additive [2]. It is also used in destruction of viruses and bacteria, inactivation of cancerous cells, as well as clean-up of oil spills [3]. TiO_2 NPs are employed for elimination of emerging contaminants [4]. Moreover, TiO_2 NPs are one of the excellent semi-conducting materials applied in solar cells because of their good chemical stability, low toxicity, low cost, and high photocatalytic activity for the degradation of organic impurity [5, 6]. Furthermore, TiO_2 NPs are widely used as photo-anode materials because of their powerful absorption of light particularly in UV range, good chemical solubility, excellent photo-corrosion resistance and low cost [7, 8]. TiO_2 is widely used as photocatalyst material due to its suitable energy band gap, which is less than 3.5 eV [9].

The recent advances in TiO_2 nanostructures and their applications have been summarized by Reghunath et al. [10]. Chen and Mao [11] have reported a review on the synthesis, properties, modifications, and applications of TiO_2 NPs. Environmental and energy applications of titanium dioxide have been discussed by Ge et al. [12]. Mao et al. [13] have completed a review on the recent progress in TiO_2 based catalysis for energy systems. In their work, Nur et al. [14] have investigated the development of TiO_2 for improved dye degradation under UV-vis irradiation. In addition, the correlation between the improved in photocatalytic activity and various surface modifications have been reported [15]. Fujishima and Honda [16] prepared TiO_2 used as photoelectrode for splitting water via photoelectrochemical water splitting.

This chapter provides recent advances in the synthesis of titanium dioxide NPs and their performance in different applications.

2. Synthesis methods of titanium dioxide nanoparticles

A number of methods have been used for the synthesis of TiO_2 NPs, which are detailed below.

2.1 Polyol method

Recently, the polyol method has been found to be a very powerful route for the fabrication of nano-oxide and chalcogenide materials [17, 18]. In addition, polyol method is a simple and low cost route for fabricating metal oxide NPs.

Thus, a number of studies have been reported on the synthesis of TiO_2 NPs by polyol method. For example, Shah and Rather [19] prepared TiO_2 NPs by polyol method using titanium (IV) butoxide, ethylene glycol, and acetone. They concluded that the mean crystallite size increased from 9.3 to 66.9 nm when calcination temperature rises from 300 to 1000°C. In addition, the obtained products showed greater stability (zeta potential of −30.8 to −37.5 mV) in aqueous solutions. Also, Sasikala et al. [20] prepared a dispersed SnO_2 on TiO_2 NPs via polyol method at calcination temperature of 500°C. They concluded that the TiO_2 containing SnO_2 showed improved photocatalytic activity compared to pure TiO_2 because of improved charge separation. Ultrafine anatase TiO_2 nanocrystals, with size of 2–5 nm, have been prepared through polyol process [21]. **Figure 1** shows the TEM images of TiO_2 nanocrystals. The samples exhibited excellent photocatalytic activities.

Furthermore, polyol method was used to synthesize TiO_2 NPs by using different mole ratios of titanium tetrachloride and polyvinylpyrrolidone [22]. The

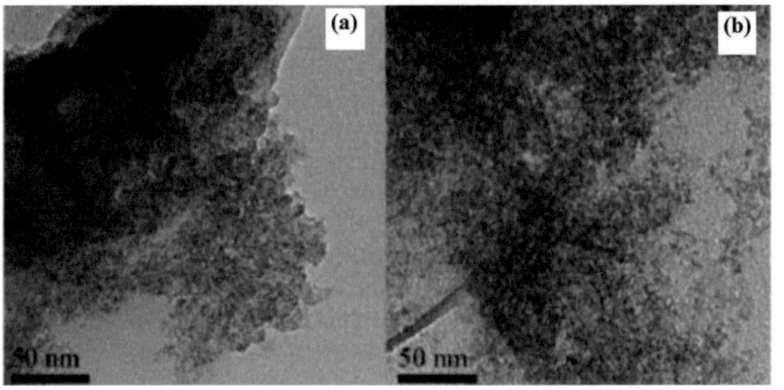

Figure 1.
TEM images of TiO_2 NPs prepared by polyol method [21].

photocatalytic performance of the prepared TiO_2 NPs attained 97.83% with a power conversion efficiency of 4.6%. Kang and co-workers [23] synthesized TiO_2 NPs, with average size of 25 nm, via polyol from titanium isopropoxide by refluxing at 270°C during 12 h. After that, the sample was heated at 600°C for 3 h. The final product showed an excellent electrochemical performance.

2.2 Hydrothermal method

Hydrothermal method is one of the most used route for nanomaterials synthesis. BiOI nanoflowers/TiO_2 nanotubes were developed for the detection of atrazine [24]. The sensing platform showed good analytical performance for detecting atrazine.

Alev et al. [25] prepared TiO_2 nanorods, with diameter of 100 nm, by hydrothermal using titanium butoxide, hydrochloric acid and deonised water. They concluded that the sensor response was 200% for 1000 ppm H_2. Additionally, TiO_2 NPs with size of 20 nm were prepared by hydrothermal method [26]. **Figure 2** shows the TEM image of TiO_2 NPs. The result obtained by UV-VIS analysis revealed that the decrease in size of TiO_2 NPs is beneficial to the blue shift of their absorption peak.

Le et al. [27] synthesized TiO_2/graphene by hydrothermal method using $TiCl_4$ as a precursor. High performance was attained for the catalysts including well dispersed TiO_2 NPs on the graphene surface with loadings ranging from 16.5% to 26%. Similarly, Yang et al. [28] prepared TiO_2 NPs by hydrothermal. The results revealed that the peptization of the precipitate favored formation of the rutile phase and highly crystalline anatase. Europium (Er) doped TiO_2 NPs were prepared by hydrothermal method for photonic application [29]. TEM analysis showed that the average particle size was about 50 nm. Indeed, the Er doping leads to a change in morphology of NPs from rodlike to triangular for Er ions increased from 1 to 3 mol%, respectively (as presented in **Figure 3**).

Ag doped TiO_2 NPs, with crystallite size of 10/13 nm, were prepared via hydrothermal at temperature of 180°C for 120 min [30]. It was revealed that the maximum photodegradation of indigo blue attained 75% after irradiation time of 150 min. Dadkhah et al. [31] prepared anatase TiO_2 NPs by hydrothermal. They achieved conversion efficiency higher than 2.61% with the influence of amine ligands as a shape controller.

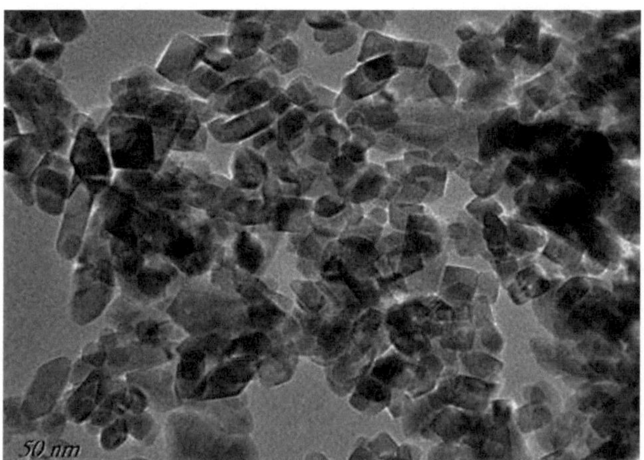

Figure 2.
TEM image of TiO_2 NPs [26].

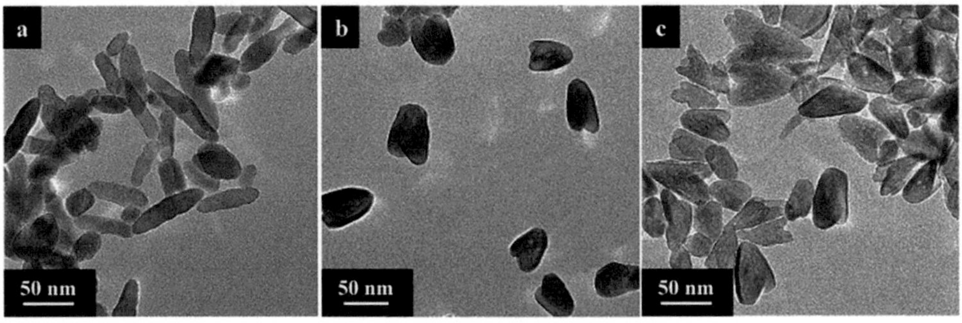

Figure 3.
TEM images of Er doped TiO_2 NPs: (a) 1 Mol %, (b) 2 Mol%, and (c) 3 Mol% Er^{2+} [29].

2.3 Sol-gel method

Sol-gel process is a powerful pathway for the synthesis of multi-component materials because of its mild synthesis conditions and low temperature. Thus, several researches have been reported on the fabrication of TiO_2 NPs by sol-gel method. For example, Sabry et al. [32] synthesized TiO_2 NPs by sol-gel process. The prepared material showed efficient photocatalytic activity of up to 68% after 180 min. Hsiung et al. [33] investigated the structure of photocatalytic active sites of TiO_2 NPs prepared by sol-gel. They concluded that the material exhibited an excellent photocatalytic activity. Additionally, TiO_2 NPs used as catalyst was prepared by sol-gel in acid at pH 3 [34]. The result showed that the material exhibited excellent reactivity for the photocatalytic reduction of nitric oxide.

Venkatachalam et al. [35] prepared alkaline earth metal (Mg^{2+} and Ba^{2+}) doped TiO_2 NPs by sol-gel method using titanium isopropoxide as precursor. **Figure 4** illustrates the SEM image of the metal-doped TiO_2 NPs, which are spherical in shape. Furthermore, the final product exhibited higher photocatalytic activity for the bisphenol.

Saravanan and Duby [36] investigated the optical and morphological properties of TiO_2 NPs synthesized via sol-gel method using titanium butoxide as a precursor. UV-Visible analyses revealed the absorbance peak in the UV region (about 380 nm) and FTIR spectrum confirmed the existence of anatase TiO_2 in the range of 400–1000 cm^{-1}. The average particle size of the TiO_2 NPs determined by dynamic light scattering (DLS) was found 131 nm. Govindaraj et al. [37] synthesized TiO_2 NPs to be used as a photo-anode by the sol-gel method. UV-Visible spectrum revealed the light absorption in the UV region with optical bandgap of 3.2 eV (see **Figure 5**).

Sinha et al. [38] studied the structural, optical, and antibacterial performance of the Mg doped TiO_2 NPs prepared by the sol-gel method. They reported that optical transmittance increases from 3 to 3.07 eV. In addition, the photoluminescence emission shows inner UV to blue resign from pure and doped TiO_2 NPs. Furthermore, Mugundan et al. [39] synthesized barium doped TiO_2 NPs by the sol-gel method. They concluded that the pure TiO_2 NPs revealed higher second harmonic generation efficiency compared to barium doped TiO_2 NPs. Nachit et al. [40] investigated the photocatalytic activity of TiO_2 NPs prepared by sol-gel process at low temperature. The mean crystallite size of TiO_2 NPs reached 30 nm at 500°C using an acid. In addition, the photocatalytic activity of TiO_2 NPs revealed that the degradation of Rhodamine B under UV light have a removal efficiency of 95% during 60 min.

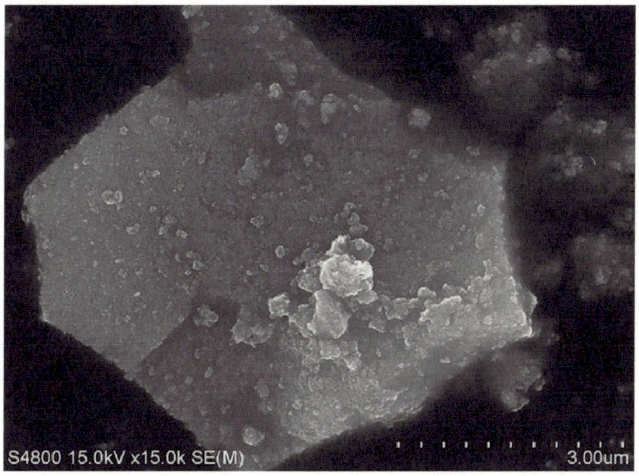

Figure 4.
SEM image of metal-doped TiO_2 NPs prepared by sol-gel [35].

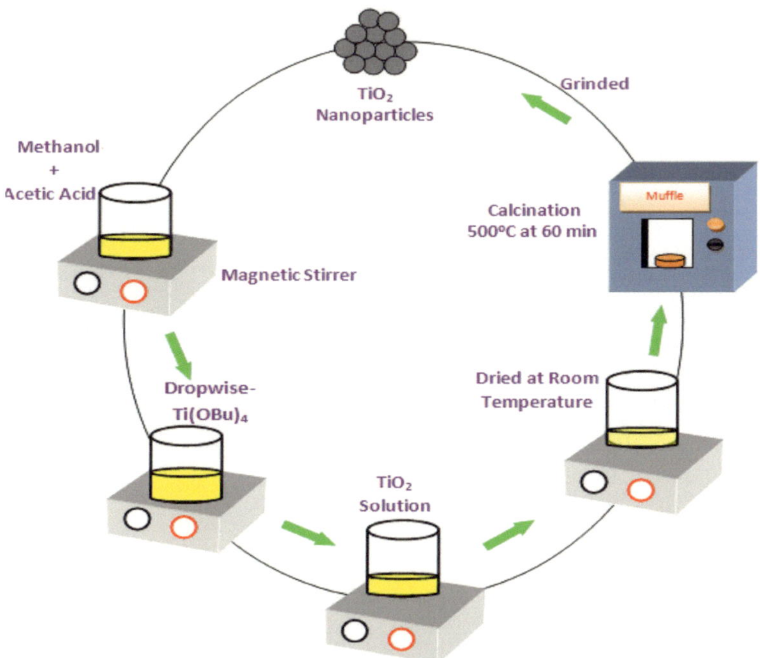

Figure 5.
Schematic diagram of the sol-gel route used for preparing TiO_2 NPs [37].

2.4 Chemical vapor deposition method

Chemical vapor deposition (CVD) is a powerful method for the synthesis of nanomaterials with enhanced performances.

Several works have been reported on the fabrication of TiO_2 NPs by CVD. For example, Liu et al. [41] prepared TiO_2 NPs by CVD method. They revealed that the gas phase hydrolysis reaction may be decomposed into two processes: (i) hydrolysis of $TiCl_4$ into $TiO(OH)_2$ and (ii) decomposition of $TiO(OH)_2$ to TiO_2. The influence of various concentrations of TiO_2 NPs on CVD grown graphene was investigated by electrical charge transport measurements and Raman spectroscopy [42]. The schematic diagram of TiO_2 doped CVD grown single layer graphene devices is presented in **Figure 6**. The obtained results showed that TiO_2 change the electronic properties besides the structure of the CVD grown graphene.

Similarly, Li et al. [43] synthesized TiO_2 NPs, with mean particle size of 22 nm, by CVD method. The TEM image of TiO_2 NPs is presented in **Figure 7**. They concluded that the TiO_2 NPs with the metal ion dopants possess elevated photocatalytic activities compared to un-doped TiO_2 NPS.

Ding et al. [44] synthesized TiO_2 NPs via CVD process. The results obtained by XPS and nitrogen ads/desorption revealed that most of TiO_2 NPs were distributed on the external surface of the support and the coating was stable. V_2O_5-TiO_2 NPs were prepared from two precursors by CVD [45]. They revealed that the CVD process was a suitable method for the single step synthesis of nanocomposite coatings. Lee et al. [46] prepared TiO_2 NPs by CVD method. The results revealed that a 60 min sample coating time gave the most highly photocatalytic activity.

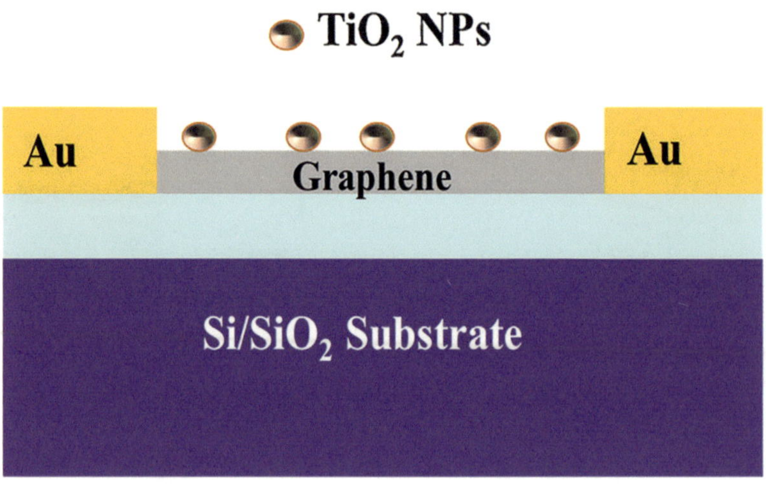

Figure 6.
Schematic diagram of TiO_2 doped CVD grown single layer graphene devices [42].

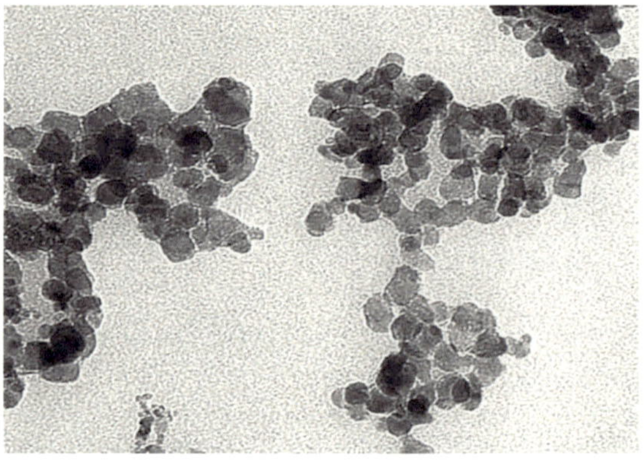

Figure 7.
TEM image of TiO_2 NPs prepared by CVD [43].

2.5 3D printing method

In the last few years, several works have been developed to fabricate 3D porous materials; principally 3D porous TiO_2 based materials.

Arango et al. [47] prepared a porous TiO_2 by 3D printing. They suggested that a large surface area could be realized for the TiO_2 via 3D printing technology.

Liu et al. [48] used 3D printing to prepare the porous Pb/TiO_2 composites applied to remove the organic contamination in the wastewater. The obtained materials exhibited high catalytic activity, good stability, and reusability against the treatment of high concentration 4-NP wastewater. The optical images of the Pb/TiO_2 scaffolds with 4, 8, 12, and 16 layers are presented in **Figure 8**.

Additionally, Aleni et al. [49] used 3D printing to fabricate a 3D dense and porous TiO_2 structure. The final products exhibited similar mechanical properties to those of porous ceramics prepared via conventional methods.

Xu et al. [50] developed 3D printing to assembly TiO_2 powders into hierarchical porous structures at macro and microscale. The schematic illustration of 3D printing process is presented in **Figure 9**. The obtained results showed that the TiO_2 structures with abundant light absorption sites and high surface area could enhance the conversion efficiency of N_2 and NH_3.

Furthermore, Wang et al. [51] synthesized TiO_2 NPs containing macrostructures by 3D printing for Arsenic (III) removal in water. They showed that 3D printing could fabricate and design macrostructures with special functions.

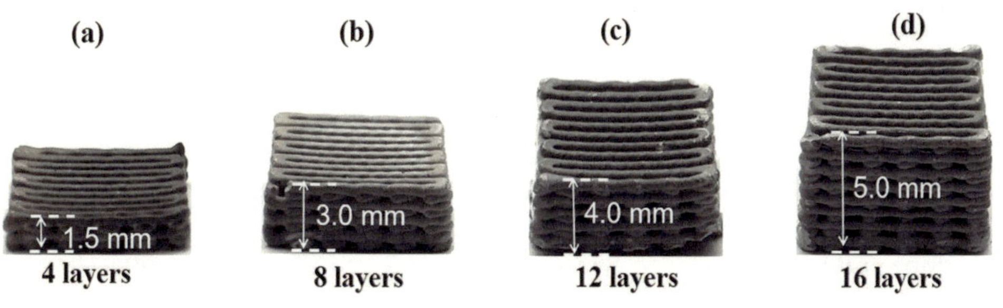

Figure 8.
Optical images of the Pb/TiO_2 scaffolds with (a): 4, (b): 8, (c): 12, and (d): 16 layers.

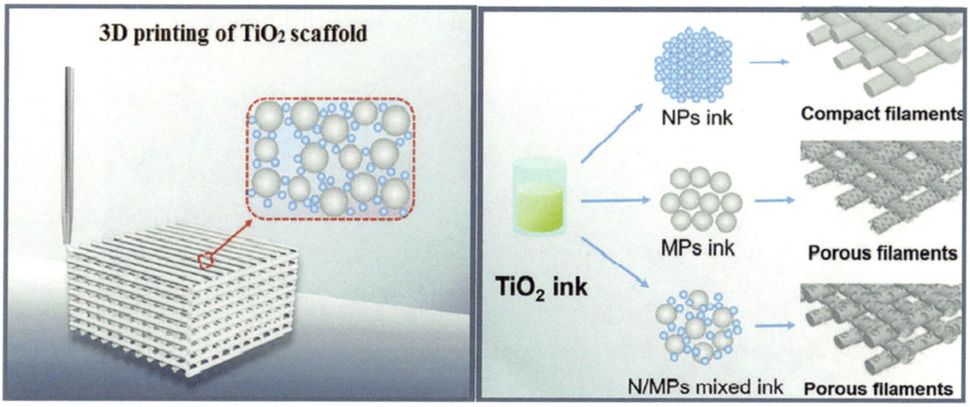

Figure 9.
Schematic illustration of 3D printing of a hierarchical porous TiO_2 [50].

2.6 Mechanical alloying

Mechanical alloying (MA) is a low cost and simple route for preparing nanostructured materials among them TiO_2 NPs. The schematic illustration of the MA is presented in **Figure 10**.

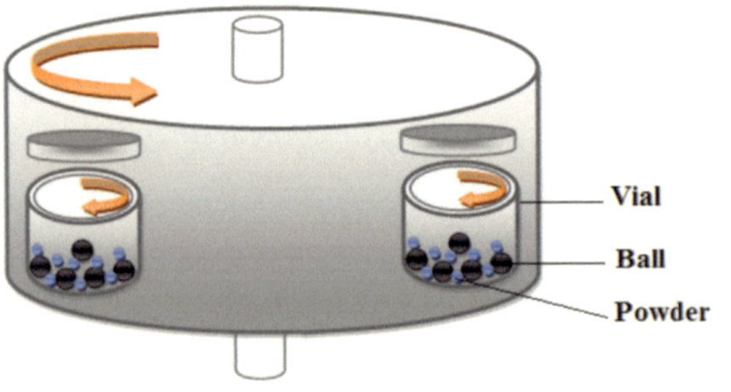

Figure 10.
Schematic illustration of the MA process.

Yao et al. [52] prepared nanostructured TiO_2 coating by mechanical alloying process. The results showed that the obtained material exhibited an excellent photocatalytic activity. Vilchez et al. [53] synthesized TiO_2 NPs by MA during 5 min. The TEM images of TiO_2 NPs are presented in **Figure 11**. The obtained material, with size in the range of 2–4 nm and specific surface area of 298 $m^2 g^{-1}$, exhibiting a good photocatalytic activity.

Kim et al. [54] prepared TiO_2 NPs by MA and heat treatment. The mean crystallite size was less than 6 nm. The UV-Visible spectrum showed that the obtained TiO_2 NPs had an elevated wavelength rage (in the range of 650 and 700 nm) compared to Ni doped TiO_2 (480–500 nm) and rutile (380–400 nm). In addition, PL spectrum exhibited a new emission peak confirming the decrease in the band gap. Furthermore, Fe (III) doped TiO_2 NPs have been synthesized via MA [55]. The final product showed excellent selectivity, stability, sensitivity, and fast response. Additionally, Eadi et al. [56] developed new Fe doped TiO_2 NPs by MA from $FeCl_3$ and TiO_2 powder. The results showed that the mean particle size was about 28 nm and the prepared material could be applied for gas sensing and photocatalytic degradation. Carniero et al. [57] investigated the effect of process parameters on the structural, optical, magnetic,

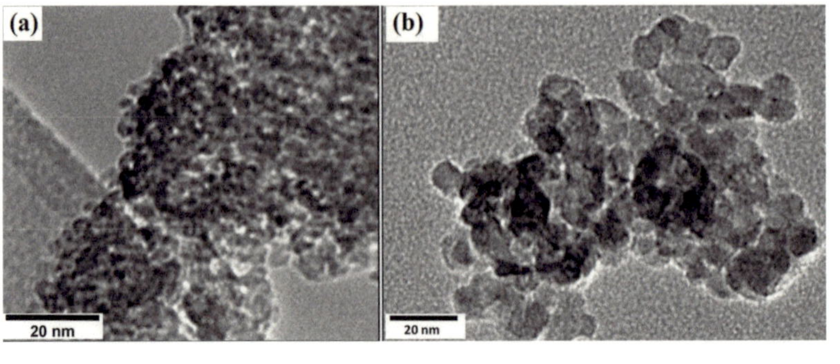

Figure 11.
TEM images of TiO_2 NPs prepared by MA [53].

and photocatalytic properties of iron doped TiO_2 NPs prepared by MA. The results showed that the incorporation of iron in the TiO_2 NPs has improved their photocatalytic activity.

2.7 Green synthesis

Green synthesis is a simple and ecofriendly method used for the preparation of nanomaterials. Abisharani et al. [58] synthesized TiO_2 NPs from titanium trychloride using Cucurbita pepo seeds extract. FTIR results showed that the existence of different functional biomolecules acted as a reducing factor for conversion of TiO_4 into TiO_2 NPs.

Isnaeni et al. [59] prepared TiO_2 NPs by green method including $TiCl_3$ hydrolysis with mango-peel extract. They revealed that the used method could be employed as an alternative to prepare phase pure anatase and rutile. Helmy et al. [60] synthesized S doped TiO_2 NPs by a novel green synthesis using Malva parviflora plant extract. They also studied their photocatalytic, antimicrobial, and antioxidant activities. The results showed that the samples exhibited good antibacterial and photocatalytic activities.

In addition, Samhitha et al. [61] studied the TiO_2 NPs prepared by various green synthesis methods for anticancer applications. Shen et al. [62] prepared Ce doped TiO_2 NPs supported on porous glass. **Figure 12** shows TEM image of TiO_2 NPs. The mean diameter was about 5 nm. This study concludes that the green method makes Ce doped TiO_2 NPs immobilized on porous glass.

Additionally, TiO_2 NPs were synthesized through green method from *Demostachaya bipinnata* extract [63]. It has been shown that the prepared TiO_2 NPs are a good candidate for controlling mosquito vectors and agricultural pest management. Nabi et al. [64] prepared TiO_2 NPs, with mean crystallite size in the range of 80–100 nm, by green method using citrus limetta extract (as presented in **Figure 13**). The results showed that the degradation activity was more than 90% within 80 min. This excellent photocatalytic activity confirms that TiO_2 NPs are ecofriendly and have powerful applications in purification of water.

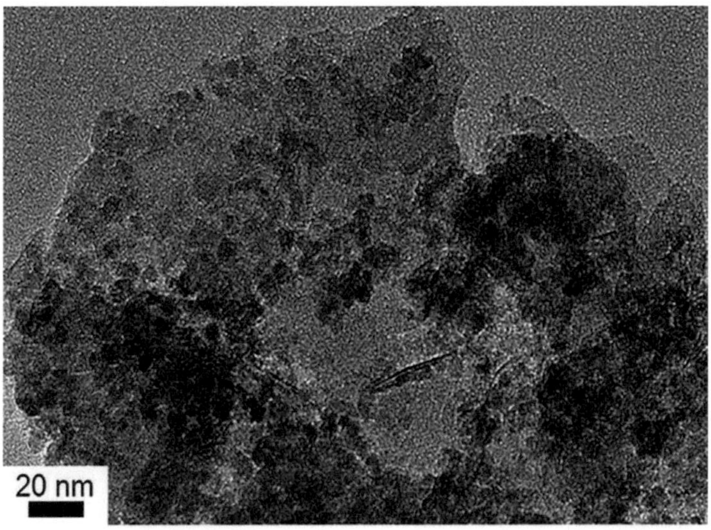

Figure 12.
TEM image of TiO_2 NPs prepared by green method [62].

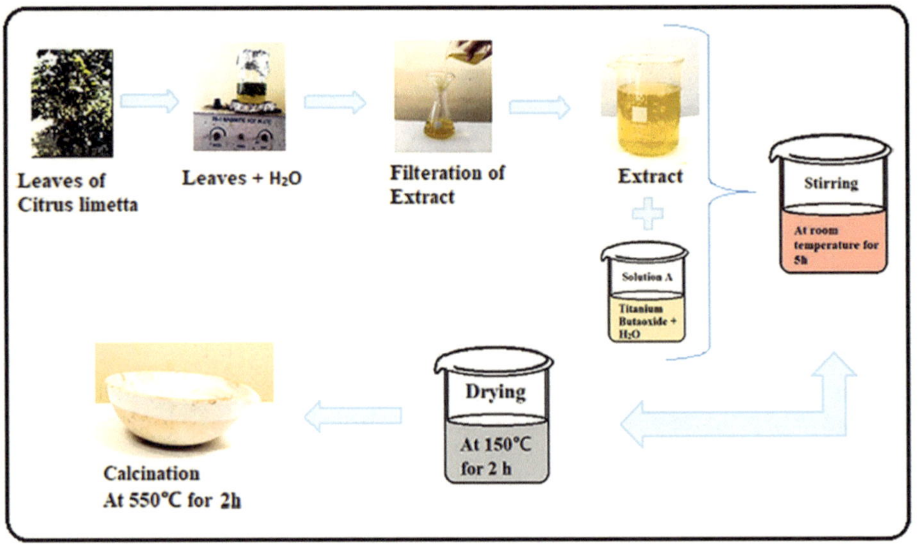

Figure 13.
Schematic diagram of the synthesis of TiO_2 NPs by green method [64].

3. Properties of titanium dioxide nanoparticles

Figure 14 shows the different crystal structures of TiO_2 [65]. As it can be seen in this figure, there are three forms (polymorphs) namely anatase, rutile, and brookite, which are classified according to their crystalline arrangements. Thus, rutile is the most stable at higher temperature, whereas anatase is the most stable at lower temperature. Furthermore, at high temperature, anatase and brookite could be transformed into rutile. Brookite in the powder or thin film forms reveals excellent stability and superior photocatalytic activity to that of anatase [66]. In addition, anatase is favored in photocatalysis because of its high photocatalytic activity between all the three polymorphs [19].

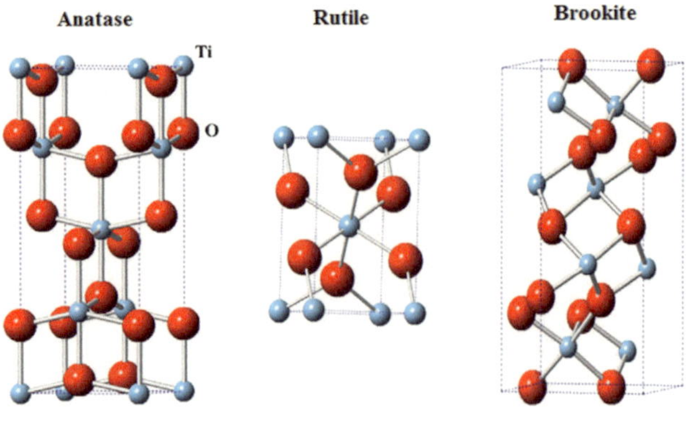

Figure 14.
Different crystal structures of TiO_2 [65].

Table 1 illustrates the different properties of titanium dioxide NPs. As it can be concluded from these values, TiO_2 NPs possess interesting physicochemical properties, which are influenced by different factors such as exposed crystal faces, morphology, and size of particles. Chen and Mao have published a review on the synthesis, properties, and applications of TiO_2 NPs [11].

Parameter	Value
Density (g/cm^3)	4.23
Crystal structure	Tetragonal
Appearance	White solid
Melting point (°C)	1870
Boiling point (°C)	2500–3000
Molecular weight (g/mol)	79.88
Chemical formula	TiO_2
Young's modulus (GPa)	244
Thermal conductivity at 800°C (W m^{-1} K^{-1})	8
Coefficient of thermal expansion (10^{-6}/K)	9
Refractive index	2.55–2.75
Mohr's hardness	5.5–7
Specific gravity	4
Size range (nm)	30–50

Table 1.
Various properties of titanium dioxide NPs [67].

Figure 15 shows the different hierarchical nanostructures of TiO_2. Four morphologies, involving 0D (quantum dots), 1D (nanotubes, nanorods, nanofibers,...), 2D (nanflakes nanosheets,...), and 3D (nanospheres, nanoflowers,...), can be obtained.

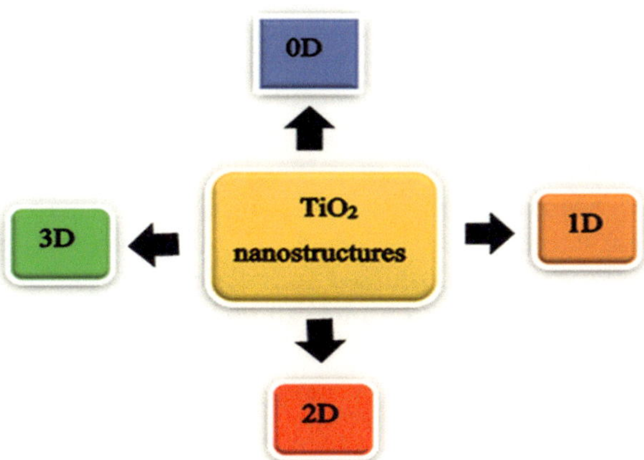

Figure 15.
Different hierarchical nanostructures of TiO_2.

4. Conclusions

In this chapter, we summarized some advances in the synthesis and properties of titanium dioxide nanoparticles. TiO_2 is basically found in three crystalline forms: brookite, anatase and rutile. Its important potential application including its use as a food additive, in cosmetics, as a pigment, semiconductor, as well as in catalysis and photocatalysis, for UV adsorption and hydrogen storage has contributed to its massive elaboration by different methods and processes. On the other hand, materials with a nanometric structure display structural, mechanical, physical, chemical, optical and electrical properties that are distinctly improved in comparison to the materials with a micrometric structure. However, each synthesis method allows favoring one or more of the above mentioned properties, allowing to promote the application of the obtained material in a specific field.

Several researches have been made on the preparation and characterization of TiO_2 NPs for various applications. Different synthesis methods have been presented to prepare titanium dioxide nanoparticles. For instance polyol process, which combines simplicity and low cost, allows to obtain TiO_2 NPs with different shapes and sizes depending on the starting reagents and operating conditions for photocatalytic activities applications. Hydrothermal is the most used method for nanomaterials synthesis and titanium dioxide can be successfully synthesized with different nanoscale shapes as sensors including dispersed TiO_2 NPs on the graphene surface. Nevertheless, the sol-gel method remains a powerful alternative for the synthesis of multi-component materials at mild and low temperature conditions leading to efficient photocatalytic activity of TiO_2. However, the Chemical vapor deposition process is suitable for the single step synthesis of nanocomposite coatings with enhanced properties. In this context, single layer graphene devices doped with TiO_2 have been obtained by CVD. This doping has shown that TiO_2 modifies the electronic properties as well as the structure of the CVD grown graphene. On the other hand, 3D porous TiO_2 based materials with high catalytic activity and good stability can be obtained through 3D printing technology. Among the simplest and most cost-effective processes for nanostructured materials synthesis, mechanical alloying is a very powerful technique for rapid elaboration of TiO_2 NPs with excellent photocatalytic activity Nevertheless, compared to conventional methods, green method has been proven to be far more efficient; low cost, and eco-friendly route to the synthesis of TiO_2 NPs.

The results obtained in this work enable a better understanding of the synthesis methods as well as the different related properties of titanium dioxide nanoparticles. However, the selection of the synthesis method is conditioned by the required properties of the titanium dioxide NPs and the cost of the final material to be obtained. This is all the more sought after for a value-added and large-scale TiO_2 elaboration, which promotes the development of more innovative applications.

XRD, SEM, and TEM are the most used techniques for the nanostructured titanium dioxide characterization. The structural, morphological, and intrinsic properties of TiO_2 NPs were also discussed and related to its performance in various applications. Titanium dioxide was a prime candidate material because of its low-cost, high-abundance, and ease of synthesis.

Acknowledgements

The authors would like to thank Ms. Maja Bozicevic, Author Service Manager, for her remarkable efforts. This work was supported by Scientific Research Projects Coordination Unit of Zonguldak Bülent Ecevit University, project n° 2022-73338635-01.

Conflict of interest

The authors declare no conflict of interest.

Author details

Mohsen Mhadhbi[1*], Houyem Abderazzak[1] and Barış Avar[2,3]

1 Laboratory of Useful Materials, National Institute of Research and Physicochemical Analysis, Ariana, Tunisia

2 Department of Metallurgical and Materials Engineering, Zonguldak Bülent Ecevit University, Zonguldak, Turkey

3 Department of Nanotechnology Engineering, Zonguldak Bülent Ecevit University, Zonguldak, Turkey

*Address all correspondence to: mhadhbi_mohsen@yahoo.fr

References

[1] Roy J. The synthesis and applications of TiO_2 nanoparticles derived from phytochemical sources. Journal of Industrial and Engineering Chemistry. 2022;**106**:1-19. DOI: 10.1016/j.jiec.2021.10.024

[2] Rowe RC, Sheskey PJ, Weller PJ. Handbook of Pharmaceutical Excipients. 4th ed. London: Pharmaceutical Press, London, United Kingdom, and the American Pharmaceutical Association; 2003

[3] Kwon S, Fan M, Cooper AT, Yang H. Photocatalytic applications of micro- and nano-TiO_2 in environmental engineering. Critical Reviews in Environmental Science and Technology. 2008;**38**(3):197-226. DOI: 10.1080/10643380701628933

[4] Arun Kumar D, Merline Shyla J, Xavier FP. Synthesis and characterization of TiO_2/SiO_2 nano composites for solar cell applications. Applied Nanoscience. 2012;**2**(4):429-436. DOI: 10.1007/s13204-012-0060-5

[5] Zhang J, Xu Q, Feng Z, Li M, Li C. Importance of the relationship between surface phases and photocatalytic activity of TiO_2. Angewandte Chemie International Edition. 2008;**47**:1766-1769. DOI: 10.1002/ange.200704788

[6] Ni M, Leung MK, Leung DY, Sumathy K. A review and recent developments in photocatalytic water-splitting using TiO_2 for hydrogen production. Renewable and Sustainable Energy Reviews. 2007;**11**:401-425. DOI: 10.1016/j.rser.2005.01.009

[7] Palmas S, Polcaro AM, Ruiz JR, Da Pozzo A, Mascia M, Vacca A. TiO_2 photoanodes for electrically enhanced water splitting. International Journal of Hydrogen Energy. 2010;**35**:6561-6570. DOI: 10.1016/j.ijhydene.2010.04.039

[8] Desai D, Wei X, Steingart DA, Banerjee S. Electrodeposition of preferentially oriented zinc for flow-assisted alkaline batteries. Journal of Power Sources. 2014;**256**:145-152. DOI: 10.1016/j.jpowsour.2014.01.026

[9] Alosfur FKM, Ridha NJ, Jumali MHH, Radiman S. Structure and optical properties of TiO_2 nanorods prepared using polyol solvothermal method. Nanotechnology. 2018;**29**(14):145707. DOI: 10.1063/1.5123095

[10] Reghunath S, Pinheiro D, Sunaja Devi KR. A review of hierarchical nanostructures of TiO_2: Advances and applications. Applied Surface Science Advances. 2021;**3**:100063. DOI: 10.1016/j.apsadv.2021.100063

[11] Chen X, Mao SS. Titanium dioxide nanomaterials: Synthesis, properties, modifications and applications. Chemical Reviews. 2007;**107**:2891-2959. DOI: 10.1021/cr0500535

[12] Ge M, Cao C, Huang J, Li S, Chen Z, Zhang K-Q, et al. A review of one-dimensional TiO_2 nanostructured materials for environmental and energy applications. Journal of Materials Chemistry A. 2016;**4**(18):6772-6801. DOI: 10.1039/C5TA09323F

[13] Mao H, Zhang F, Meng D, Dai L, Qian Y, Pang H. Review on synthesis of porous TiO_2-based catalysts for energy conversion systems. Ceramics International. 2021;**47**(18):25177-25200. DOI: 10.1016/j.ceramint.2021.06.039

[14] Nur ASM, Sultana M, Mondal A, Islam S, Robel FN, Islam A, et al. A

review on the development of elemental and codoped TiO_2 photocatalysts for enhanced dye degradation under UV–vis irradiation. Journal of Water Process Engineering. 2022;**47**:102728. DOI: 10.1016/j.jwpe.2022.102728

[15] Low J, Cheng B, Yu J. Surface modification and enhanced photocatalytic CO_2 reduction performance of TiO_2: A review. Applied Surface Science. 2017;**392**:658-686. DOI: 10.1016/j.apsusc.2016.09.093

[16] Fujishima A, Honda K. Electrochemical photolysis of water at a semiconductor electrode. Nature. 1972;**238**:37-38. DOI: 10.1038/238037a0

[17] Toneguzzo P, Viau G, Acher O, Fievet-Vincent F, Fievet F. Monodisperse ferromagnetic particles for microwave applications. Advanced Materials. 1998;**10**:1032-1035. DOI: 10.1002/(SICI)1521-4095(199809)10:13<1032::AID-ADMA1032>3.0.CO;2-M

[18] Grisaru H, Palchik O, Gedanken A, Palchik V, Slifkin MA, Weiss AM. Microwave-assisted polyol synthesis of $CuInTe_2$ and $CuInSe_2$ nanoparticles. Inorganic Chemistry. 2003;**42**:7148-7155. DOI: 10.1021/ic0342853

[19] Shah AH, Rather MA. Effect of calcination temperature on the crystallite size, particle size and zeta potential of TiO_2 nanoparticles synthesized via polyol-mediated method. Materials Today: Proceedings. 2021;**44**:482-488. DOI: 10.1016/j.matpr.2020.10.199

[20] Sasikala R, Shirole A, Sudarsan V, Sakuntala T, Sudakar C, Naik R, et al. Highly dispersed phase of SnO_2 on TiO_2 nanoparticles synthesized by polyol-mediated route: Photocatalytic activity for hydrogen generation. International Journal of Hydrogen Energy. 2009;**34**:3621-3630. DOI: 10.1016/j.ijhydene.2009.02.085

[21] Wang Y, Zhang L, Li S, Jena P. Polyol-mediated synthesis of ultrafine TiO_2 nanocrystals and tailored physiochemical properties by Ni doping. Journal of Physical Chemistry C. 2009;**113**(21):9210-9217. DOI: 10.1021/jp902306h

[22] Pratheep P, Vijayakumar E, Subramania A. Polyol thermolysis synthesis of TiO_2 nanoparticles and its paste formulation to fabricate photoanode for dye-sensitized solar cells. Applied Physics A: Materials Science & Processing. 2015;**119**:497-502. DOI: 10.1007/s00339-014-8980-4

[23] Kang JW, Kim DH, Mathew V, Gim JH, Yu I, Woo CH, et al. Polyol-mediated synthesis of TiO2 nanoparticles. Defect and Diffusion Forum. 2011;**312-315**:160-165. Trans Tech Publications, Ltd. DOI: 10.4028/www.scientific.net/ddf.312-315.160

[24] Fan L, Liang G, Zhang C, Fan L, Yan W, Guo Y, et al. Visible-light-driven photoelectrochemical sensing platform based on BiOI nanoflowers/TiO_2 nanotubes for detection of atrazine in environmental samples. Journal of Hazardous Materials. 2021;**409**:124894. DOI: 10.1016/j.jhazmat.2020.124894

[25] Alev O, Şennik E, Kilinç N, Oztürk ZZ. Gas sensor application of hydrothermally growth TiO_2 nanorods. Procedia Engineering. 2015;**120**:1162-1165. DOI: 10.1016/j.proeng.2015.08.747

[26] Xiaoming F. Synthesis and optical absorpition properies of anatase TiO_2 nanoparticles via a hydrothermal hydrolysis method. Rare Metal Materials and Engineering. 2015;**44**(5):1067-1070. DOI: 10.1016/S1875-5372(15)30068-0

[27] Le T-LT, Le T-HT, Van KN, Van Bui H, Le TG, Vo V. Controlled growth of TiO_2 nanoparticles on graphene by hydrothermal method for visible-light photocatalysis. Journal of Science: Advanced Materials and Devices. 2021;**6**(4):516-527. DOI: 10.1016/j.jsamd.2021.07.003

[28] Yang J, Mei S, Ferreira MF. Hydrothermal synthesis of nanosized titania powders: Influence of peptization and peptizing agents on the crystalline phases and phase transitions. Journal of the American Ceramic Society. 2000;**83**:1361-1368. DOI: 10.1111/j.1151-2916.2000.tb01394.x

[29] Jeon S, Braun PV. Hydrothermal synthesis of Er-doped luminescent TiO_2 nanoparticles. Chemistry of Materials. 2003;**15**(6):1256-1263. DOI: 10.1021/cm0207402

[30] Avciata O, Benli Y, Gorduk S, Koyun O. Ag doped TiO_2 nanoparticles prepared by hydrothermal method and coating of the nanoparticles on the ceramic pellets for photocatalytic study: Surface properties and photoactivity. Journal of Engineering Technology and Applied Sciences. 2016;**1**:34-50. DOI: 10.30931/jetas.281381

[31] Dadkhah M, Salavati-Niasari M, Mir N. Synthesis and characterization of TiO_2 nanoparticles by using new shape controllers and its application in dye sensitized solar cells. Journal of Industrial and Engineering Chemistry. 2014;**20**(6):4039-4044. DOI: 10.1016/j.jiec.2014.01.003

[32] Sabry RS, Al-Haidarie YK, Kudhier MA. Synthesis and photocatalytic activity of TiO_2 nanoparticles prepared by sol–gel method. Journal of Sol-Gel Science and Technology. 2016;**78**:299-306. DOI: 10.1007/s10971-015-3949-0

[33] Hsiung TL, Paul Wang H, Lin HP. Chemical structure of photocatalytic active sites in nanosize TiO_2. Journal of Physics and Chemistry of Solids. 2008;**69**:383-385. DOI: 10.1016/j.jpcs.2007.07.009

[34] Karami A. Synthesis of TiO_2 nano powder by the sol-gel method and its use as a photocatalyst. Journal of the Iranian Chemical Society. 2010;**7**(Suppl. 2):S154-S160. DOI: 10.1007/BF03246194

[35] Venkatachalam N, Palanichamy M, Murugesan V. Sol–gel preparation and characterization of alkaline earth metal doped nano TiO_2: Efficient photocatalytic degradation of 4-chlorophenol. Journal of Molecular Catalysis A. 2007;**273**(1-2):177-185. DOI: 10.1016/j.molcata.2007.03.077

[36] Saravanan S, Dubey RS. Optical and morphological studies of TiO_2 nanoparticles prepared by sol–gel method. Materials Today: Proceedings. 2021;**47**:1811-1814. DOI: 10.1016/j.matpr.2021.03.207

[37] Govindaraj R, Senthil Pandiyan M, Ramasamy P, Mukhopadhyay S. Sol–gel synthesized mesoporous anatase titanium dioxide nanoparticles for dye sensitized solar cell (DSSC) applications. Bulletin of Materials Science. 2015;**38**:291-296. DOI: 10.1007/s12034-015-0874-3

[38] Sinha S, Patel V, Verma H, Akhani TK, Rathore MS. Structural, optical and antibacterial performance of the Mg doped TiO_2 nanoparticles synthesized using sol–gel method. Materials Today: Proceedings. 2022;**67**(Part 5):694-700. DOI: 10.1016/j.matpr.2022.06.543

[39] Mugundan S, Praveen P, Sridhar S, Prabu S, Lawrence Mary K, Ubaidullah M, et al. Sol-gel synthesized

barium doped TiO_2 nanoparticles for solar photocatalytic application. Inorganic Chemistry Communications. 2022;**139**:109340. DOI: 10.1016/j.inoche.2022.109340

[40] Nachit W, Ait Ahsaine H, Ramzi Z, Touhtouh S, Goncharova I, Benkhouja K. Photocatalytic activity of anatase-brookite TiO_2 nanoparticles synthesized by sol gel method at low temperature. Optical Materials. 2022;**129**:112256. DOI: 10.1016/j.optmat.2022.112256

[41] Liu W, Tang H, Liu D. Combining density functional theory and CFD-PBM model to predict TiO_2 nanoparticle evolution during chemical vapor deposition. Chemical Engineering Journal. 2022;**454**:140174. DOI: 10.1016/j.cej.2022.140174

[42] Singh AK, Chaudhary V, Singh AK, Sinha SRP. Effect of TiO_2 nanoparticles on electrical properties of chemical vapor deposition grown single layer grapheme. Synthetic Metals. 2019;**256**:116155. DOI: 10.1016/j.synthmet.2019.116155

[43] Li W, Ismat Shah S, Huang C-P, Jung O, Ni C. Metallorganic chemical vapor deposition and characterization of TiO_2 nanoparticles. Materials Science and Engineering B. 2002;**96**(3):247-253. DOI: 10.1016/S0921-5107(02)00352-5

[44] Ding Z, Xijun H, Lu GQ, Yue P-L, Paul F. Greenfield novel silica gel supported TiO_2 photocatalyst synthesized by CVD method. Langmuir. 2000;**16**(15):6216-6222. DOI: 10.1021/la000119l

[45] Song MY, Park Y-K, Jurng J. Direct coating of V_2O_5/TiO_2 nanoparticles onto glass beads by chemical vapor deposition. Powder Technology. 2012;**231**:135-140. DOI: 10.1016/j.powtec.2012.07.043

[46] Lee H, Song MY, Jurng J, Park Y-K. The synthesis and coating process of TiO_2 nanoparticles using CVD process. Powder Technology. 2011;**214**(1):64-68. DOI: 10.1016/j.powtec.2011.07.036

[47] Torres Arango MA, Kwakye-Ackah D, Agarwal S, Gupta RK, Sierros KA. Environmentally friendly engineering and three-dimensional printing of TiO_2 hierarchical mesoporous cellular architectures. ACS Sustainable Chemistry & Engineering. 2017;**5**:10421-10429. DOI: 10.1021/acssuschemeng.7b02450

[48] Liu T, Sun Y, Jiang B, Guo W, Qin W, Xie Y, et al. Pd nanoparticle-decorated 3D-printed hierarchically porous TiO_2 scaffolds for the efficient reduction of a highly concentrated 4-nitrophenol solution. ACS Applied Materials & Interfaces. 2020;**12**:28100-28109. DOI: 10.1021/acsami.0c03959

[49] Aleni AH, Kretzschmar N, Jansson A, Ituarte IF, St-Pierre L. 3D printing of dense and porous TiO_2 structures. Ceramics International. 2020;**46**:16725-16732. DOI: 10.1016/j.ceramint.2020.03.248

[50] Xu C, Liu T, Guo W, Sun Y, Liang C, Cao K, et al. 3D printing of powder-based inks into functional hierarchical porous TiO_2 materials. Advanced Engineering Materials. 2020;**22**:1901088. DOI: 10.1002/adem.201901088

[51] Wang D, Zhi T, Liu L, Yan L, Yan W, Tang Y, et al. 3D printing of TiO_2 nano particles containing macrostructures for As(III) removal in water. Science of the Total Environment. 2022;**815**:152754. DOI: 10.1016/j.scitotenv.2021.152754

[52] Yao Q, Sun J, Zhu Y, Zhang H, Tong W. TiO_2 coating prepared by mechanical alloying treatment for photocatalytic degradation. Surface

Engineering. 2019;**35**(11):927-932. DOI: 10.1080/02670844.2018.1554738

[53] Vílchez A, Rodriguez-Abreu C, Esquena J, et al. Mechanochemical synthesis of TiO_2 nanoparticles and their self-organization at interfaces to produce emulsion-templated photocatalytic porous polymers. Journal of Inorganic and Organometallic Polymers. 2021;**31**:1912-1930. DOI: 10.1007/s10904-021-01885-7

[54] Kim DH, Park HS, Kim S-J, Lee KS. Synthesis of novel TiO_2 by mechanical alloying and heat treatment-derived nanocomposite of TiO_2 and $NiTiO_3$. Catalysis Letters. 2006;**106**(1-2):29-33. DOI: 10.1007/s10562-005-9186-3

[55] Darbandi M, Panahi P, Asadpour-Zeynali K. Fe (III) doped TiO_2 nanoparticles prepared by high energy ball milling as booster for non-enzymatic, mediator-free and sensitive electrochemical sensor. Microchemical Journal. 2022;**183**:108093. DOI: 10.1016/j.microc.2022.108093

[56] Eadi SB, Kim S, Jeong SW, Jeon HW. Novel preparation of Fe doped TiO_2 nanoparticles and their application for gas sensor and photocatalytic degradation. Advances in Materials Science and Engineering. 2017;**2017**:2191659, 6 pages. DOI: 10.1155/2017/2191659

[57] Carneiro JO, Azevedo S, Fernandes F, Freitas E, Pereira M, Tavares CJ, et al. Synthesis of iron-doped TiO_2 nanoparticles by ball-milling process: The influence of process parameters on the structural, optical, magnetic, and photocatalytic properties. Journal of Materials Science. 2014;**49**:7476-7488. DOI: 10.1007/s10853-014-8453-3

[58] Abisharani JM, Devikala S, Kumar RD, Arthanareeswari M, Kamaraj P. Green synthesis of TiO_2 nanoparticles using Cucurbita pepo seeds extract. Materials Today: Proceedings. 2019;**14**:302-307. DOI: 10.1016/j.matpr.2019.04.151

[59] Isnaeni IN, Indriyati D, Sumiarsa D, Primadona I. Green synthesis of different TiO_2 nanoparticle phases using mango-peel extract. Materials Letters. 2021;**294**:129792. DOI: 10.1016/j.matlet.2021.129792

[60] Helmy ET, Abouellef EM, Soliman UA, Pan JH. Novel green synthesis of S-doped TiO_2 nanoparticles using Malva parviflora plant extract and their photocatalytic, antimicrobial and antioxidant activities under sunlight illumination. Chemosphere. 2021;**271**:129524. DOI: 10.1016/j.chemosphere.2020.129524

[61] Shiva Samhitha S, Raghavendra G, Camila Quezada P, Bindu H. Green synthesized TiO_2 nanoparticles for anticancer applications: Mini review. Materials Today: Proceedings. 2022;**54**:765-770. DOI: 10.1016/j.matpr.2021.11.073

[62] Shen C, Pang K, Le D, Luo G. Green synthesis and enhanced photocatalytic activity of Ce-doped TiO_2 nanoparticles supported on porous glass. Particuology. 2017;**34**:103-109. DOI: 10.1016/j.partic.2017.01.007

[63] Shyam-Sundar N, Karthi S, Senthil-Nathan S, Narayanan KR, Santoshkumar B, Sivanesh H, et al. Eco-friendly biosynthesis of TiO_2 nanoparticles using Desmostachya bipinnata extract: Larvicidal and pupicidal potential against Aedes aegypti and Spodoptera litura and acute toxicity in non-target organisms. Science of the Total Environment. 2023;**858**:159512. DOI: 10.1016/j.scitotenv.2022.159512

[64] Nabi G, Majid A, Riaz A, Alharbi T, Kamran MA, Al-Habardi M. Green synthesis of spherical TiO_2 nanoparticles using Citrus Limetta extract: Excellent photocatalytic water decontamination agent for RhB dye. Inorganic Chemistry Communications. 2021;**129**:108618. DOI: 10.1016/j.inoche.2021.108618

[65] Etacheri V, Di Valentin C, Schneider J, Bahnemann D, Pillai SC. Visible-light activation of TiO_2 photocatalysts: Advances in theory and experiments. Journal of Photochemistry and Photobiology C Photochemistry Reviews. 2015;**25**:1-29. DOI: 10.1016/j.jphotochemrev.2015.08.003

[66] Tay Q, Liu X, Tang Y, Jiang Z, Sum TC, Chen Z. Enhanced photocatalytic hydro- gen production with synergistic two-phase anatase/brookite TiO_2 nanostructures. Journal of Physical Chemistry C. 2013;**117**:14973-14982. DOI: 10.1021/jp4040979

[67] El-Mahallawi IS, Shash AY, Amer AE. Nanoreinforced cast Al-Si alloys with Al_2O_3, TiO_2 and ZrO_2 nanoparticles. Metals. 2015;**5**:802-821. DOI: 10.3390/met5020802

Section 2

Environmental and Energy Applications of TiO_2

Chapter 6

Modification Strategies of Titanium Dioxide

Jingyi Wang, Hui Xiao and Huaxin Wang

Abstract

Titanium dioxide (TiO_2) is a standard white pigment. However, when TiO_2 is exposed to ultraviolet light, it will catalyze the degradation of the surrounding organic matrix. Surface coating of TiO_2 is an effective method for reducing the catalytic effect of TiO_2. It can also improve the dispersion of TiO_2 in an organic matrix. This review critically introduces recent results on the surface coating of TiO_2. First, the main features of TiO_2, including processes, structure, and final properties, are described briefly. Second, this chapter reports and discusses different surface coating methods for TiO_2 with inorganic oxides and organic matter. Inorganic oxides, such as Al_2O_3, SiO_2, and ZrO_2, would form a continuous dense film and block the defects of the TiO_2 lattice. They can give TiO_2 excellent weather resistance. The organic matter available for surface treatment includes the surfactant, the coupling agent, and the macromolecule. They can improve the affinity of TiO_2 with various organic matrices. Surfactant treatment is relatively simple. Coupling agents can give TiO_2 more novel properties, such as thermal stability. Macromolecules can increase the volume of TiO_2 particles through steric hindrance and improve the dispersion of TiO_2 in an organic matrix. However, coating TiO_2 in a single matter is challenging to meet the increasing performance requirements. Therefore, it is necessary to study further the effect of co-coating with different inorganic oxides and organic matter on the structure and properties of TiO_2.

Keywords: titanium dioxide, inorganic coating, organic coating, structure, pigmentary properties

1. Introduction

Since the discovery of titanium in 1791, TiO_2 has been used commercially for over 100 years. Compared with other white pigments, such as ZnS, $BaSO_4$, ZnO, etc., TiO_2 is a nontoxic, stable pigment [1]. It shows high hiding power, refractive index, whiteness, and other excellent physical properties (**Table 1**) [2]. As a result, TiO_2 is widely used in printing ink, plastics, paper, coating materials, and cosmetics. It is also an indispensable raw material for the light industry, electronic industry, and other fields [3]. TiO_2 is produced by the sulfate and chloride process. The quality of TiO_2 produced by the two processes is different (**Table 2**) [2, 4]. Upmarket TiO_2 is mainly produced by the chloride process [5].

Pigment	Relative density	Refractive index	Lightening power		Covering power/%
			Reynolds number	Relative value/%	
Rutile TiO_2	4.20	2.76	1650	100	100
Anatase TiO_2	3.91	2.55	1270	77	78
ZnS	4.00	2.37	660	40	39
$BaSO_4$	4.50	1.64	—	—	—
Sb_2O_3	5.67	2.09	280	17	15
ZnO	5.60	2.02	200	12	14
Lithopone	4.20	1.84	260	16	—
Lead white (basic lead carbonate)	6.10	2.00	159	9	12

Table 1.
Technical index comparison of white pigment [2].

	Sulfate process	Chloride process
Raw ore	(1) Titanium concentrate: low price, stable, can be obtained directly from mining; (2) acid-soluble titanium slag: relatively high price, good quality, need to be chemically processed	(1) Titanium concentrate/white titanium: low price, stable, high process technology requirements; (2) rutile: relatively high price, low process technology requirements; (3) Titanium chloride slag and artificial rutile: higher price, low process technology
Auxiliary raw materials	H_2SO_4	Cl_2
Price	Low	High
Type	Anatase, middle-end rutile	High-end rutile
H_2SO_4 recovery/%	13	75
Flow	Long and complex	Short and simple process
Technology	Mature	Domestic immaturity
Control accuracy	Low	High
Quality	Coverage and yellowing resistance are weaker than chloride methods but cheaper and less used in specific areas such as papermaking and chemical fiber	High purity, good comprehensive performance, high price
Energy consumption	(1) Pressure on environmental protection, but recycling of waste by-products can be improved; (2) large consumption of coal, natural gas, steam, water, and electricity	(1) Less three wastes, small pressure for environmental protection, but the treatment of ferric chloride in solid waste is difficult; (2) consumption is relatively small
Government policy	Restricted	Supported

Table 2.
Comparison of different processes [2, 4].

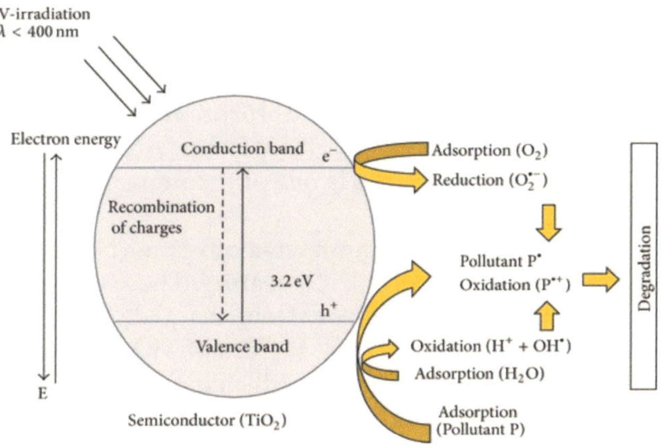

Figure 1.
The main photocatalytic process of TiO_2 [9].

TiO_2 shows a regular lattice structure. There are three crystalline forms of TiO_2 in nature: anatase, rutile, and brookite [6]. Anatase has a tetragonal crystal system. However, it will slowly transform into rutile after heating at about 610°C [7] and completely transform into rutile at 915°C. The latter also has a tetragonal crystal structure, each unit contains six atoms, and its oxygen atoms are densely packed, so rutile shows the highest stability [8]. Compared with anatase, rutile exhibits higher density, hardness, refractive index, and dielectric constant.

TiO_2 has excellent physical and chemical properties, however, TiO_2 surface has a photocatalytic active site (**Figure 1**). After absorbing ultraviolet light energy, electron–hole pairs (the charge carrier) are generated [10]. The valence band hole (h^+) is highly oxidizing while the conduction band electron (e^-) is highly reducing [11]. The h^+ oxidizes H_2O or OH^- ion to the hydroxyl radical ($OH^•$), the e^- reduces adsorbed oxygen (O_2) species to superoxide ($O_2^•$) and then undergoes a series of reactions to give the $OH^•$ radical. These radicals will react with surrounding organic substances, resulting in the decomposition of the organic matrix [12]. What's worse, TiO_2 particles are easy to agglomerate due to the high special surface area, causing poor dispersion in the organic matrix [13].

To overcome the drawbacks of TiO_2 mentioned above, one can use coatings. The coating of TiO_2 by inorganic oxides, such as alumina, silica, and zirconia [14], can effectively inhibit the oxidative degradation of the organic matrix, finally improving the light and weather resistance [15, 16]. The poor dispersion of TiO_2 can also be effectively solved by coating [17]. Therefore, it is of great social significance and economic value to study the coating of TiO_2 to improve the physical stability and dispersion, extending new applications of TiO_2. In this chapter, we introduce the modification strategies of TiO_2 to the readers. To fully describe the modification mechanism, processes and properties of modified TiO_2 will be discussed.

2. Inorganic modification of TiO_2

The purpose of the inorganic modification is to coat TiO_2 with a layer of the inorganic hydrated oxide film. This film can block and cover the lattice defects of TiO_2 and reduce the connecting possibility between organic matrix and active groups of TiO_2. Such films comprise alumina, silica, zirconia, etc.

2.1 Alumina

Alumina (Al_2O_3) is a suitable electron acceptor, which can annihilate the photoelectrons generated by TiO_2 after ultraviolet absorption and excitation, inhibiting subsequent active groups' generation [18]. In addition, Al_2O_3 can also reflect ultraviolet from natural light [19]. Thus, Al_2O_3 is one of the most used materials for the inorganic coating of TiO_2.

A variety of chemicals, such as sodium metaaluminate ($NaAlO_2$) and aluminum sulfate ($Al_2(SO_4)_3$), have been used for TiO_2 coating. These metal salts are added into TiO_2 suspension at various pH, and the positively charged OH-Al hydrolyzed by soluble salt is adsorbed and wrapped on the surface of TiO_2 particles to form hydrated alumina. The structure of hydrated alumina will change at different pH values. It shows an amorphous structure at pH 5, a floccular false boehmite structure at pH 8–10, and a flaky gibbsite structure at pH above 10.

Zhang et al. [20] reported the preparation of compact amorphous Al_2O_3 film on the TiO_2 under the molar ratio $NaAlO_2/TiO_2$ of 1/22 at 80°C in pH 5. After being coated by Al_2O_3 films, the whiteness and brightness of the modified TiO_2 samples increased with the increase of the Al_2O_3 loading, while the relative light scattering index depended on the alumina loading.

Dong et al. [21] synthesized alumina-coated rutile TiO_2 samples using the chemical liquid deposition method under various pH and aging temperatures. The results showed that this film-coating process should mainly be attributed to chemical bonding and physical adsorption (**Figure 2a**). The higher aging temperature was in favor of

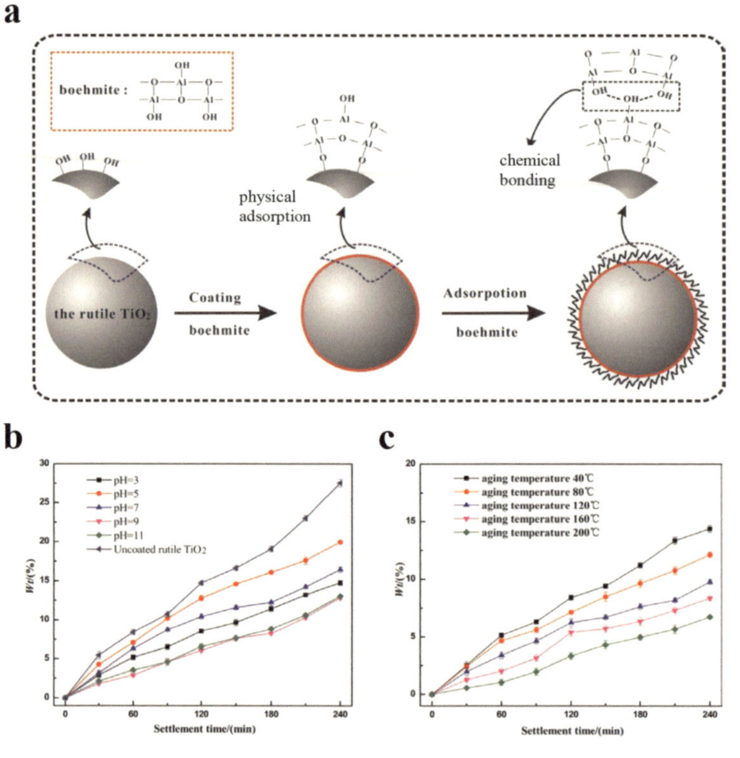

Figure 2.
(a) Schematic diagram of physical adsorption and chemical bonding of Al_2O_3 coated rutile; dispersion stability of Al_2O_3-coated rutile TiO_2 samples at different pH values (b) and aging temperatures (c) [21].

the elevation of the boehmite content of the coating film, causing the enhancement of dispersion stability. It contributed to the increase of steric hindrance and electrostatic repulsion. The coated TiO_2 exhibited well dispersion stability at pH 9 (**Figure 2b**) and aging temperature 200°C (**Figure 2c**), respectively.

Wu et al. [22] discussed the mechanism of the film-coating process of hydrated alumina on TiO_2 particles in an aqueous solution. The effects of temperature, pH value, and $Al_2(SO_4)_3$ solution were investigated. It is found that TiO_2 particles promote the hydrolysis of $Al_2(SO_4)_3$ in both acidic and basic solutions and adsorb positively charged OH-Al species in slurries. When the OH-Al species or TiO_2 particles have enough energy to cross the repulsion threshold, the hydroxyl groups on the surface of the TiO_2 particles will condense with the OH-Al species, leading to the coating of OH-Al species on the surface of the TiO_2 particles. As a result, the Al_2O_3 film is formed.

2.2 Silica

Silica coating shows a similar function as alumina. Compared with alumina, silica film gives more chemical stability to TiO_2. TiO_2 suspension is added to water-soluble silicon compound in base condition. Silicon is deposited on TiO_2 particles as $Si(OH)_4$ through physical adsorption and chemical bonding between $Si(OH)_4$ and TiO_2. The deposited $Si(OH)_4$ is further condensed into a silica gel, finally realizing the coating of TiO_2 particles (**Figure 3**).

Liu et al. [23] prepared SiO_2-coated TiO_2 powders by the chemical deposition method starting from rutile TiO_2 and Na_2SiO_3. The evolution of island-like and uniform coating layers depended on the ratio of Na_2SiO_3 to TiO_2, reaction temperature, and pH. The result showed that the whiteness and brightness of the TiO_2 product increased with the loading of SiO_2.

Lin et al. [24] studied the surface characteristics of hydrous silica-coated TiO_2 particles. Different analytical techniques were used to characterize the silica oxide coatings on TiO_2 particles. Analyses showed that hydrous silica is continuously coated on the surface of TiO_2 particles. The hydrous silica film coating can improve the durability of pigment weather and dispersion properties.

SiO_2 can be easily deposited on TiO_2 surfaces. However, SiO_2 coating layers with a lower polarity cannot significantly enhance the dispersibility of TiO_2 in a polar solvent. Moreover, the hydrogen bond interaction between the hydrated SiO_2 will lead to thixotropy. Al_2O_3 coating layers with many –OH groups not only improve the dispersibility of TiO_2 powders in polar solvents but also provide abundant active sites for further organic modification. However, Al_2O_3 coating layers tend to anchor loosely at TiO_2 surfaces. Therefore, various reports are about the formation of binary Al_2O_3/SiO_2 films on the TiO_2 surface.

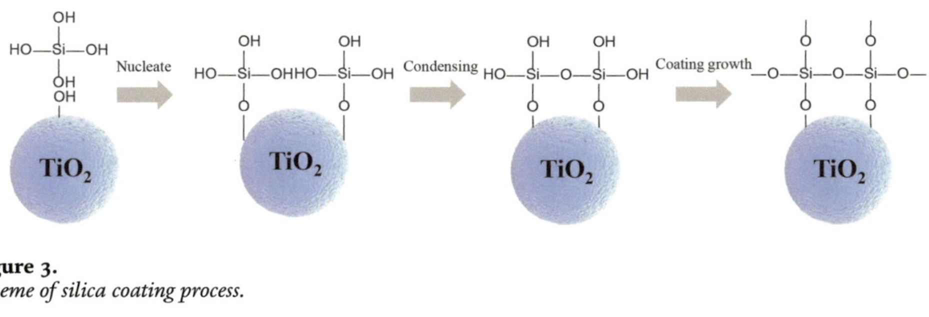

Figure 3.
Scheme of silica coating process.

Zhang et al. [25] prepared binary Al_2O_3/SiO_2-coated rutile TiO_2 composites by a liquid-phase deposition method starting from $Na_2SiO_3{\cdot}9H_2O$ and $NaAlO_2$. The formation of continuous and dense binary Al_2O_3/SiO_2 coating layers depended on the pH value of the reaction solution and the alumina loading. The coated TiO_2 particle had a high dispersibility in water. Compared with SiO_2-coated TiO_2 samples, the whiteness and brightness of the binary Al_2O_3/SiO_2-coated TiO_2 particles were higher.

To improve the dispersion and reduce the photocatalytic activity of TiO_2, Godnjavec et al. [26] modified TiO_2 by the SiO_2/Al_2O_3 films on the surface of particles and incorporated modified TiO_2 into the polyacrylic coating. The results showed that surface treatment of TiO_2 with SiO_2/Al_2O_3 could improve the dispersion of TiO_2 in the polyacrylic matrix, and the UV protection of the clear polyacrylic composite coating was enhanced.

2.3 Zirconia

Zirconia (ZrO_2) has a high refractive index (2.170) and weak ultraviolet light absorption. Therefore, the ZrO_2 coating considerably reduces UV absorption causing higher photostability [27] and increasing the glossiness of TiO_2 particles. This coating can increase the amount of hydroxyl groups on the surfaces of the TiO_2 particles, which improves the dispersibility of TiO_2 powders in aqueous media and provides more active sites for the subsequent organic modification.

The TiO_2 powders are dispersed in distilled water with ultrasonic treatment to obtain TiO_2 suspension, and the zirconium salt solution is added as follows. The zirconium salt hydrolyzes rapidly, and the zirconia nanoparticles grow and form aggregates on the surface of TiO_2 through Zr–O–Ti bonds. The zirconia nanoparticles will grow and form a continuous and dense film.

Zhang et al. [28] reported that the ZrO_2-coated rutile TiO_2 could be prepared by the chemical liquid deposition method starting from rutile TiO_2 and $ZrOCl_2$. The formation of zirconia coating depended on pH value of reaction solution and the mole ratio of $ZrOCl_2$ to TiO_2. When the pH value reached to 9 with a mole ratio of $ZrOCl_2$ to TiO_2 of 1:51, the zirconia aggregates with an average particle size of about 4 nm coated on the surface of the TiO_2 particles (**Figure 4a, b**). Compared with the exposed rutile TiO_2, the dispersibility, whiteness, brightness, and relative light scattering index of the ZrO_2-coated TiO_2 were significantly improved.

Li et al. [29] prepared ZrO_2-coated TiO_2 by a precipitation method. The $Zr(SO_4)_2$ solution was added to TiO_2 suspension at the pH of 5.2 at 50°C. The mass ratio of ZrO_2 to TiO_2 was 1.0%, and a dilute NaOH solution was used to adjust the pH value. The results showed that supersaturation of the $Zr(SO_4)_2$ solution is one of the key factors influencing the type of nucleation in the zirconia coating. Lower supersaturation benefits the heterogeneous nucleation of zirconia on the surface of TiO_2 particles, while higher supersaturation leads to the homogeneous nucleation of zirconia itself. A suitable ZrO_2 content is about 1.0 wt.%, and this thick and continuous film gives better pigmentary properties.

To sum up, the function of the inorganic oxide-coated film of TiO_2 is to form a barrier, reducing the photoactivity of TiO_2 and the production of free radicals on the surface of TiO_2. As a result, the coated TiO_2 has good pigmentary properties, including weather and light resistance. However, using a single inorganic oxide coating is often not sufficient to meet the requirements of several applications. So, it is one of the

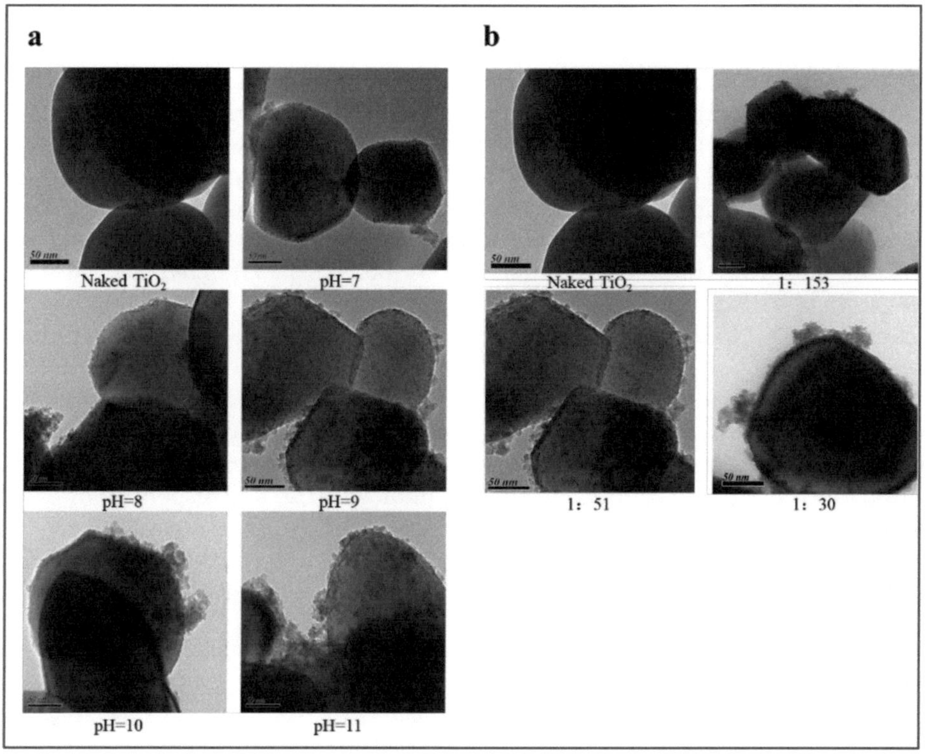

Figure 4.
TEM micrographs of bare rutile TiO_2 and ZrO-coated TiO_2 at T = 80°C, $ZrOCl_2$:TiO_2 = 1:51 with different pH (a) and with a different molar ratio of $ZrOCl_2$ to TiO_2 (b) [28].

essential directions to study further the co-coating of various inorganic oxides and the regulation and process of coating structure.

3. Organic modification of TiO_2

Modifying TiO_2 by organic agents is realized by coating them with organic substances such as surfactants, coupling agents and polymers. It can improve the affinity of TiO_2 particles with organic matrices, resulting in better dispersion of TiO_2. Thus, the pigmentary properties of TiO_2, such as tint-reducing power, hiding power, and whiteness, are shown.

In the modification, there are two mechanisms: physical adsorption of organic agents on the surface of TiO_2 and chemical bonding between TiO_2 and organic agents [30]. The principle of physical adsorption is that the hydrophilic group of the organic coating agent is adsorbed on the surface of TiO_2 particles. In contrast, the oleophilic group is outwardly affinity to the surrounding polymer matrix. Therefore, the polymer chains can penetrate the TiO_2 aggregates and separate the TiO_2 particles, finally improving the dispersion of TiO_2. For chemical bonding, the hydroxyl groups on the surface of TiO_2 particles act as active sites, which will react with organic coating agents and form covalent bonds. As a result, the TiO_2 particles change from hydrophilic to hydrophobic. Several kinds of organic agents can be used for the surface modification of TiO_2, including surfactants, coupling agents and polymers.

3.1 Surfactants

Surfactants can be divided into cationic, anionic, and nonionic surfactants. One can use surfactants singly or together to modify TiO_2 particles to evaluate the performance of TiO_2. Li et al. [31] chose anionic sodium dodecyl sulfate (SDS) and nonionic nonylphenol ethoxylate (NPEO, Tergitol NP-9) to study the effect of surfactants on the behaviors of TiO_2 in aqueous solution. The results showed that both surfactants could be adsorbed onto the surface of nano-TiO_2 but that only SDS can significantly decrease the zeta potential of TiO_2. Both surfactants reduced the aggregation of TiO_2 and retarded the aggregate sedimentation at surfactant concentrations ≥0.015% (w:v). In addition, SDS exerted a more substantial reductive effect than NP-9.

Wei et al. [32] used different surfactants, such as cetyltrimethylammonium bromide (CTAB), sodium dodecylbenzene sulfonate (SDBS), and diethanolamine (DEA), to modify TiO_2 particles. The crystal type of TiO_2 has no noticeable change with the addition of different surfactants, but the morphology, size, and dispersion of the TiO_2 particles have changed to some extent. Among the three surfactants, CTAB is beneficial in reducing TiO_2 particle size and improving TiO_2 dispersion and agglomeration. And this CTAB-coated TiO_2 had the greatest photostability in methyl orange degradation.

Wittmar et al. [33] prepared modified TiO_2 particles by adding a cationic imidazolium salt solution. It was found that an increase in the alkyl chain length was beneficial, leading to a narrowing of the particle size distribution and a decrease of the agglomerate size in dispersion. The smallest average nanoparticle sizes in dispersion were around 30 nm.

Zhang et al. [34] discussed the influence of different surfactants on the thermal stability, weather fastness, and pigmentary properties of TiO_2 particles. The results are collected in **Table 3**. Compared with neopentyl glycol (NPG),

Coating agent	Dosage /%	Hue			Oil absorption /%	290°C evaporation /%
		L (Whiteness)	a (Red and green saturation)	b (Blue and green saturation)		
TMP	0.3	97.64	−0.41	2.39	15.98	0.41
	0.5	97.64	−0.43	2.29	14.72	0.43
	0.7	97.18	−0.52	2.23	13.95	0.52
	0.9	96.85	−0.53	2.14	13.49	0.54
NPG	0.3	97.54	−0.40	2.46	16.26	0.49
	0.5	97.49	−0.46	2.29	16.01	0.52
	0.7	97.21	−0.53	2.9	15.69	0.56
	0.9	96.89	−0.56	2.10	15.33	0.60
PEG	0.3	97.59	−0.42	2.49	16.86	0.44
	0.5	97.53	−0.44	2.39	16.56	0.49
	0.7	97.26	−0.57	2.23	16.01	0.53
	0.9	7.00	−0.61	2.10	15.49	0.59

Coating agent	Dosage /%	Hue			Oil absorption /%	290°C evaporation /%
		L (Whiteness)	a (Red and green saturation)	b (Blue and green saturation)		
TME	0.3	97.62	−0.44	2.42	17.37	0.43
	0.5	97.59	−0.49	2.31	17.04	0.49
	0.7	97.35	−0.55	2.26	16.55	0.52
	0.9	96.92	−0.59	2.09	16.01	0.61
Raw TiO_2	/	97.01	−0.50	2.22	14.29	0.44

Table 3.
Routine index of TiO_2 *[34].*

polyethylene glycol (PEG) and trimethylolethane (TME), trimethylolpropane (TMP) can bring the highest whiteness (97.64 of L) to TiO_2 particles only in the content of 0.3 wt%.

3.2 Coupling agent

Coupling agents are amphoteric structural compounds, which can be divided into silane coupling agents, titanate coupling agents, aluminate coupling agents, etc. One end group of the coupling agent can react with the hydroxyl group on the surface of TiO_2 particles to form a strong chemical bond. The other can react with the polymeric matrix. Consequently, two kinds of materials of different polarity, TiO_2, and a polymer, are closely combined to give the composite material excellent comprehensive performance.

Silane coupling agents were first developed and used widely to modify TiO_2. In the reaction, organic silicon is adsorbed on TiO_2, and the molecule part reacts with the hydroxyl group on the surface of TiO_2 to prevent the aggregation of particles. Wang et al. [35] reported the modification of TiO_2 by three kinds of silane coupling agents (KH550, KH570, and HDTMS). The chemical structure and the reaction mechanism with TiO_2 are shown in **Figure 5(a)**. The results showed that TiO_2 modified by silane coupling agent had small particle size, improved hydrophobicity, and low surface energy (**Figure 5(b, c)**). Furthermore, compared with raw TiO_2 and KH550 coated TiO_2, HDTMS-coated TiO_2 and KH 570 coated TiO_2 had excellent dispersion stability as white pigments in blue light curing inks (**Figure 5(d)**).

Sabzi et al. [36] used aminopropyl trimethoxyl silane (APS) as a coupling agent to modify TiO_2. The results showed that silane coupling agents could significantly improve the dispersion of TiO_2 in polyurethane composite and the mechanical properties of composite. Xuan et al. [37] reported the modification of TiO_2 by vinyltrimethoxyl silane (A171) and the reinforcement of modified TiO_2 on wheat straw fiber/polypropylene composite. The modified TiO_2 could effectively improve the tensile, flexural, and impact resistance as well as the UV light stability of the composite. However, the thermal stability of the coupling agents is poor. This leads to the degradation of the organic layer on the surface of TiO_2. Finally, the color and whiteness of TiO_2 are changed.

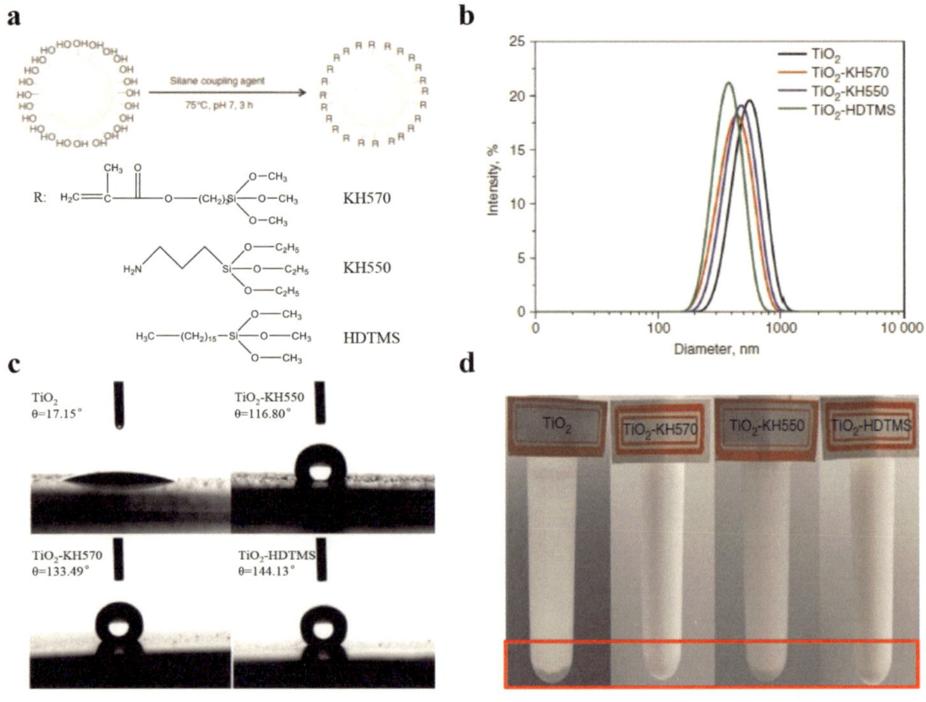

Figure 5.
(a) The scheme of reaction between silane coupling agents with different chemical structures and TiO_2 particle, (b) particle size distribution of raw and modified TiO_2, (c) contact angle of raw and modified TiO_2-water interface, (d) dispersion of raw and modified TiO_2 [35].

3.3 Polymer

The above two organic treatment methods depend on the reaction of small molecular modifiers with the surface groups of TiO_2. In contrast, a modification with a macromolecule uses the polymer to coat the TiO_2 particles directly or the reactive monomer to polymerize on the surface of TiO_2 particles. In the coating with polymers, there is no interaction between the polymeric groups and TiO_2, but the polymer induces a steric hindrance [38]. As a result, the dispersion of TiO_2 in the subsequent polymer matrix is improved. TiO_2 shows good pigmentary properties. The reaction mechanism and classification of polymer coating modification are collected in **Table 4**.

Man et al. [40] used the microcapsule method to modify TiO_2. The *in situ* polymerization of acrylic monomer on the surface of TiO_2 particles obtained the core-shell structure of modified TiO_2. This core-shell structure TiO_2 showed improved dispersion in organic media and excellent UV shielding ability. Olad et al. [41] used polyaniline (PANI) to modify TiO_2 through *in situ* polymerization. The results showed that PANI was successfully implanted on the surface of TiO_2, effectively inhibiting the aggregation of TiO_2 nanoparticles.

In the "Anchor positioning" coating method, the polymers used are named hyper-dispersant, which Schofield first proposed in the 1980s. Compared with the structure of traditional dispersants, such as surfactant SDS, with the hydrophilic and lipophile groups, hyper-dispersants have two completely different groups: anchoring group and

Method	Mechanism	Classification
Microcapsule method	The continuous and dense polymer capsule is formed by *in situ* polymerization with active monomer or adsorbing polymers directly using the Van der Waals force	1. Active monomers are adsorbed on the surface of TiO_2 and then polymerized. 2. Surface active points of TiO_2 are stimulated and initiate the polymerization of the monomer. 3. Polymer chains are adsorbed directly to form a dense film on the surface of TiO_2 by using Van der Waals force
Surface grafting modification	The surface of inorganic particles was pretreated first, and graft polymerization was initiated. There are two different pretreatments: coupling agent pretreatment and surfactant pretreatment	1. Surface coupling reaction with polymerizable organic monomers. 2. Introducing free radical-producing compounds to graft polymerizable organic monomers. 3. The free radicals of the particle itself capture the polymer chain to achieve graft polymerization
"Anchor positioning" coating	The functional group of the polymer can anchor on the surface of TiO_2, and the solvent chain of the polymer extends in a nonaqueous system to provide steric stability	Terminal group anchoring method of macromolecules

Table 4.
Modification by polymer coating [39].

solvent group. The anchoring groups are anchored on the particles' surface by single-point or multipoint anchoring or co-anchoring with a surface synergist (**Figure 6**). At the same time, the solvent chain is extended in a nonaqueous system to provide steric stability. Therefore, the particles are stably dispersed (**Figure 7**). So, the hyper-dispersants have a unique dispersion effect on the nonaqueous system.

There are a lot of different anchoring groups, and solvent chains can be designed to synthesize hyper-dispersant (**Tables 5** and **6**). Thus, hyper-dispersants with different effects can be designed and synthesized by selecting different anchoring groups and solvent chains.

Schaller et al. [43] modified TiO_2 with poly(acrylic acid)-polystyrene block copolymer hyper-dispersants. It is proved that the end group of polymers will form some bond interactions with TiO_2 particles, which improves the stability between polymer and TiO_2 particles and then improves the dispersion of TiO_2 in water.

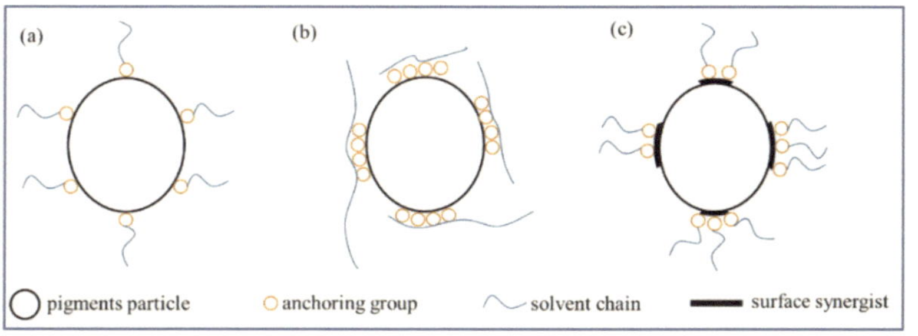

Figure 6.
The anchorage form of hyper-dispersant on particle surface: (a) single-point, (b) multipoint, and (c) co-anchoring with a surface synergist.

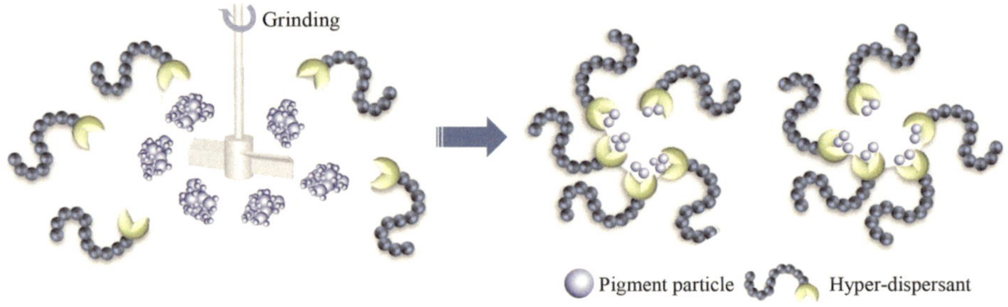

Figure 7.
The scheme of particle dispersion by using hyper-dispersant.

Anchoring group	Electronegativity	Section width of anchoring group (nm)
–OH	3.9	0.22
$-NH_2$	3.7	0.36
–SH	2.6	0.37
$-SO_3^-$	4.33	0.58
$-PO_4^{2-}$	4.86	0.60
–COOH	4.1	0.52
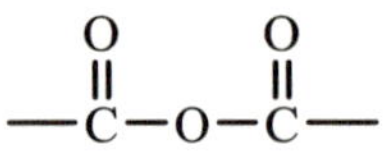	4.88	0.74
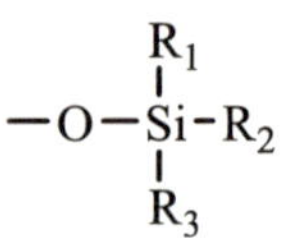	4.15	0.22
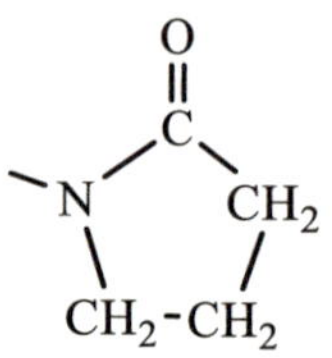	3.8	0.65

Table 5.
Electronegativity and section width of the anchoring group [42].

Zhang et al. [44] synthesized three hyper-dispersants: nonterminated, carboxyl-terminated, and polyethylene imine-grafted poly(hydroxyl carboxylic acid) ester. It is found that polyethylene imine-grafted hyper-dispersant has the best dispersion performance in nano-TiO_2/resin solution dispersion systems.

A novel acrylic polyester hyper-dispersant containing methacrylic acid (MAA), butyl acrylate (BA), and 3-pentadecylphenyl acrylate (PDPA) was polymerized by Liu et al. [45]. This hyper-dispersant was used to disperse TiO_2 in a nonpolar solvent system. The results showed that the viscosity and particle size of suspensions were affected by monomer ratio and molecular weight. The optimum monomer ratio and molecular weight were MAA: BA: PDPA = 1:10:1.2 (wt%) and 6000, respectively. Liu et al. [46] further reported the effects of acrylic polyester hyper-dispersant on the

Polymer	Solvent chain structural unit	Flexibility	Solubility parameter /$J^{0.5}m^{-1.5}$
Polyethylene glycol	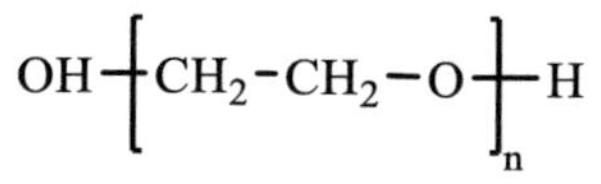	Good	1.88~2.49
Polyvinyl alcohol	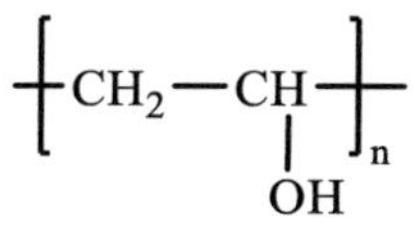	Good	2.58~2.91
Polyacrylamide	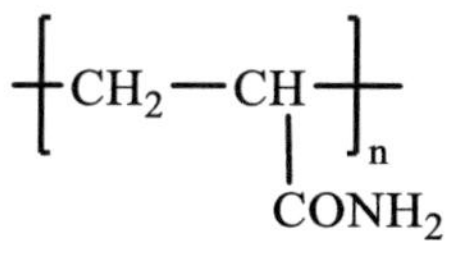	Good	2.65
Polyacrylic acid	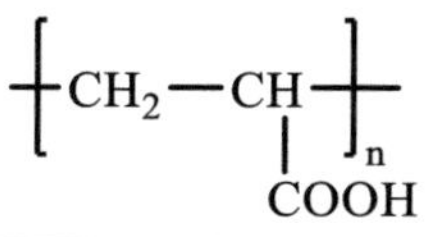	Good	2.44
Polymethylacrylic acid	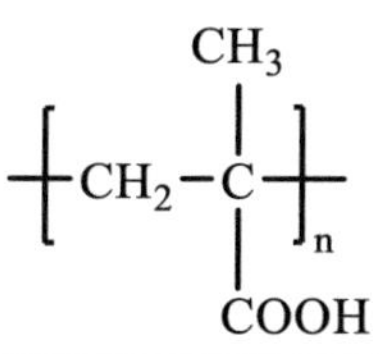	General	1.86~2.64
Polymaleic Anhydride	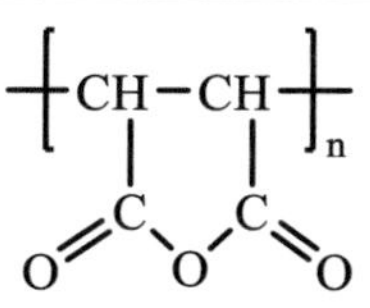	Bad	1.58~2.83
Polyvinylpyrrolidone	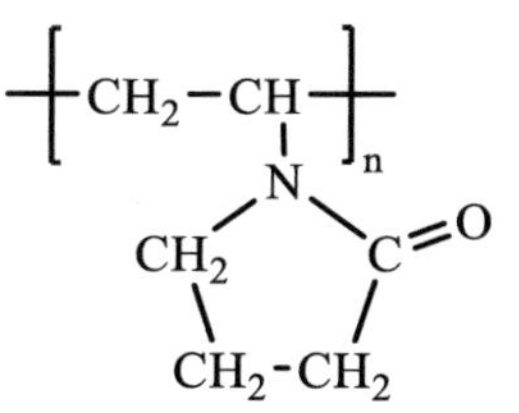	Very bad	2.30~2.62

Table 6.
Solubility parameters of some polymer structural units [42].

dispersion of TiO_2 in different organic solvents. The results showed that acrylic polyester hyper-dispersant adsorption onto TiO_2 is spontaneous and physical.

4. Conclusion and prospects

As a critical pigment, TiO_2 has a considerable application market. However, due to the crystal defects, there is a photocatalytic active site on the surface of TiO_2. After absorbing ultraviolet light, free radicals are produced, causing organic compound degradation in the surrounding TiO_2. Besides, TiO_2 particles are easy to agglomerate and are dispersed poorly in the organic matrix due to the high specific surface area. The surface inorganic/organic modification of TiO_2 is an excellent choice to overcome the drawbacks of TiO_2. The coating films on the surface of TiO_2 can effectively inhibit the oxidative degradation of the organic matrix and improve the dispersion of TiO_2, finally improving pigmentary properties of TiO_2, such as whiteness, hiding power, light resistance, and weather resistance. With the development of the economy, the demands for applying TiO_2, such as high weather resistance, light resistance, and dispersion stability, are gradually increased. Thus, the coating treatments of TiO_2 are an essential strategic development direction in the future.

Acknowledgements

The authors gratefully acknowledge the support by the Scientific Foundation of the Liming Vocational University (grant no. LMPT 202101, LZB202101).

Author details

Jingyi Wang*, Hui Xiao and Huaxin Wang
School of New Materials and Shoes and Clothing Engineering, Liming Vocational University, Quanzhou, China

*Address all correspondence to: jingleewong@gmail.com

References

[1] Chen C, Chong KC, Pan Y, et al. Revisiting carbazole: Origin, impurity, and properties. ACS Materials Letters. 2021;**3**:1081-1087. DOI: 10.1021/acsmaterialslett.1c00138

[2] Gázquez MJ, Bolívar JP, García-Tenorio G-BR, et al. A review of the production cycle of titanium dioxide pigment. Materials Sciences and Applications. 2014;**5**:441-458. DOI: 10.4236/msa.2014.57048

[3] Likodimos V. Photonic crystal-assisted visible light activated TiO_2 photocatalysis. Applied Catalysis B: Environmental. 2018;**230**:269-303. DOI: 10.1016/j.apcatb.2018.02.039

[4] Gupta SM, Tripathi M. A review of TiO_2 nanoparticles. Chinese Science Bulletin. 2011;**56**:1639-1657. DOI: 10.1007/s11434-011-4476-1

[5] Liu X, Chen Y, Rao S, et al. One step preparation of highly dispersed TiO_2 nanoparticles. Chemical Research in Chinese Universities. 2015;**31**: 688-692. DOI: 10.1007/s40242-015-5300-2

[6] Pang CL, Lindsay R, Thornton G. Structure of clean and adsorbate-covered single-crystal rutile TiO_2 surfaces. Chemical Reviews. 2013;**113**:3887-3948. DOI: 10.1021/cr300409r

[7] Gu D-E, Yang B-C, Hu Y-D. A novel method for preparing V-doped titanium dioxide thin film photocatalysts with high photocatalytic activity under visible light irradiation. Catalysis Letters. 2007; **118**:254-259. DOI: 10.1007/s10562-007-9179-5

[8] Thomas AG, Syres KL. Adsorption of organic molecules on rutile TiO_2 and anatase TiO_2 single crystal surfaces. Chemical Society Reviews. 2012;**41**: 4207-4217. DOI: 10.1039/c2cs35057b

[9] Haider AJ, Jameel ZN, Al-Hussaini IHM. Review on: Titanium dioxide applications. Energy Procedia. 2019;**157**: 17-29. DOI: 10.1016/j.egypro.2018.11.159

[10] Haider AJ, Alhaddad RM, Yahya KZ. Nanostructure dopants TiO_2 films for gas sensing. Iraqi Journal of Applied Physics. 2011;**7**:27-31. DOI: 10.1007/s10562-007-9179-5

[11] Haider AJ, AL-Anbari RH, Kadhim GR, et al. Exploring potential environmental applications of TiO_2 nanoparticles. Energy Procedia. 2017; **119**:332-345. DOI: 10.1016/j.egypro.2017.07.117

[12] Ahmad MK, Aziz AFA, Soon CF, et al. Rutile phased titanium dioxide (TiO_2) nanorod/nanoflower based waste water treatment device. In: Jablonski R, Szewczyk R, editors. Recent Global Research and Education: Technological Challenges. 2017. pp. 483-490

[13] Liu Y, Zhang Y, Ge C, et al. Evolution mechanism of alumina coating layer on rutile TiO_2 powders and the pigmentary properties. Applied Surface Science. 2009;**255**:7427-7433. DOI: 10.1016/j.apsusc.2009.04.013

[14] Palanisamy B, Babu CM, Sundaravel B, et al. Sol-gel synthesis of mesoporous mixed Fe_2O_3/TiO_2 photocatalyst: Application for degradation of 4-chlorophenol. Journal of Hazardous Materials. 2013;**252**: 233-242. DOI: 10.1016/j.jhazmat.2013.02.060

[15] Dong X, Sun Z, Liu Y, et al. Insights into effects and mechanism of

pre-dispersant on surface morphologies of silica or alumina coated rutile TiO_2 particles. Chemical Physics Letters. 2018;**699**:55-63. DOI: 10.1016/j.cplett.2018.03.043

[16] Vargas MA, Rodríguez-Páez JE. Amorphous TiO_2 nanoparticles: Synthesis and antibacterial capacity. Journal of Non-Crystalline Solids. 2017; **459**:192-205. DOI: 10.1016/j.jnoncrysol.2017.01.018

[17] Hamieh T, Toufaily J, Alloul H. Physicochemical properties of the dispersion of titanium dioxide in organic media by using zetametry technique. Journal of Dispersion Science and Technology. 2008;**29**:1181-1188. DOI: 10.1080/01932690701856626

[18] Yu Y, Zhu Y, Guo J, et al. Suppression of TiO_2 photocatalytic activity by low-temperature pulsed CVD-grown SnO_2 protective layer. Industrial and Engineering Chemistry Research. 2018;**57**:8679-8688. DOI: 10.1021/acs.iecr.8b00270

[19] Xiong S, Tang Y, Ng HS, et al. Specific surface area of titanium dioxide (TiO_2) particles influences cyto- and photo-toxicity. Toxicology. 2013;**304**: 132-140. DOI: 10.1016/j.tox.2012.12.015

[20] Zhang Y, Liu Y, Ge C, et al. Evolution mechanism of alumina nanofilms on rutile TiO_2 starting from sodium metaaluminate and the pigmentary properties. Powder Technology. 2009;**192**:171-177. DOI: 10.1016/j.powtec.2008.12.009

[21] Dong X, Sun Z, Jiang L, et al. Investigation on the film-coating mechanism of alumina-coated rutile TiO_2 and its dispersion stability. Advanced Powder Technology. 2017;**28**: 1982-1988. DOI: 10.1016/j.apt.2017.05.001

[22] Wu HX, Wang TJ, Jin Y. Film-coating process of hydrated alumina on TiO_2 particles. Industrial and Engineering Chemistry Research. 2006; **45**:1337-1342. DOI: 10.1021/ie0510167

[23] Liu Y, Ge C, Ren M, et al. Effects of coating parameters on the morphology of SiO_2-coated TiO_2 and the pigmentary properties. Applied Surface Science. 2008;**254**:2809-2819. DOI: 10.1016/j.apsusc.2007.10.021

[24] Lin Y-L, Wang T-J, Jin Y. Surface characteristics of hydrous silica-coated TiO_2 particles. Powder Technology. 2002;**123**:194-198. DOI: 10.1016/S0032-5910(01)00470-3

[25] Zhang Y, Yin H, Wang A, et al. Deposition and characterization of binary Al_2O_3/SiO_2 coating layers on the surfaces of rutile TiO_2 and the pigmentary properties. Applied Surface Science. 2010;**257**:1351-1360. DOI: 10.1016/j.apsusc.2010.08.071

[26] Godnjavec J, Zabret J, Znoj B, et al. Investigation of surface modification of rutile TiO2 nanoparticles with SiO_2/Al_2O_3 on the properties of polyacrylic composite coating. Progress in Organic Coating. 2014;**77**:47-52. DOI: 10.1016/j.porgcoat.2013.08.001

[27] Wei B-X, Zhao L, Wang T-J, et al. Photo-stability of TiO_2 particles coated with several transition metal oxides and its measurement by rhodamine-B degradation. Advanced Powder Technology. 2013;**24**:708-713. DOI: 10.1016/j.apt.2012.12.009

[28] Zhang Y, Yin H, Wang A, et al. Evolution of zirconia coating layer on rutile TiO_2 surface and the pigmentary property. Journal of Physics and Chemistry of Solids. 2010;**71**:1458-1466. DOI: 10.1016/j.jpcs.2010.07.013

[29] Li J, Sun T-C, Wang Y, et al. Preparation and film-growing mechanism of hydrous zirconia coated on TiO_2. International Journal of Minerals Metallurgy and Materials. 2010;**17**:660-667. DOI: 10.1007/s12613-010-0371-z

[30] Liang Y, Ding H. Mineral-TiO_2 composites: Preparation and application in papermaking, paints and plastics. Journal of Alloys and Compounds. 2020; **844**:1-23. DOI: 10.1016/j.jallcom.2020.156139

[31] Li X, Yoneda M, Shimada Y, et al. Effect of surfactants on the aggregation and stability of TiO_2 nanomaterial in environmental aqueous matrices. Science of the Total Environment. 2017; **574**:176-182. DOI: 10.1016/j.scitotenv.2016.09.065

[32] Wei J, Wen X, Zhu F. Influence of surfactant on the morphology and photocatalytic activity of anatase TiO_2 by solvothermal synthesis. Journal of Nanomaterials. 2018;**2018**:3086269. DOI: 10.1155/2018/3086269

[33] Wittmar A, Gajda M, Gautam D, et al. Influence of the cation alkyl chain length of imidazolium-based room temperature ionic liquids on the dispersibility of TiO_2 nanopowders. Journal of Nanoparticle Research. 2013; **15**:1463-1467. DOI: 10.1007/s11051-013-1463-2

[34] Zhang X, Guo H, Zhong Z, et al. Effect of organic coating agent on properties of titanium dioxide products for Masterbatch. China Plastics. 2020;**34**: 32-36. DOI: 10.19491/j.issn.1001-9278.2020.10.006

[35] Wang L, Jiang X, Wang C, et al. Titanium dioxide grafted with silane coupling agents and its use in blue light curing ink. Coloration Technology. 2020; **136**:15-22. DOI: 10.1111/cote.12434

[36] Sabzi M, Mirabedini SM, Zohuriaan-Mehr J, et al. Surface modification of TiO_2 nano-particles with silane coupling agent and investigation of its effect on the properties of polyurethane composite coating. Progress in Organic Coating. 2009;**65**:222-228. DOI: 10.1016/j.porgcoat.2008.11.006

[37] Xuan L, Han G, Wang D, et al. Effect of surface-modified TiO_2 nanoparticles on the anti-ultraviolet aging performance of foamed wheat straw fiber/polypropylene composites. Materials (Basel, Switzerland). 2017;**10**: 456-469. DOI: 10.3390/ma10050456

[38] Pastrana-Martinez LM, Morales-Torres S, Kontos AG, et al. TiO_2, surface modified TiO_2 and graphene oxide-TiO_2 photocatalysts for degradation of water pollutants under near-UV/Vis and visible light. Chemical Engineering Journal. 2013;**224**:17-23. DOI: 10.1016/j.cej.2012.11.040

[39] Kango S, Kalia S, Celli A, et al. Surface modification of inorganic nanoparticles for development of organic–inorganic nanocomposites—A review. Progress in Polymer Science. 2013;**38**:1232-1261. DOI: 10.1016/j.progpolymsci.2013.02.003

[40] Man Y, Mu L, Wang Y, et al. Synthesis and characterization of rutile titanium dioxide/polyacrylate nanocomposites for applications in ultraviolet light-shielding materials. Polymer Composites. 2015;**36**:8-16. DOI: 10.1002/pc.22903

[41] Olad A, Behboudi S, Entezami AA. Effect of polyaniline as a surface modifier of TiO_2 nanoparticles on the properties of polyvinyl chloride/TiO_2 nanocomposites. Chinese Journal of

Polymer Science. 2013;**31**:481-494. DOI: 10.1007/s10118-013-1236-5

[42] Wang S. Structural Design and Synthesis of Hyper-Dispersant Agent and It's Application in Plastics Modification. Hefei, Anhui, China: Hefei University of Technology; 2008

[43] Schaller C, Schauer T, Dirnberger K, et al. Synthesis and properties of hydrophobically modified water-borne polymers for pigment stabilization. Progress in Organic Coating. 1999;**35**: 63-67. DOI: 10.1016/s0300-9440(99) 00034-x

[44] Zhang B-H, Yang Q, Qian F, et al. Synthesis and properties of hyper-dispersants. Journal of Shanghai University (English Edition). 2009;**13**: 283-286. DOI: 10.1007/s11741-009-0406-1

[45] Liu W, Cheng L, Yao X, et al. Synthesis and properties of novel acrylic polyester hyper-dispersant. Journal of Coatings Technology and Research. 2016;**13**:763-768. DOI: 10.1007/s11998-016-9787-3

[46] Liu W, Cheng L, Yao X, et al. Adsorption studies for the acrylic polyester hyper-dispersant on titanium dioxide. Journal of Dispersion Science and Technology. 2017;**38**:33-39. DOI: 10.1080/01932691.2015.1115361

Chapter 7

Tuning the Magnetic and Photocatalytic Properties of Wide Bandgap Metal Oxide Semiconductors for Environmental Remediation

Ganeshraja Ayyakannu Sundaram, Rajkumar Kanniah and Vaithinathan Karthikeyan

Abstract

The review focuses on recent developments towards preparing room temperature ferromagnetic metal oxide semiconductors for better photocatalytic performance. Here we reported the combined study of photocatalytic and ferromagnetic properties at room temperature on metal oxides, particularly TiO_2, which is rapidly an emerging field in the development of magnetism and environmental remediation. Even after decades of research in this area, the exact mechanism of the combination of ferromagnetism and photocatalysis in these materials has been not understood completely. However, some of the critical factors were hinted about the contribution to magnetism. Many reports demonstrated that oxygen vacancy and various metal doping plays a primary role in the room temperature ferromagnetism and photocatalysis in wide-band-gap metal oxides. However, it is not easy to understand the direct correlation between magnetism, oxygen vacancies, dopant concentration, and photocatalysis. This review primarily aims to encompass the recent progress of metal oxide for understanding magnetism and photocatalyst under visible light.

Keywords: metal oxide, titania, ferromagnetism, photocatalysts, semiconductors

1. Introduction

The optical, magnetic, and photocatalytic properties of wide bandgap metal oxide semiconductors (MOS) are easily tunable by adjusting the defect concentration, attaining great attention in the scientific research community [1, 2]. The position of the defect levels significantly influences the photons of various absorption and emission energies, and the intensity of intrinsic magnetism is also affected by the number of unpaired electron spins created by the defect levels in MOS compounds [3]. Therefore, tuning the magnetic properties of the MOS nanoparticles by defect engineering could be directly correlated with the optical as-well-as photocatalytic properties [1, 2]. The tuning of the absorption spectra by the defects of varying charge states helps prepare light-emitting diodes, optic-magnetic-based devices, or optically writable oxides by the d^0-magnetism various wavelengths of light [4, 5]. The nature of MOS and their recent research on n-type and p-type models were remarkable in many applications [6].

The MOS nanoparticles with a unique combination of magnetic and charge transport properties such as TiO_2, ZnO, and SnO_2 are attracting substantial attention from the academic and industrial community. From all these various MOS materials, TiO_2 gains special attention due to its solid photocatalytic behavior and several other advantages like low cost, chemical and thermal stability, innocuity, and high refractive index [7, 8]. However, this wide-bandgap TiO_2 semiconductor is activated to perform photocatalysis only under irradiation of ultraviolet (UV) light, which needs to improve for practical applications. Many investigations have been reported and strategies to enhance TiO_2 photo-absorption capability [9–13]. Various strategies to improve photo-absorption, doping, co-doing, surface grafting, the combination of surface grafting and doping are efficient and established routes [14–18]. Suppose MOS nanoparticles are sitting in the core. In that case, the structure of MOS composite nanomaterials could be divided into four forms: core-shell, matrix-dispersed, Janus, and shell-core-shell structures, as shown in **Figure 1**.

For example, metal-doped TiO_2 nanoparticles improve the bandgap from the range of wide to mid-level electronic states, which imparts enhancement in charge migration or produces a strong redshift in the photo-absorption spectrum. More

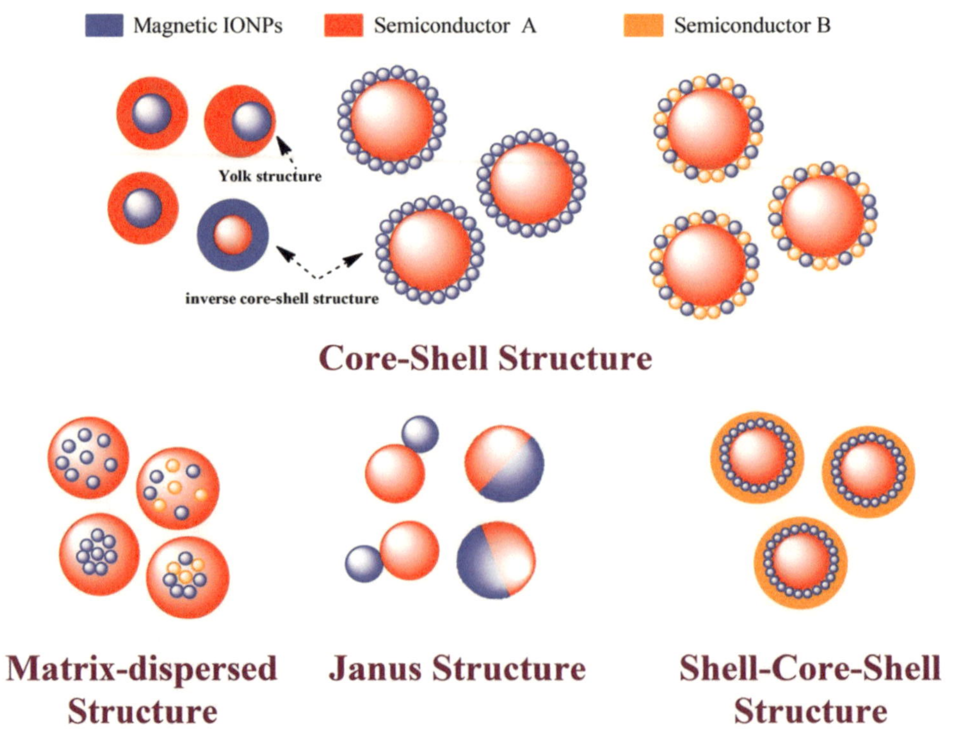

Figure 1.
Various structures of magnetic MOS composite materials. Blue spheres indicate the magnetic MOS nanoparticles, and the non-magnetic matrix and secondary materials are shown in another color [19].

emphasis has been explained in recent years on the $[Sn_xTi_{1-x}O_2]$ system by coupling TiO_2 with SnO_2 oxide. It is highly acceptable that these new nanocomposites exhibit high photocatalytic activity compared to pure TiO_2 [20]. The simple hydrothermal synthesis route will produce SnO_2-TiO_2 nanocomposites; however, a small variation in the synthesis condition could lead to the formation of distinct secondary phases [21]. Cao *et al.* reported that annealing temperature strongly influences the Sn^{4+} ions doping into TiO_2 lattice, depends on temperature, which may substitute in lattice and exist as secondary phases like $SnCl_x$ or SnO_2 [22]. Sn-doped TiO_2 nanoparticles showed significant enhancement in performance as components of active visible light photocatalyst [23, 24], lithium-ion batteries [25], antibacterial activity [26], dye-sensitized solar cells [27], photo-electrochemical conversion [28] and water splitting [29] has been reported. It is important to find a reliable way to synthesize Sn-doped TiO_2 nanostructures, as TiO_2 and SnO_2 are environmentally benign, highly stable, and strong oxide materials [30, 31]. We developed a simple hydrothermal method to synthesize Sn-TiO_2 nanocrystals with sufficient oxygen vacancies, in this nanocrystal with different concentrations of Sn observed ferromagnetism and excellent photocatalytic activity [32, 33]. Wang *et al.* reported that Sn doping and Sn-Fe co-doping in TiO_2 showed a strong red-shift in the optical absorption spectrum [34]. The reason for this shift in absorption spectrum in the Sn-doped TiO_2 system comes from the most of the Sn 5 s states are located at the bottom of the conduction band where Ti 3d states are present and mixed with them.

The combination of non-transition metal and non-metal co-doping improves the visible-light activities of MOS materials. The non-metal doping in TiO_2 can make the new extra valance band and non-transition metal doping create the additional charge carrier traps, which improve the separation efficiency of photo-generated electron–hole pairs, reducing the bandgap width, and broadening the photo-absorption limit [35, 36]. Therefore, the combination of metal and non-metal co-doping will be applied to drastically enhance the visible-light photocatalytic performance of TiO_2. Among the various non-metals, nitrogen is an effective and promising candidate because N doping modifies the charge transport properties of TiO_2 along with which also induces the oxygen-defect sites, therefore improving the photocatalytic performance [37]. The substitutional nitrogen doping on TiO_2 showed an effective reduction in the bandgap width [38]. The nitrogen atoms were successfully substituted by either titanium or oxygen vacant atomic sites in the lattice of TiO_2 lattice. Asahi *et al.* reported that nitrogen atoms successfully replaced the oxygen lattice sites and reduced the bandgap width by mixing N 2*p* and O 2*p* states [39]. Wang *et al.* have studied that the TiO_2 nanocrystals were compacted closely together to form the solid TiO_2. By doping nitrogen, some extra impurity levels were distributed on the surface of the TiO_2 [40], as shown in **Figure 2a**. The solid TiO_2 with a close packing structure creates the difficulty of nitrogen doping into the bulk structure of TiO_2 and makes the diffusion of nitrogen difficult. However, the addition of the dodecyl tri-methyl ammonium bromide (CTAB) to TiO_2 nanocrystals produces a loose packing mesoporous structure, which is conducive for TiO_2 to take up ammonia into the interspaces.

Compared to undoped mesoporous TiO_2, the nitrogen-doped mesoporous TiO_2 with uniform distribution from the inside out produced successive energy levels from the bulk to the surface (**Figure 2b**). This subsequent impurity energy-band level formed by nitrogen doping are located above the valence band and successfully reduces the bandgap of the mesoporous TiO_2, which is the primary attribution for the improved photocatalytic activity throughout the visible-light range. Zhuang *et al.* have reported that the facile sol–gel method prepared Sn and N co-doped TiO_2 (SNT)

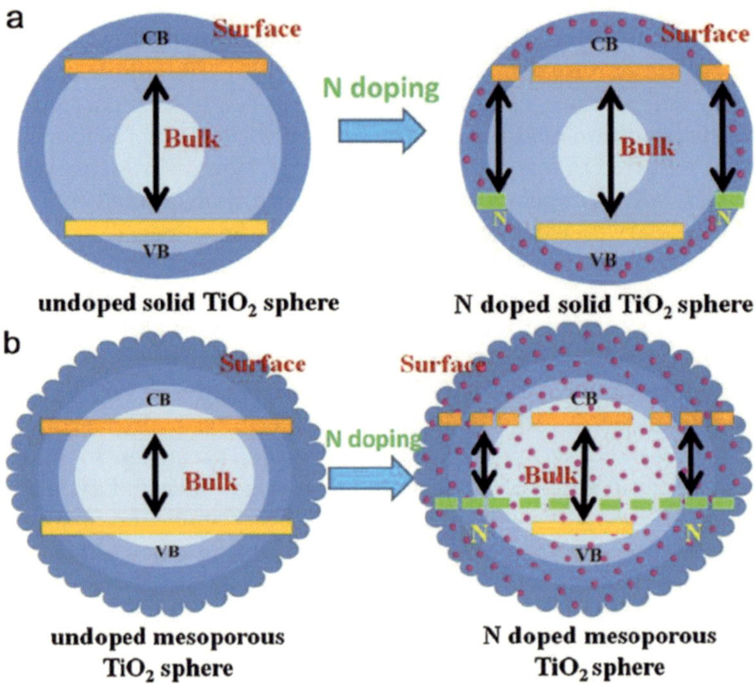

Figure 2.
Schematic diagrams depicting the band structures of (a) solid and (b) mesoporous TiO_2 before and after doping on N [41, 42].

photocatalysts. The post-nitridation treatment enhances the photocatalytic performance of co-doped TiO_2 under visible light or simulated solar light irradiation [43]. However, more studies are required to clearly understand the effect of doping on the physical, chemical and catalytic properties of SNT microspheres.

2. Diluted magnetic semiconductors

Diluted magnetic semiconductors (DMS), referred to as doping of magnetic impurities in bulk semiconductors, also called "semi-magnetic semiconductors", have been studied. This concept has had a particular interest in the research community for the past few years because ferromagnetism in diluted magnetic semiconductors (DMS) has been another important subject that can manipulate the carrier-associated charge and spin-based parameters [44, 45]. Especially, DMS with room temperature ferromagnetic oxides gained particular attention in the applications of magnetic fluids, biomedical, magnetic resonance imaging, catalysis, and environmental remediation [46, 47]. Wang *et al*. developed a facile method to synthesize ZnO crystals with Zn vacancies, and these doped Zn vacancies created p-type conductivity, room-temperature ferromagnetism, and excellent photocatalytic performance [48]. The recent development of ferromagnetic ordering in photo-induced transition metal-doped TiO_2 nanoparticles can be justified by creating defects in the samples [15]. However, the actual role of dopants (e.g. transition metals) at the room temperature ferromagnetism in TiO_2 nanoparticles is still an unclarified problem [49]. In one of our recent papers, our group proposed a new model for combined mechanics of ferromagnetism and their photocatalytic activity in wide-band-gap metal oxide-associated nanocomposites [32]. The study of ferromagnetism and photocatalytic activity on synthesized metal oxide-based nanocomposites suggesting a significant role of oxygen vacancies present on the surface and improved charge carrier concentration on magnetism and photocatalytic performance [50]. Charanpahari *et al*. reported room-temperature ferromagnetic nanocomposites showing better photocatalytic performance compared to commercially available diamagnetic photocatalysts under visible light irradiation [51]. Doping and co-doping have the advantage of high activity in semiconductor nanocomposites, which imparts the concept of magnetic photocatalysts with charge carrier and separation function was raised [51, 52]. Hence, in the research of photocatalytic activity today researchers are focusing on the development of photocatalyst possessing ferromagnetic property and visible-light activity.

DMS with room temperature ferromagnetism has been extensively studied for the applications of spin-based field-effect transistors, spin-based light-emitting diodes (LEDs), and non-volatile memory devices [53, 54]. In DMS materials are due to the coupling of magnetic ordering with one of the other types of ferroic ordering parameters like ferroelasticity or ferroelectricity, which are very interesting from the standpoint of device applications in fields such as spintronic and magneto-optics. Therefore, DMS offering certainly promising immense opportunities for new next-generation applications [55]. Theoretical and experimental studies on these metal oxides have shown improved ferromagnetism by the presence of defects or lightweight doping elements like C, N, and Li [56]. The addition of light elements in DMS can develop magnetism and significantly stabilizes the intrinsic defects in the oxide materials [56]. In these systems, the improved ferromagnetism is mainly attributed to the following mechanisms (i) the concentration of the oxygen vacancies (V_O) and defects sites and (ii) the substitution of an oxygen atom with the doping element and associated formation of spin-polarized states in the bandgap and (iii) the change of titanium oxidation state (Ti^{3+}) in the occurrence of ferromagnetic order. Therefore, defect engineering is a powerful tool to tune or improve the functional properties of the metal oxides like their electronic band structure, charge carrier transport, and catalytic performance [48]. The photocatalytic performance of TiO_2 significantly depends on their electrical and optical properties, which are primarily determined and altered by the crystal structure, optimized concentration of dopants, and defects [57].

Figure 3(A) showing the schematic diagram of the magnetic orientation of Fe doped TiO_2 nanoparticles, which are annealed under vacuum. It shows the possible paramagnetic species, their distribution in the nanoparticles lattice, surface, and interfacial boundary, and the potential interaction with ferromagnetic or antiferromagnetic species. The red circles inside the nanoparticles representing the magnetic polaron and overlapped magnetic polarons form BMPs. Along with BMPs, coupled F+ centres on the surface and interface also contribute towards ferromagnetism. However, F^{2+} without any electrons and F Centre with two trapped electrons are not likely to contribute towards ferromagnetism [58]. In vacuum annealed pristine TiO_2 nanoparticles, the total magnetization is contributed from the surface and interfacial oxygen vacancies, i.e. $M_{total} = M_{surface} + M_{interface}$. However, an extra BMP factor is added in the Fe doped vacuum annealed TiO_2 nanoparticles; therefore, the total magnetization is written as $M_{total} = M_{BMP} + M_{surface} + M_{interface}$. These observations of paramagnetic behavior in Fe doped TiO_2 nanoparticles suggest that the density of oxygen vacancies is possibly insufficient to generate solid ferromagnetic coupling with the nearest lattice site of Fe^{3+} ions. To improve the magnetization in pure and 2%

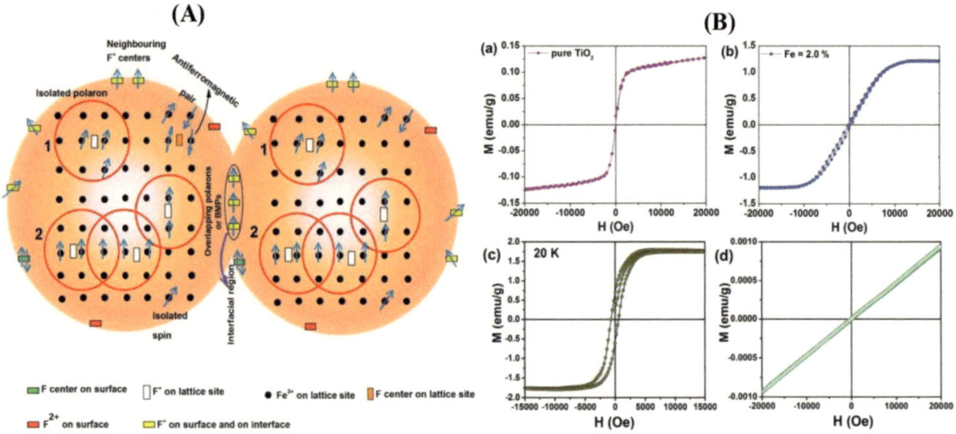

Figure 3.
(A) Diagram represents various possible magnetic species, their distribution, and interaction [58]. (B) M–H curves of vacuum annealed nanoparticles of (a) pristine TiO_2 and, (b) 2% Fe doped TiO_2 at room temperature, (c) 2% Fe doped TiO_2 at 20 K and, (d) paramagnetic M–H curve of vacuum annealed 2% Fe doped TiO_2 after reheating in the air at 450°C [58].

Fe doped TiO_2, vacuum annealed at 200°C for 3 h, generating donor carrier or oxygen vacancies. M–H measurements are carried out after the annealing on the samples, and as plotted in **Figure 3(B)**, initially diamagnetic pristine TiO_2 and paramagnetic Fe doped TiO_2 nanoparticles both have exhibited ferromagnetism. The observed ferromagnetism in pure TiO_2 nanoparticles could be attained from either Ti^{3+} ions or the presence of oxygen vacancies on the lattice site or the surface. Even though pristine and Fe doped TiO_2 showed ferromagnetically, the saturation magnetization of pure TiO_2 is less than that of Fe doped TiO_2 nanoparticles. The enhanced magnetization in Fe doped samples could be due to the extra magnetic interaction generated by both Fe dopants and defects in the ferromagnetic exchange coupling. The ferromagnetism is again switched back to paramagnetic for reheated vacuum annealed Fe doped TiO_2 in the air at 450°C samples as shown in **Figure 3(B)d**. The above results support that the oxygen vacancies possibly play the driving role in switching the magnetic ordering from paramagnetic to ferromagnetic and then back to paramagnetic in Fe doped TiO_2 nanoparticles. Just simple doping of Fe may not be sufficient to induce ferromagnetic solid exchange interaction. Only, when a high concentration of oxygen vacancies and Fe doping combining may participate in ferromagnetic exchange interaction.

Irradiation of various energy ion beams is one of the sophisticated techniques for incorporating the defects (i.e., vacancies, interstitials, etc.) into transition metal-doped metal oxide semiconductor matrix materials. Many researchers have studied that ion beam irradiation could improve the structural complexity of the ZnO nanoparticles by dissolving the secondary impurity phases, helps in substitutional incorporation of Mn^{2+} at the Zn^{2+} site (Mn and Zn) and improves the ferromagnetic property of the samples [59–61]. To avoid the segregation of nano-dimensional doped transition metal or its oxide clusters and to induce intrinsic structural defects in the host material in a controlled fashion, irradiation of a low energy ion beam using inert gases such as Xe or Ar is the best option which also eradicates the complexities arising from the chemical reactivity of the ion beams [60]. A multilayer coating and high-temperature calcination, thus affecting the photocatalytic efficiency, often influence the magnetic properties [62]. Therefore, a novel and facile approach to the low-cost preparation of the ferromagnetic and photocatalytic TiO_2 nanocomposite at relatively low temperatures is highly recommended. We have reported several research articles related to the photocatalytic performance and magnetic properties of TiO_2-based photocatalysts such as various metal (Sn, Cu and, Fe) oxide coupled TiO_2 [32], Sn doped TiO_2 [33], Fe_2O_3 coupled. Doped TiO_2 [63], nickel(II)-imidazole doped TiO_2 [64], hierarchical Sn and N co-doped TiO_2 [65] and hierarchical AgCl loaded Sn doped TiO_2 [66].

3. Visible light photocatalysts

Progressive research towards solar power-based energy conversion, wastewater treatment, and efficient photocatalysts attracting great attention [67–70]. Photocatalytic and photovoltaic solar cells convert solar-based light energy into chemical reaction and electrical power generation. Consequently, improving the stabilizations of photo-induced charge carrier transportation is the critical factor for light-harvesting systems. TiO_2-based materials are widely used in environmental and energy-related applications like photocatalysis, photovoltaics, artificial photosynthesis, and spintronic, which have been often foreseen. For better performance, TiO_2 is usually employed as nanocrystals or nanostructures [71–73]. However, the efficiency of photocatalytic activity of TiO_2 needs to improve to induce charge carrier activity using visible light or sunlight. Noble metal (Pt, Pd, Rh, and Au) doped and modified TiO_2 photocatalysts have been attracted great attention towards efficiency enhancement [74–76]. Especially in this context of an investigation, Ag-loaded TiO_2 that is Ag cluster-incorporated AgBr nanoparticles [77], Ag nanoparticles and CuO nanoclusters [78], and Ag/AgCl [79] in TiO_2 photocatalysts are undoubtedly intriguing to attain high performance [80]. The interfacial heterojunction between TiO_2 and SnO_2 particles can have a synergetic effect on photo activity [24]. Furthermore, any agglomeration in TiO_2/Ag/AgCl system due to the nature of the materials process used can influence the observed photocatalytic activity given that Ag/AgCl is a plasmonic system.

Therefore to improve the photocatalytic performance of metal oxide nanoparticles by expanding the range of photo-response and increasing the efficiency of electron–hole carrier separation, the hierarchical assembly of nanoscale building photocatalytic blocks with a tunable dimensionality and structural complexity offers a practical strategy towards the realization of multi-functionality of nanomaterials [81]. In general, hierarchical heterostructures are formed by connecting two different low-dimensional nanostructure materials; this type of structure provides the ultrahigh specific surface area and a network system consisting of parallel connective paths and provides interconnection of various functional components [82].

Liu et al., in their work, explained the photocatalytic mechanisms operating in the Fe(III)-$Fe_xTi_{1-x}O_2$ system as illustrated in **Figure 4**. are discussed [17, 18]. They are owing to the wide bandgap of pristine TiO_2, which is inactive under the illumination of visible or sunlight. However, by the selected surface grafting and bulk doping of Fe(III) ions, which have band energy levels identical to TiO_2, the visible-light absorption of TiO_2 is drastically improved by the bulk-doped Fe(III) ions. The QE was unaffected because of the efficient transfer of electrons between doped Fe(III) and surface Fe(III). Moreover, a good interface junction between surface-grafted and bulk-doped Fe(III) ions is needed for efficient charge carrier transfer. Notably, the visible-light activity reaction was markedly reduced by introducing a thin layer between the

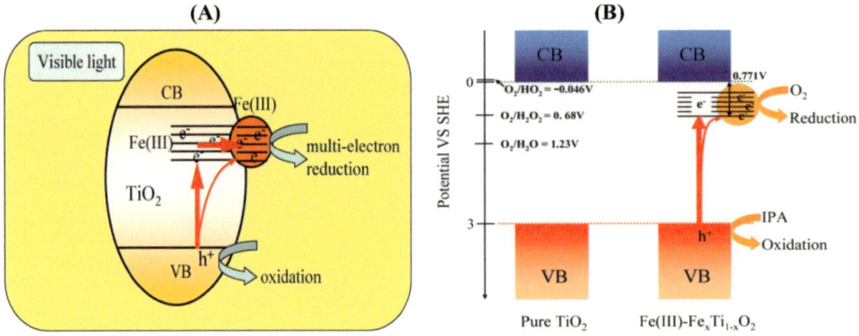

Figure 4.
(A) Proposed photocatalysis process. (B) Change in bandgap and photo-activity by Fe doping [17, 18].

surface Fe(III) ions and doped TiO_2. The photo-generated charge carriers are effectively transferred to the surface of Fe(III) doped TiO_2, which acts as an efficient co-catalyst for multi-electron reduction reactions. In photocatalysis by Fe(III) doped TiO_2, holes with high oxidation potential are kept in the deep level of the valance band and effectively decompose the organic compounds. Therefore, efficient visible-light photocatalysts with high R is achieved.

The conceptual ferromagnetic photocatalysts show a better charge carrier separation function to take advantage of high activity in the couple, doped, surface modified, or co-doped semiconductor nanocomposites. However, furthermore development in these TiO_2-based photocatalysts requires other strategies to improve photocatalytic efficiency. In today's research, one of the effective strategies is AgCl nanoparticles loaded in Sn-doped TiO_2 microsphere to enhance the visible-light activity have become an essential outcome in the photocatalytic and photovoltaic applications [83, 84].

4. Ferromagnetic TiO_2-based photocatalyst

In our previous reports, we worked on various concentrations of Sn doping to improve the structural, electronic, magnetic, and photocatalytic properties of TiO_2 nanoparticles [32, 33, 85, 86]. Significantly, the study of room temperature photocatalytic and ferromagnetic performance in the Sn-doped TiO_2 nanoparticles is one of the most emerging and fascinating fields in environmental remediation. Adding various concentrations of $SnCl_4$ in $Ti(NO_3)_4$ aqueous solutions produced any one of the anatase, a mixture of anatase-rutile and rutile phases of TiO_2 nanoparticles with the added Sn atoms, which are synthesized using the facile hydrothermal method. To study the photocatalytic performance of the synthesized Sn-TiO_2 nanoparticles, both methyl orange (MO) and *R*PhOH (where PhOH is phenol group and *R* is 3-NH_2, H, and 4-Cl) in water were chosen as model pollutants under both the illumination of visible light and UV light irradiation. Light irradiation showed a significant relationship between the Hammett substitution constant (σ) of RPhOH and the photocatalytic degradation efficiency of Sn-TiO_2 nanoparticles. The concentration of Sn doping significantly affected the structural, electronic, magnetic and, photocatalytic properties of the TiO_2 nanoparticles. Even after decade-long research, the actual mechanism of ferromagnetism combined with photocatalytic behavior in these materials is still not understood. However, hints about some of the critical factors that contribute to magnetism have been revealed. It is believed that oxygen vacancies, phase changes, and doping level play a significant role in the RTFM of semiconductor oxides; however, demonstration of a direct correlation between the magnetism, dopant concentration, oxygen vacancies, and photocatalytic activity has been strenuous. Because of these reasons, in this work, we made an effort to investigate the essential role of Sn^{4+} ions on the above properties of TiO_2 nanoparticles.

In another report, we first follow the facile hydrothermal synthesis route for preparing ST microspheres, followed by nitriding treatment by flowing an ammonia gas to successfully fabricate hierarchical SNT microspheres with V_O [64]. The fabricated as-prepared samples are characterized by the conventional analytical techniques and ^{119}Sn Mössbauer spectroscopy to understand the structure, magnetism, and photocatalytic performance. The main objective of this study is to improve the photocatalytic performance and RTFM of TiO_2 by the co-doping of Sn and N atoms. As compared to pristine and Sn doped TiO_2 nanoparticles, SNT microspheres showed significant absorption of visible light for photocatalytic activity is observed. Then we have further studied the photocatalytic movement of Rhodamine B (RhB) degradation under the illumination of visible light irradiation on pristine TiO_2, P25, ST, and SNT microspheres and observed vigorous photocatalytic activity in SNT microspheres. However, until now, no one reported magnetic studies on the SNT microspheres. Suppose, if the photocatalysts exhibit RTFM, the phenomenon may insist on the electrons trapped in V_O or structural defects. In this aspect, we can believe that this study can be implemented in the various other types of facile designing semiconductors to obtain an insight into the role of the visible light photocatalytic performance, RTFM behavior, and combined performance enhancement. In addition, we also studied the photovoltaic performance of ST and SNT microspheres in the applications of Perovskite solar cells. The combined mental and non-metal doped TiO_2 nanoparticles with other structural defect sites represent a new kind of semiconductor materials and provide novel opportunities for TiO_2-based materials.

For the first time, we have reported a facile hydrothermal synthesis route to successfully fabricate hierarchical AgCl in Sn-TiO_2 (AST) microspheres using post-calcination treated with different temperature samples [66, 87]. The primary objective of this study is to modify Sn doped TiO_2 by loading AgCl nanoparticles to enhance photocatalytic performance. Improved visible light absorption capability was observed in the AST microspheres compared to Sn-TiO_2, AgCl, Ag/AgCl, and commercial Degussa P25 photocatalysts. To check the photocatalytic performance of the as-synthesized AST microspheres, the rhodamine B (RhB) and 3-nitrophenol aqueous solutions were used as the model systems under visible light ($\lambda \geq 420$ nm). The obtained results indicate that the hierarchical AST microsphere photocatalysts showed a higher photodegradation rate than Ag/AgCl, AgCl, Sn-TiO_2, and the commercial TiO_2 (P25) materials. However, the study on various concentrations of AgCl in the AST microsphere is crucial to understand the optimized amount needed to obtain the best photocatalytic performance. To the best of our knowledge, for the first time, we reported the facile preparation route, high visible-light photocatalytic performance in hierarchical AST microspheres, and the magnetic behavior of these photocatalysts characterized by the ^{119}Sn Mössbauer technique. The new semiconductor family of noble metal halide and metal-doped TiO_2 nanoparticles opens up novel opportunities for TiO_2-based materials.

We have option $[Fe(III)(bipy)_2Cl_2)]^+[Fe(III)Cl_4]^-$ ionic salt-like complex as precursor complex [73]. The aqueous solution of precursor complex could behave like electrolytes. While the reduction potential from free Fe(III) to free Fe(II) is 0.77 V, that of photo-reduction from $[Fe^{III}Cl_4]^-$ to $[Fe^{II}Cl_3]^-$ is 0.34 V which indicates that photoreduction of the $[FeCl_4]^-$ ion is easier than the normal chemical reduction of free ferric ions [12]. Hence chosen iron(III) complex interacts with n-type TiO_2 semiconductors.

It reduces Fe(III) to Fe(II) *via* interfacial electron transfer dynamics under dark (poor efficiency), near-UV (good efficiency), and visible light (moderate efficiency) irradiation systems. At the same time, the precursor complex is adsorbed on the TiO_2 surface to form a surface complex; it acts as a co-catalyst for the reduction of Fe(III) to Fe(II) with TiO_2. However, there are no reports on the study of photosensitized *via* IFET dynamics between Fe(III)-bipy complex (bipy without -OH or -COOH groups) and titania semiconductor interface until now. Hence, we report the near-UV and visible-light-induced IFET process on $[Fe^{III}(bipy)_2Cl_2][Fe^{III}Cl_4]$ (precursor complex) with TiO_2 NPs, and the photochemical product was mainly characterized by electronic absorption, Fe K-edge X-ray absorption fine structure (XAFS), electron paramagnetic resonance (EPR) and ^{57}Fe Mössbauer spectroscopes method. In addition, electron transfer was confirmed by cyclic voltammetric and photoluminescence measurements. However, the following factors control the IFET reaction, those are (i) the presence of TiO_2 nanoparticles, (ii) the irradiation time-lapse, (iii) light source with various wavelengths ($380 \leq \lambda \leq 520$ nm), and (iv) different types of TiO_2 nanoparticles.

In one of our works, nickel(II)-imidazole-anatase nanocomposites prepared by a simple adsorption method showed room-temperature ferromagnetism and good photocatalytic performance, which were designed by mixing of $[Ni(1\text{-}MeIm)_6]Cl_2H_2O$ complex and anatase TiO_2 starting materials in an aqueous medium [64]. Various conventional techniques as adsorption already elucidated the deposition of the surface species. We observed the ferromagnetic behavior in the composite sample under the vibrating sample magnetometer at room temperature. This Ni-dopedTiO_2 nanocomposite has good visible light absorption ability than pristine TiO_2. To understand and evaluate the adsorption and photocatalytic activity of the Ni-doped TiO_2 nanocomposite, selected methylene blue (MB) as an organic pollutant illuminating under visible light irradiation. We first reported the Ni(II)-imidazole complex deposited on the anatase (TiO_2) semiconductor with good photocatalytic and magnetic properties prepared by a simple adsorption method. The research of metal oxide-based photocatalysis is expected to open up a general method for synthesizing other transition metal-loaded metal oxide semiconductor photocatalysts.

In all of our previous reports covers the studies related to Mössbauer spectroscopic, photocatalytic and magnetic investigations of Sn and Fe doped TiO_2 nanocomposites [32, 33, 63–66, 73, 85–87]. Using the facile hydrothermal synthesizing route, we prepared Sn-based TiO_2. For structural and magnetic characterization, Mössbauer spectroscopy has unique advantages to mature into one of the classical techniques for Sn or Fe-based TiO_2 nanoparticles. Mössbauer spectroscopic results provided a strong understanding and evidence of the relationship between the structural, photocatalytic, and magnetic properties of Sn or Fe-based TiO_2 nanoparticles. The Sn or Fe-doped TiO_2 nanocomposites have promising applications in photocatalysis for water purification by degrading organic pollutants using efficient visible light absorption to produce strong stability and high photocatalytic activity. This review helps in the fundamental understanding of structural and magnetic properties of Sn or Fe-doped TiO_2 nanocomposites and their contribution towards environmental remediation by visible-light photocatalysis.

5. Conclusion

This review mainly highlighted the importance of the development of wide bandgap metal oxide nanoparticles for photocatalyst applications. Several researchers are primarily focused on developing a room-temperature ferromagnetic TiO_2 as the photocatalyst, which has a high potentiality to absorb visible light from the solar spectrum. However, there are certain limitations in pristine TiO_2 nanoparticles: their high photo-generated holes and electrons recombination rate, and they require UV light for photocatalysis. These problems can be overcome by introducing metallic or non-metallic dopants or creating oxygen vacancies and defect sites into TiO_2. The two successful approaches that have been discussed are the doping and grafting of TiO_2 nanoparticles with either anionic or cationic elements and coupling TiO_2 nanoparticles with other semiconductors. Further study is needed to understand the use of novel ferromagnetic metal oxide-based photocatalyst for large-scale applications.

Acknowledgements

Dr. ASG is thankful to National College (Autonomous), Tiruchirappalli, Tamil Nadu for financial support through a college minor research project scheme (No. NCT/SEC/010/2022-2023/19-07-2022).

Author details

Ganeshraja Ayyakannu Sundaram[1,2]*, Rajkumar Kanniah[3] and Vaithinathan Karthikeyan[4]

1 Department of Chemistry, SRM Institute of Science and Technology, Chennai, India

2 Mössbauer Effect Data Center, Dalian Institute of Chemical Physics, Chinese Academy of Sciences, Dalian, China

3 Department of Chemistry, Central University of Tamil Nadu, Thiruvarur, India

4 James Watt School of Engineering, University of Glasgow, United Kingdom

*Address all correspondence to: asgchem84@gmail.com

References

[1] Fischer DK, de Fraga KR, Choi CWS. Ionic liquid/TiO_2 nanoparticles doped with nonexpensive metals: New active catalyst for phenol photodegradation. RSC Advances. 2022;**12**:2473-2484

[2] Song H, Lee JD, Kim SKR. Correlated visible-light absorption and intrinsic magnetism of $SrTiO_3$ due to oxygen deficiency: Bulk or surface effect? Inorganic Chemistry. 2015;**54**:3759-3765

[3] Fan CM, Peng Y, Zhu Q, Lin L, Wang RX, Xu AW. Synproportionation reaction for the fabrication of Sn^{2+} self-doped SnO_{2-x} nanocrystals with tunable band structure and highly efficient visible light photocatalytic activity. Journal of Physical Chemistry C. 2013;**117**:24157-24166

[4] Kan D, Terashima T, Kanda R, Masuno A, Tanaka K, Chu S, et al. Blue-light emission at room temperature from Ar^+-irradiated $SrTiO_3$. Nature Materials. 2004;**4**:816-819

[5] Sun S, Wu P, Xing P. d^0 ferromagnetism in undoped n and p-type In_2O_3 films. Applied Physics Letters. 2012;**101**:132417

[6] Lou C, Lei G, Liu X, Xie J, Li Z, Zheng W, et al. Design and optimization strategies of metal oxide semiconductor nanostructures for advanced formaldehyde sensors. Coordination Chemistry Reviews. 2022;**452**:214280

[7] Xiang Y, Li Y, Zhang X, Zhou A, Jing N, Xu Q. Hybrid Cu_xO–TiO_2 porous hollow nanospheres: Preparation, characterization and photocatalytic properties. RSC Advances. 2017;**7**:31619-31627

[8] Gao D, Wu X, Wang P, Xu Y, Yu H, Yu J. Simultaneous realization of direct photoinduced deposition and improved H_2-evolution performance of Sn-Nano particle modified TiO_2 Photocatalyst. ACS Sustainable Chemistry & Engineering. 2019;**7**:10084-10094

[9] Ehsan MF, Khan R, He T. Visible-light photoreduction of CO_2 to CH_4 over Zn Te-Modifited TiO_2 coral-like nanostructures. ChemPhysChem. 2017;**18**:3203-3210

[10] Han F, Kamabala VSR, Srinivasan M, Rajarathnam D, Naidu R. Tailored titanium dioxide photocatalysts for the degradation of organic dyes in wastewater treatment. A review. Applied Catalysis A: General. 2009;**359**:25-40

[11] Liu G, Wang LZ, Yang HG, Cheng HM, Lu GQ. Titania-based photocatalysts-crystal growth, doping and heterostructuring. Journal of Materials Chemistry. 2010;**20**:831-843

[12] Zhang H, Chen G, Bahnemann DW. Photoelectrocatalytic materials for environmental applications. Journal of Materials Chemistry. 2009;**19**:5089-5121

[13] Leung DYC, Fu X, Wang C, Ni M, Leung MKH, Wang X, et al. Hydrogen production over titania-based photocatalysts. ChemSusChem. 2006;**36**:681-694

[14] El-sheikh SM, Zhang G, El-hosainy HM, Ismail AA, Shea KEO, Falaras P, et al. High performance Sulfur, Nitrogen and Carbon doped mesoporous anatase – brookite TiO_2 photocatalyst for the removal of microcystin-LR under visible light irradiation. Journal of Hazardous Materials. 2014;**280**:723-733

[15] Anbalagan K. UV-sensitized generation of phase pure cobalt-doped

Anatase: $Co_xTi_{1-x}O_{2-\delta}$ nanocrystals with ferromagnetic behavior using Nano-TiO_2/cis-[Co^{III}(en)$_2$($MeNH_2$) Cl]$^{2+}$. Journal of Physical Chemistry C. 2011;**115**:3821-3832

[16] Irie H, Kamiya K, Shibanuma T, Miura S, Tryk DA, Yokoyama T, et al. Visible light-sensitive Cu(II)-grafted TiO_2 photocatalysts: Activities and X-ray absorption fine structure analyses. Journal of Physical Chemistry C. 2009;**113**:10761-10766

[17] Liu M, Qiu X, Miyauchi M, Hashimoto K. Energy-level matching of Fe(III) ions grafted at surface and doped in bulk for efficient visible-light Photocatalysts. Journal of the American Chemical Society. 2013;**135**:10064-10072

[18] Dubey M, Kumar R, Srivastava KS, Joshi M. Visible light induced photodegradation of chlorinated organic pollutants using highly efficient magnetic Fe_3O_4/TiO_2 nanocomposite. Optik. 2021;**243**:167309

[19] Wei W, Jiang C, Roy VA. L, recent progress in magnetic iron oxide – Semiconductor composite nanomaterials as promising photocatalysts. Nanoscale. 2015;**7**:38-58

[20] Abdel-Messih MF, Ahmed MA, El-sayed AS. Photocatalytic decolorization of Rhodamine B dye using novel mesoporous SnO_2–TiO_2 nano mixed oxides prepared by sol–gel method. Journal of Photochemistry and Photobiology A: Chemistry. 2013;**260**:1-8

[21] Mourão HAJL, Avansi WJ, Ribeiro C. Hydrothermal synthesis of Ti oxide nanostructures and TiO_2:SnO_2 heterostructures applied to the photodegradation of rhodamine B. Materials Chemistry and Physics. 2012;**135**:524-532

[22] Cao Y, He T, Zhao L, Wang E, Yang W, Cao Y. Structure and phase transition behavior of Sn^{4+} – doped TiO_2 nanoparticles. Journal of Physical Chemistry C. 2009;**113**:18121-18124

[23] Boppana VBR, Lobo RF. Photocatalytic degradation of organic molecules on mesoporous visible-light-active Sn (II) -doped titania. Journal of Catalysis. 2011;**281**:156-168

[24] Li J, Xu X, Liu X, Yu C, Yan D, Sun Z, et al. Sn doped TiO_2 nanotube with oxygen vacancy for highly efficient visible light photocatalysis. Journal of Alloys and Compounds. 2016;**679**:454-462

[25] Lübke M, Johnson I, Makwana NM, Brett D, Shearing P, Liu Z, et al. High power TiO_2 and high capacity Sn-doped TiO_2 nanomaterial anodes for lithium-ion batteries. Journal of Power Sources. 2015;**294**:94-102

[26] Dhanapandian S, Arunachalam A, Manoharan C. Highly oriented and physical properties of sprayed anatase Sn-doped TiO_2 thin films with an enhanced antibacterial activity. Applied Nanoscience. 2016;**6**:387-397

[27] Duan Y, Fu N, Liu Q, Fang Y, Zhou X, Zhang J, et al. Sn-doped TiO_2 photoanode for dye-sensitized solar cells. Journal of Physical Chemistry C. 2012;**116**:8888-8893

[28] Xu M, Da P, Wu H, Zhao D, Zheng G. Controlled Sn-doping in TiO_2 nanowire photoanodes with enhanced Photoelectrochemical conversion. Nano Letters. 2012;**12**:1503-1508

[29] Asefa BAA, Pan CJ, Su WN, Chen HM, Rick J, Hwang BJ. Facile one-pot controlled synthesis of Sn and C codoped single crystal TiO_2 nanowire arrays for highly efficient

photoelectrochemical water splitting. Applied Catalysis B: Environmental. 2015;**163**:478-486

[30] Chang S, Chen S, Huang Y. Synthesis, structural correlations, and photocatalytic properties of TiO_2 nanotube/SnO_2–Pd nanoparticle Heterostructures. Journal of Physical Chemistry C. 2011;**115**:1600-1607

[31] Banerjee S, Dionysiou DD, Pillai S. Self-cleaning applications of TiO_2 by photoinduced hydrophilicity and photocatalysis. Applied Catalysis B: Environmental. 2015;**176**:396-428

[32] Sundaram A, Samy A, Rajkumar K, Wang Y, Wang Y, Wang J, et al. Simple hydrothermal synthesis of metal oxides coupled nanocomposites: Structural, optical, magnetic and photocatalytic studies. Applied Surface Science. 2015;**353**:553-563

[33] Ganeshraja AS, Thirumurugan S, Rajkumar K, Zhu K, Wang Y, Anbalagan K, et al. Effects of structural, optical and ferromagnetic states on the photocatalytic activities of Sn–TiO_2 nanocrystals. RSC Advances. 2016;**6**:409-421

[34] Wang Y, Zhang Y, Yu F, Jin C, Liu X, Ma J, et al. Correlation investigation on the visible-light-driven photocatalytic activity and coordination structure of rutile Sn-Fe-TiO_2 nanocrystallites for methylene blue degradation. Catalysis Today. 2015;**258**:112-119

[35] Xu H, Ouyang S, Liu L, Reunchan P, Umezawa N, Ye J. Recent advances in TiO_2-based photocatalysis. Journal of Materials Chemistry A. 2014;**2**:12642-12661

[36] Sang L, Zhao Y, Burda C. TiO_2 nanoparticles as functional building blocks. Chemical Reviews. 2014;**114**:9283-9318

[37] Zhang Y, Zhu W, Cui X, Yao W, Duan T. One-step hydrothermal synthesis of iron and nitrogen co-doped TiO_2 nanotubes with enhanced visible-light photocatalytic activity. CrystEngComm. 2015;**17**:8368-8376

[38] Irie H, Washizuka S, Yoshino N, Hashimoto KY. Visible-light induced hydrophilicity on nitrogen-substituted titanium dioxide films. Chemical Communications. 2003;**11**:1298-1299

[39] Asahi R, Morikawa T, Ohwaki T, Aoki K, Taga Y. Visible-light photocatalysis in nitrogen-doped titanium oxides. Science. 2001;**293**:269-272

[40] Wang W, Tadé MO, Shao Z. Nitrogen-doped simple and complex oxides for photocatalysis: A review. Progress in Materials Science. 2018;**92**:33-63

[41] Li X, Liu P, Mao Y, Xing M, Zhang J. Preparation of homogeneous nitrogen-doped mesoporous TiO_2 spheres with enhanced visible-light photocatalysis. Applied Catalysis B: Environmental. 2015;**164**:352-359

[42] Pu X, Hu Y, Cui S, Cheng L, Jiao Z. Preparation of N-doped and oxygen-deficient TiO_2 microspheres via a novel electron beam-assisted method. Solid State Sciences. 2017;**70**:66-73

[43] Zhuang H, Zhang Y, Chu Z, Long J, An X, Zhang H, et al. Synergy of metal and nonmetal dopants for visible-light photocatalysis: A case-study of Sn and N co-doped TiO_2. Physical Chemistry Chemical Physics. 2016;**18**:9636-9644

[44] Phokha S, Pinitsoontorn S, Maensiri S. Structure and magnetic properties of monodisperse Fe^{3+} –doped CeO_2 Nanospheres. Nano-Micro Letters. 2013;**3**:223-233

[45] Dakhel AA. Microstructural, optical and magnetic properties of TiO_2:Fe:M (M = Ga, Zn) dilute magnetic semiconductor nanoparticles: a comparative study. Applied Physics A: Materials Science & Processing. 2021;**127**:440

[46] Lu A, Salabas EL, Schüth F. Magnetic nanoparticles: Synthesis, protection, functionalization, and application. Angewandte Chemie. 2007;**46**:1222-1244

[47] Gupta A, Zhang R, Kumar P, Kumar V, Kumar A. Nano-structured dilute magnetic semiconductors for efficient Spintronics at room temperature. Magnetochemistry. 2020;**6**:15

[48] Wang S, Pan L, Song J, Mi W, Zou J, Wang L, et al. Titanium-defected undoped anatase TiO_2 with p-type conductivity, room-temperature ferromagnetism, and remarkable photocatalytic performance. Journal of the American Chemical Society. 2015;**137**:2975-2983

[49] Chetri P, Basyach P, Choudhury A. Exploring the structural and magnetic properties of TiO_2/SnO_2 core/shell nanocomposite: An experimental and density functional study. Journal of Solid State Chemistry. 2014;**220**:124-131

[50] Cheng C, Amini A, Zhu C, Xu Z, Song H, Wang N. Enhanced photocatalytic performance of TiO_2-ZnO hybrid nanostructures. Scientific Reports. 2014;**4**:1-5

[51] Charanpahari A, Ghugal SG, Umare SS, Sasikala R. Mineralization of malachite green dye over visible light responsive bismuth doped TiO_2-ZrO_2 ferromagnetic nanocomposites. New Journal of Chemistry. 2015;**39**:3629-3638

[52] Khang NC, Khanh N, Anh NH, Nga D, Minh N. The origin of visible light photocatalytic activity of N-doped and weak ferromagnetism of Fe-doped TiO_2 anatase. Advances in Natural Sciences: Nanoscience and Nanotechnology. 2011;**2**:015008

[53] Na C, Park S, Kim SJ, Woo H, Kim HJ, Chung J, et al. Chemical synthesis of CoO – ZnO: Co hetero-nanostructures and their ferromagnetism at room temperature. CrystEngComm. 2012;**14**:5390-5393

[54] Alivov Y, Singh V, Ding Y, Cerkovnik LJ, Nagpal P. Doping of wide-bandgap titanium-dioxide nanotubes: Optical, electronic and magnetic properties. Nanoscale. 2014;**6**:10839-10849

[55] Thakare VP, Game OS, Ogale SB. Ferromagnetism in metal oxide systems: Interfaces, dopants, and defects. Journal of Materials Chemistry C. 2013;**1**:1545-1557

[56] Rahman G. Nitrogen-induced ferromagnetism in BaO. RSC Advances. 2015;**5**:33674-33680

[57] Liu G, Yang HG, Pan J, Yang YQ, Lu GQ, Cheng H. Titanium dioxide crystals with tailored facets. Chemical Reviews. 2014;**114**(19):9559-9612

[58] Choudhury B, Verma R, Choudhury A. Oxygen defect assisted paramagnetic to ferromagnetic conversion in Fe doped TiO_2 nanoparticles. RSC Advances. 2014;**4**:29314-29323

[59] Neogi SK, Midya N, Pramanik P, Banerjee A, Bhattacharyya A, Taki GS, et al. Correlation between defect and magnetism of low energy Ar^{+9} implanted and un-implanted $Zn_{0.95}Mn_{0.05}O$ thin films suitable for electronic application.

Journal of Magnetism and Magnetic Materials. 2016;**408**:217-227

[60] Kumar S, Asokan K, Singh R, Chatterjee S, Kanjilal D, Ghosh AK. Investigations on structural and optical properties of ZnO and ZnO:Co nanoparticles under dense electronic excitations. RSC Advances. 2014;**4**:62123-62131

[61] Borges R, Silva R, Magalhaes S, Cruz M, Godinho M. Magnetism in Ar-implanted ZnO. Journal of Physics Condensed Matter. 2007;**19**:476207

[62] Dong H, Zeng G, Tang L, Fan C, Zhang C, He X, et al. An overview on limitations of TiO_2-based particles for photocatalytic degradation of organic pollutants and the corresponding counter measures. Water Research. 2015;**79**:128-146

[63] Ganeshraja AS, Rajkumar K, Zhu K, Li X, Thirumurugan S, Xu W, et al. Facile synthesis of iron oxide coupled and doped titania nanocomposites: Tuning of physicochemical and photocatalytic properties. RSC Advances. 2016;**6**:72791-72802

[64] Ganeshraja AS, Thirumurugan S, Rajkumar K, Wang J, Anbalagan K. Ferromagnetic nickel(II) imidazole-anatase framework: An enhanced photocatalytic performance. Journal of Alloys and Compounds. 2017;**706**:485-494

[65] Ganeshraja AS, Yang M, Nomura K, Maniarasu S, Veerappan G, Liu T, et al. ^{119}Sn Mössbauer and ferromagnetic studies on hierarchical tin- and nitrogen-Codoped TiO_2 microspheres with efficient photocatalytic performance. Journal of Physical Chemistry C. 2017;**121**:6662-6673

[66] Ganeshraja AS, Zhu K, Nomura K, Wang J. Hierarchical assembly of AgCl@ Sn-TiO_2 microspheres with enhanced visible light photocatalytic performance. Applied Surface Science. 2018;**441**:678-687

[67] Long R, Li Y, Liu Y, Chen S, Zheng X, Gao C, et al. Isolation of Cu atoms in Pd lattice: Forming highly selective sites. Journal of the American Chemical Society. 2017;**139**:4486-4492

[68] Zhang P, Li J, Lv L, Zhao Y, Qu L. Vertically aligned graphene sheets membrane for highly efficient solar thermal generation of clean water. ACS Nano. 2017;**11**:5087-5093

[69] Zhou X, Liu N, Schmuki P. Photocatalysis with TiO_2 nanotubes: "Colorful" reactivity and designing site-specific photocatalytic Centers into TiO_2 nanotubes. ACS Catalysis. 2017;**7**:3210-3235

[70] Zhang X, Li Z, Xu S, Yaowen Ruan Y. Carbon quantum dot-sensitized hollow TiO_2 spheres for high-performance visible light photocatalysis. New Journal of Chemistry. 2021;**45**:8693-8700

[71] Kou J, Lu C, Wang J, Chen Y, Xu Z, Varma R. Selectivity enhancement in heterogeneous photocatalytic transformations. Chemical Reviews. 2017;**117**:1445-1514

[72] Mattioli G, Bonapasta AA, Bovi D, Giannozzi P. Photocatalytic and photovoltaic properties of TiO_2 nanoparticles investigated by ab initio simulations. Journal of Physical Chemistry C. 2014;**118**:29928-29942

[73] Ganeshraja AS, Yang M, Xu W, Anbalagan K, Wang J. Photoinduced interfacial electron transfer in 2, 2'-Bipyridyl Iron (III) complex-TiO_2 nanoparticles in aqueous medium. ChemistrySelect. 2017;**2**:10648-10653

[74] Wang F, Jiang Y, Lawes DJ, Ball GE, Zhou C, Liu Z, et al. Analysis of the promoted activity and molecular mechanism of hydrogen production over fine Au–Pt alloyed TiO_2 photocatalysts. ACS Catalysis. 2015;**5**:3924-3931

[75] Seh ZW, Liu S, Low M, Zhang S, Liu Z, Milayah A, et al. Janus Au-TiO_2 photocatalysts with strong localization of plasmonic near-fields for efficient visible-light hydrogen generation. Advanced Materials. 2012;**24**:2310-2314

[76] Fontelles-carceller O, Muñoz-Batista MJ, Rodríguez-castellón E, Conesa JC, Fernández-garcía M, Kubacka A. Measuring and interpreting quantum efficiency for hydrogen photo-production using Pt-titania catalysts. Journal of Catalysis. 2017;**347**:157-169

[77] Hayashido Y, Naya S, Tada H. Local electric field-enhanced plasmonic photocatalyst: Formation of Ag cluster-incorporated AgBr nanoparticles on TiO_2. Journal of Physical Chemistry C. 2016;**120**:19663-19669

[78] Méndez-Medrano MG, Kowalska E, Lehoux A, Herissan A, Ohtani B, Bahena D, et al. Surface modification of TiO_2 with Ag nanoparticles and CuO nanoclusters for application in photocatalysis. Journal of Physical Chemistry C. 2016;**120**:5143-5154

[79] Yang L, Wang F, Shu C, Liu P, Zhang W, Hu S. An in-situ synthesis of Ag/AgCl/TiO_2/hierarchical porous magnesian material and its photocatalytic performance. Scientific Reports. 2016;**6**:1-7

[80] Shah ZH, Wang J, Ge Y, Wang C, Mao W, Zhang S, et al. Highly enhanced plasmonic photocatalytic activity of Ag/Agcl/TiO_2 by CuO co-catalyst. Journal of Materials Chemistry. 2015;**3**:3568-3575

[81] Zhu L, Hong M, Ho GW. Hierarchical assembly of SnO_2/ZnO nanostructures for enhanced photocatalytic performance. Scientific Reports. 2015;**5**:1-11

[82] Her Y, Yeh B, Huang S. Vapor–solid growth of p-Te/n-SnO_2 hierarchical Heterostructures and their enhanced room-temperature gas sensing properties. ACS Applied Materials & Interfaces. 2014;**6**:9150-9159

[83] Ingram DB, Christopher P, Bauer JL, Linic S. Predictive model for the design of plasmonic metal/semiconductor composite photocatalysts. ACS Catalysis. 2011;**1**:1441-1447

[84] Saliba M, Zhang W, Burlakov VM, Stranks SD, Sun Y, Ball JM, et al. Plasmonic-induced photon recycling in metal halide perovskite solar cells. Advanced Functional Materials. 2015;**25**:5038-5046

[85] Ganeshraja AS, Kiyoshi G, Wang J. ^{119}Sn Mossbauer studies on ferromagnetic and photocatalytic Sn – TiO_2 nanocrystals. Hyperfine Interactions. 2016;**237**:139

[86] Vázquez-Robaina O, Cabrera AF, Cruz AF, Torres CER. Observation of room-temperature ferromagnetism induced by high-pressure hydrogenation of anatase TiO_2. Journal of Physical Chemistry C. 2021;**125**(26):14366-14377

[87] Sundaram AG, Maniarsu S, Vijendra RP, Ganapathy V, Karthikeyan V, Nomura K, et al. Hierarchical Sn and AgCl co-doped TiO_2 microspheres as electron transport layer for enhanced perovskite solar cell performance. Catalysis Today. 2020;**355**:333-339

Chapter 8

TiO_2 Nanocoatings on Natural Fibers by DC Reactive Magnetron Sputtering

Helena Cristina Vasconcelos, Telmo Eleutério, Maria Gabriela Meirelles and Susana Sério

Abstract

The surface functionalization of natural fibers, mainly using TiO_2 films, shows a growing interest in its application as yarns in fabrics that require advanced properties, allowing the use of their excellent physical and chemical properties in the textile area. The DC magnetron sputtering technique is a potential method for depositing TiO_2 films onto natural fibers, allowing for the creation of advanced and competitive properties compared to synthetic fibers. Different crystalline phases of TiO_2 have been shown to be effective in photocatalytic applications. Reactive discharges like the Ar/O2 gas mixture can be used to deposit TiO_2 films with desired characteristics, and controlling deposition parameters can further manipulate the properties of the coatings. Analytical techniques such as XRD, XPS, and SEM/EDS can be used to study the surface properties of TiO_2 films. XRD determines crystal structure, XPS provides information on chemical composition, and SEM/EDS examines morphology and elemental composition.

Keywords: TiO_2, DC reactive magnetron sputtering, natural fibers, *Hedychium gardnerianum*, X-ray diffraction (XRD), X-ray photoelectron spectroscopy (XPS)

1. Introduction

In the last few years natural fibers of natural origin, especially those of lignocellulosic composition [1], attracted a great attention due its interesting properties as high tensile strength and rigidity, and to the advantages of being recyclable, lightweight, biocompatible, and extremely low costing, when compared to the synthetic ones. Among others, ramie, jute, kenaf, etc. reinforced composites have been highly emphasized [2]. The use of natural fibers fit into the concept of circular economy, which seeks to reduce, reuse, recover, and recycle products. In addition, there are economic and functional advantages in the use of natural fibers compared to the most common artificial fibers, made of carbon, glass, or polymeric resins [3], namely due to their low production cost and high abundance.

Sustainability is becoming a concern in the development of new materials, mainly due to problems related to the use of scarce resources and waste management.

On the other hand, there is a kind of activation energy for the creation of new products and functionalities, enabling new commercial paradigms or complementing the existing ones.

Cellulose fibers can be obtained from many plants and represent one of the most abundant organic materials on earth. Invasive plants, such as ginger lily (*Hedychium gardnerianum*) [4], are abundant in various countries and are therefore an ecological source of fibers for scientific and industrial applications as an alternative to the traditional glass, polymer, or carbon fibers [5]. Invasive species are a threat to ecosystems and the survival of many endemic species, and their monitoring, control, or eradication is crucial, to prevent the modification of ecological processes and the loss of biodiversity. The ginger lily plant can be found in large quantities at Azores Islands (Portugal), and they are mainly considered as waste. However, its biological nature gives them specific characteristics less good for high-tech applications, namely the ease way since they absorb water. This causes dimensional changes and swelling on the fibers, mainly because its main composition of cellulose, which is structurally a linear polymeric chain with OH groups, and thus highly hydrophilic.

Nowadays, titanium dioxide (TiO_2) is one of the most effective photocatalysis [6] demonstrating high efficiency of decomposition and detoxification of several toxins and pollutants [7]. However, there is a huge disadvantage that involves the removal of TiO_2 catalysts after their applications, in the case of catalysts based on particles in suspension. In general, water purification reactors employ photocatalyst particles (powder type) that have higher photocatalytic activity due to less mass transfer limitations between the treated contaminants and the photocatalyst. However, powdered photocatalysts need to be filtered and separated after water treatment, which is a tedious and expensive process. So, to commercialize the process as a full-scale technology, it is critical to increase the photocatalytic activity of TiO_2 and manufacture devices with TiO_2 immobilized on a specific support. This strategy can bring a great benefit. Therefore, several attempts have been already used to immobilize TiO_2 on different supporting materials [8] and shapes. From these, glass substrates [9, 10], glass spheres [11], fiberglass [12], activated carbon, zeolite, and ceramics [13] stainless steel [14] and polyamide fibers [15], can be emphasized. Moreover, photocatalytic fiber is an emerging solution to immobilize catalyst powders [16, 17]. The natural fibers are nowadays preferable due to their multiple advantages in terms of environmental sustainability. Inspired by these remarkable characteristics, fibers have found a great interest as supporting substrates [18]. Cotton and ginger lily fibers, have booth cellulose in its main composition and so have abundant hydroxyl groups (OH) [18, 19] to link photocatalysts through hydrogen bonds and van der Waals forces. Wool fiber, instead, possesses plenty of disulfide bonds (-S-S-), carboxyl groups (-COOH), and amino groups (-NH_2) [16]. Moreover, natural fibers are considered desirable for TiO_2 immobilization platforms on account of their intrinsic porous structure [20], large specific surface area, and flexibility [16]. Its flexible form can be adapted to different spaces and purification devices. In addition, they can be cut to any size, rolled up, etc., to meet the function's requirements. For example, cotton fibers were proved to be easily installed inside photoreactors [21].

Since TiO_2 can impart antibacterial [22] and self-cleaning [23] properties to the fibers it becomes clear that is of great interest for the textile industry. The multiple fiber-related substrates involving fibers, yarns, and fabrics with different structures can be used as support substrates for photocatalyst proposes [16]. Important fiber properties are good adhesion of TiO_2 which demands improvement of binding efficiency with fibers to keep the necessary high specific surface area to enhance the absorption affinity.

There are many methods used for the synthesis of TiO_2 [24]. TiO_2 nanoparticles/films immobilization was prepared successfully in a variety of natural fibers, by the sol-gel method [25], microwave-assisted liquid phase deposition process [26], and DC-reactive magnetron sputtering [18].

A promising method to prepare photocatalytic immobilized TiO_2-based thin films is by DC-reactive magnetron sputtering. This technique enables large-area deposition with high uniformity, yet it is essential to understand why the film properties exhibit after deposition. This is because the properties of the coatings obtained are highly dependent on the selected parameters, and so it is necessary to establish the ideal ones that satisfy any film application and understand the basic processes that control the properties of that films. These include, for example, the type of species deposited, their energy, and consequently the effects that their bombardment will have on the surface of a growing film, etc. In addition, the influence of substrate temperature on the nature of the film is also to be considered, among other factors, namely, for instance, the substrate position relative to the target, discharge pressure, and the gas mixture. In fact, bombardment can result in a series of surface effects, namely displacement of lattice atoms, creation of defects that can lead to increased atomic mobility, surface heating that can promote crystallization of nanoparticles, etc. These effects will consequently affect the internal stress, crystal size, morphology, and roughness of the deposited films. Quite often the physical structure of the thin film is directly responsible for the expected film property. For instance, in photocatalytic TiO_2 films, deposition of crystalline or amorphous TiO_2 is of crucial importance for their functionality [22]. So, understanding the relationship between deposition parameters that will affect film properties is therefore important for defining procedures.

2. Basics of DC reactive magnetron sputtering

2.1 Sputtering

Sputter deposition is a physical vapor deposition (PVD) process [27] based on the ejection of atoms from a solid target as a result of collisions with energetic particles. However, if the collisions are due to the impact of positive ions, the process in known as cathodic sputtering.

Sputtering allows the deposition of thin films of a variety of materials, including metals and certain compounds such as oxides and nitrides and any type of substrate can be applied for deposition. Simultaneous deposition from various sources permits to develop complex compositions.

It has advantages over other deposition methods when the intended film is a compound (e.g., oxide) or an alloy, avoiding non-stoichiometric films, separation of phases of the constituent elements, and even differences in the desired composition. Coatings deposited generally have good adhesion and exceptional coverage.

The objective is to remove material from a target and bring it to the substrate (e.g., the fiber to be coated). This is achieved by means of ion bombardment in the plasma, usually by Ar^+ ions. Further, the ions that reach the target with enough energy can eject atoms from the target that are then dropped on the fiber (or other substrate) surface placed nearby to the target. The process basically consists of three distinct steps that occur altogether (**Figure 1**):

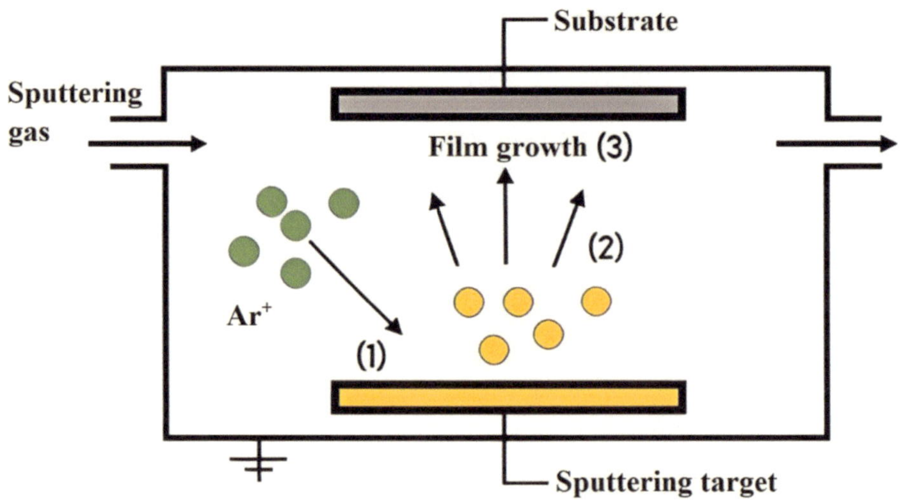

Figure 1.
Schematic diagram of magnetron sputtering. Adapted from [28].

1. Production of atoms to deposit from a target material source,

2. Transference of the atoms through the plasma to the substrate,

3. Deposition of atoms on the substrate and film growth.

The widespread use of sputtering is explained by the many advantages of this technique, mainly due to its simplicity of operate and the quality of the thin films through the stoichiometry control in complex compositions, excellent film adhesion to the substrate, uniform deposition over a large area and tailor of the film thickness. Moreover, by the change of deposition parameters such as oxygen partial pressure, working pressure, and sputtering power is possible to achieve desired film parameters, for example, microstructure, composition, step coverage, among others.

2.2 Chamber preparation and DC reactive plasma

The process begins by creating vacuum inside the chamber and thus the air is pumped out. The chamber is then filled with argon, an inert gas, reaching a pressure between 1 and 10 Pa. When a DC voltage is applied between the electrodes (with a gas in between them), a plasma is formed. The applied voltage is high enough to enable that a large quantity of inert gas atoms turns into ions; electrons acquire enough kinetic energy to ionize gas atoms (break gas atoms) and thus, the plasma is formed [29]. The ions and electrons are then accelerated towards opposite electrodes. Plasma is thus a partially ionized gas. Depending on the mean free path in the gas, the accelerated particles can collide with inert gas atoms and give rise to scattering occurrences at a rate that can change with the pressure and nature of the gas. Moreover, these scattering occurrences can lead to ionization of further gas atoms. The probability of ionization (α) occurs will depend basically on the threshold voltage to initiate the breakdown of the gas (trigger the gas discharge) which must surpass the ionization potential of the gas species and can be calculated by [30]:

$$\alpha = \frac{1}{\lambda} \exp\left(-\frac{V_i}{eE\lambda}\right) \tag{1}$$

where λ is the mean free path of the sputtering ion, V_i is the ionization potential of the gas in electron volts, e is the electron charge, and $E = V/d$ is the electric field between electrodes [30].

The breakdown voltage of a gas, which is the voltage required break a sustained plasma, is established by Paschen's law, which is a function of the product of electrode gap spacing and chamber pressure, according to:

$$V_b = A \frac{pd}{ln\,(pd) + B} \tag{2}$$

where V_b is the breakdown voltage, d is the gap electrode distance (cm), p is the pressure (torr), and A and B constants depending on the gas mixture inside the chamber. Paschen's law relationship the breakdown voltage versus the product of the pressure and the gap electrode distance (pd) as shown in **Figure 2** [31] and predict a minimum breakdown voltage for any gas.

Once the V_b is achieved plasma becomes self-sustaining and plasma reaches a steady state, exhibit enough energy to be used in sputtering.

2.3 Principle of sputter deposition

Typical gases used in the sputtering process are from the group of noble gases because they tend not to react with the target material. Argon (Ar) gas is the most common one in this process. Positively charged argon ions from the plasma (Ar^+ ions) are accelerated by an electrical potential difference toward the negatively biased target (cathode), where the target material, for example, Ti, is placed and hits it.

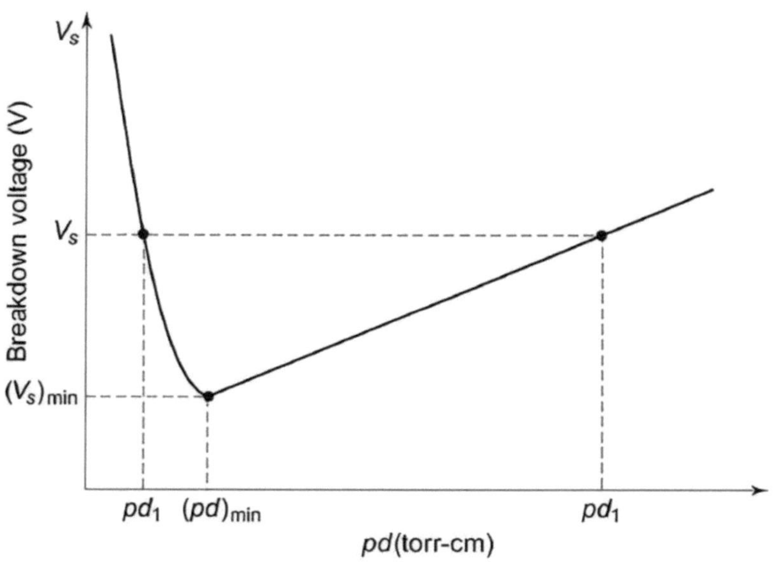

Figure 2.
The breakdown voltage versus gas pressure curve [31].

With the impact energy, atoms are ejected from the target and diffuse through the vacuum chamber until they are deposited on the substrate to form a thin film (**Figure 1**). This atom ejection is known as sputtering. From a physical point of view, the principles of sputtering are based on a simple momentum transfer model, which allows understanding how atoms are ejected from the surface of a material due to successive collisions. The collision of particles and the transfer of momentum are important aspects of the DC sputtering process. In a plasma, there are various types of particles, such as electrons, ions, and neutral atoms or molecules. When these particles collide with each other or with the target material, momentum is transferred between them.

Because of the bombardment of the target, beyond ejected or sputtered atoms, additional events can occur as shown in **Figure 3**, including the followed briefly underlined: secondary electrons, reflected ions at the target surface, ion implantation in the structural atomic network, lattice defects, and structural rearrangement by trapping ion species.

The mass of the energetic ions is key to the energy and momentum transferred to the film atom during the collision. From the physics laws of the conservation (of energy and momentum), the energy transferred in a collision of an incident particle (i) and a target particle (t) is given by:

$$\frac{E_t}{E_i} = \frac{4m_i m_t}{(m_i + m_t)^2} \cos^2\theta \tag{3}$$

where E and m, are, respectively, the energy and the mass. θ is the angle of incidence as measured from a line across the two centers of masses, as shown in **Figure 4**. When the ejected particles reach the substrate, they deposit onto its surface due to the momentum transfer that occurs during the collision. The amount of momentum transferred during the collision depends on the mass and velocity of the particles involved. In general, heavier particles transfer more momentum than lighter particles, and faster particles transfer more momentum than slower particles.

The transfer of momentum is an important factor in determining the quality and properties of the deposited thin film. If the momentum transfer is too low, the deposited film may be porous and have a low density. On the other hand, if the

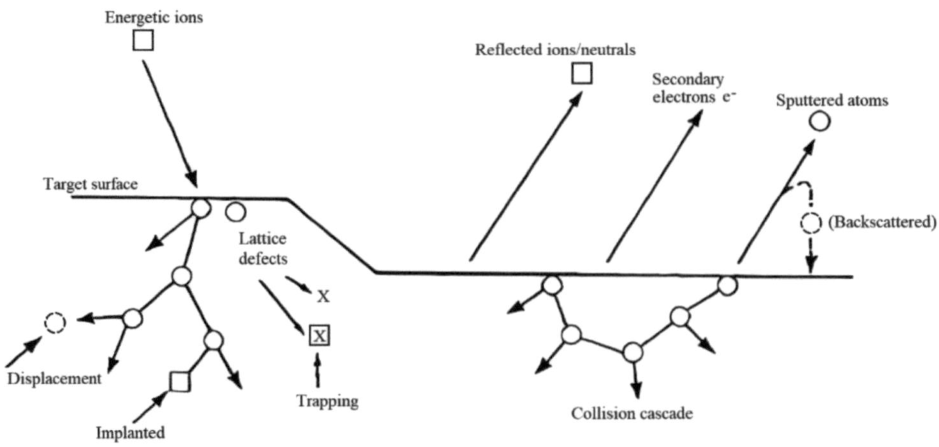

Figure 3.
Events that may occur on the target surface being bombarded with energetic ions. Adapted from [27].

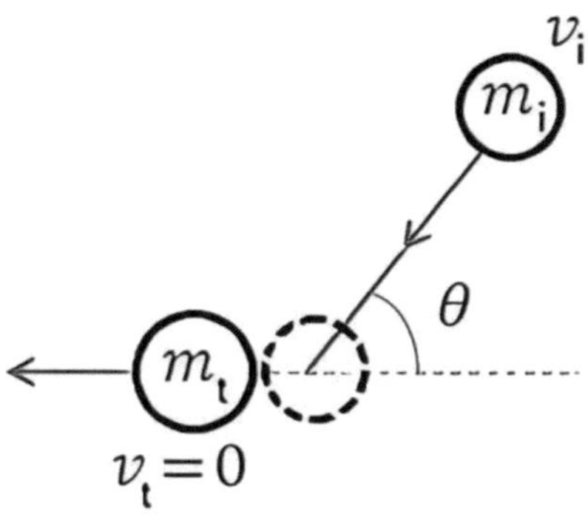

Figure 4.
Collision of particles and the transfer of momentum. Adapted from [32].

momentum transfer is too high, the deposited film may be dense but have high levels of residual stress. Therefore, it is important to carefully control the parameters of the DC sputtering process, such as the gas pressure, target material, and substrate temperature, to optimize the momentum transfer and achieve the desired properties of the deposited thin film. The efficiency of the momentum transfer is the highest, $\left(\frac{E_t}{E_i}\right)_{max}$, when $\cos\theta = 1$ and $m_i = m_t$, that is, is desired that the atomic weight of the sputtering gas could be identical to that of one of the target.

Ejected atoms must be able to diffuse freely toward the substrate with desirable little opposition to their movement, which explains the necessity of the sputtering to be done in vacuum conditions. To achieve this, a low pressure within the chamber and a suitable large DC voltage applied between the electrodes, in other words, between the target and the substrate, give rise to a glow discharge that allow accelerate the positive ions to the target. Therefore, ions can retain their high energies. Besides atom-gas collisions can be prevent after ejection from the target. Still, the initial kinetic energy of the atoms transported through the plasma can be lost by collisions within the plasma, failing the energy needed to deposit themselves on the substrate. Thereby, not all atoms ejected from the target reach the substrate, many are projected in different directions and deposit on any surface they encounter. The atoms that can reach the substrate thereby form a layer called a thin film. So, sputtering is also described by its yield, which is the ratio of the number of atoms ejected to the number of incident energetic ions and depends on the chemical bonding of the target atoms and the energy transferred by impact [27].

The sputtering yield (Y) is an important parameter that characterizes the efficiency of the sputtering process since it determines the rate at which atoms are ejected from the target material and deposited onto the substrate. Therefore, the sputtering yield plays a crucial role in the fabrication and processing of thin films, coatings, and surface modifications using sputtering techniques. Y is defined as the number of atoms or molecules sputtered from the target per incident particle. Y is zero for ion energies below the threshold energy of sputtering, Φ. This means that particles with energy below this threshold are not able to cause sputtering. Mathematically, we can express this as $Y = 0$ for $E < \Phi$, where Y is the sputtering yield and E is the ion energy. Y is a function of the ion energy and the target material properties, with a threshold energy below which sputtering does not occur and a power law relationship above the threshold energy.

The sputtering yield depends on various factors, such as the energy and flux of the incident ions, the target material properties, and the surface conditions. By controlling these factors, the sputtering yield can be optimized to achieve desired properties, such as film thickness, composition, morphology, and adhesion. The sputtering yield also affects the overall efficiency and quality of the sputtering process, as well as the cost and environmental impact.

Reactive sputtering is a widely used technique for depositing compound films on substrates. In reactive sputtering, a target material is bombarded with ions in the presence of a reactive gas such as oxygen, nitrogen, or hydrogen. The sputtered species react with the reactive gas to form a compound film on the substrate surface.

One of the advantages of reactive sputtering is that it allows for precise control of the stoichiometry of the deposited film by adjusting the flow rate of the reactive gas. This makes it possible to deposit films with desired properties such as optical, electronic, magnetic, or mechanical properties.

Another advantage of reactive sputtering is that it can be used to deposit films on complex substrates, fibers, nanoparticles, films and materials with irregular surfaces, porous materials, etc. This is because the sputtered species have high kinetic energies, which enable them to penetrate the pores and irregularities of the substrate surface. As a result, the deposited film can conformally coat the entire surface of the substrate, including its complex features, such as corners, edges, and high aspect ratio structures.

2.4 Magnetron sputtering

The sputtering process is a relatively simple technology, but it still requires additional support systems, such as efficient cooling of the substrate because the electrons that are repelled by the negative cathode can reach the substrate heating it; and the use of magnets to confine the electron paths towards the cathode surface (magnetron sputtering) to increase the plasma efficiency and therefore the deposition rate. This allows the plasma thus located/confined to improve deposition rates due to the greater number of ions colliding with the target and reduces the temperature of the substrate as less electrons collide with it.

The presence of magnets behind the cathode creates a magnetic field close to the surface of the target. These magnets are positioned to produce a magnetic field near the target is a such way that magnetic field lines are parallel to the cathode surface and perpendicular to the electric field lines (**Figure 5**). This arrangement allows to concentrate the electrons close to the target, as shown in **Figure 5a**, instead of them circulating randomly dispersed around it, while the ion trajectories are not influenced by the deflection due to their greater mass. The combined action of the electric (E) and magnetic (B) fields near the target generates the $E \times B$ drift phenomenon. The trajectories of electrons, of charge q and velocity v, captured in this drift are forced to bend and follow helical trajectories around the magnetic field lines (**Figure 5b**), and to follow them because of the Lorenz force (F_L). The Lorenz force acting on a charged particle is given by the following equation:

$$\vec{F}_L = q\left(\vec{E} + \vec{v}\text{x}\vec{B}\right) \tag{4}$$

where q is the charge of the particle, $\vec{E}$ is the electric field, $\vec{v}$ is the velocity of the particle, and $\vec{B}$ is the magnetic field. The term $\vec{v}$ x $\vec{B}$ represents the cross-product of

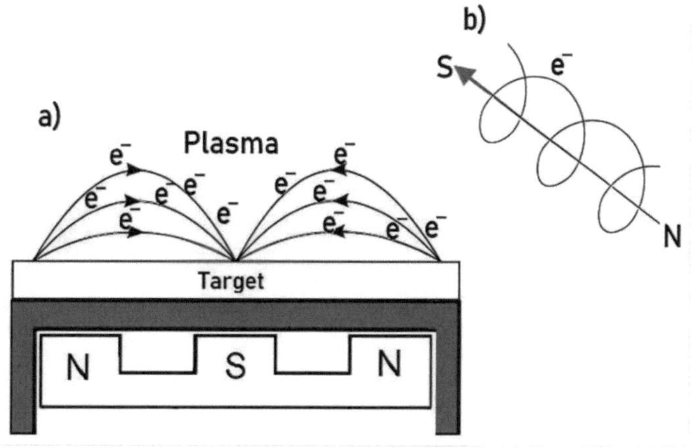

Figure 5.
(a) Layout of the DC magnetron sputtering system near the target; (b) helical electron trajectory around the magnetic field line due to the Lorenz force.

the velocity and magnetic field vectors. In the case of DC sputtering, the magnetic field is typically generated by a permanent magnet. The electric field is created by the potential difference between the cathode and anode. As the electrons move toward the anode, they experience a Lorentz force that causes them to follow a curved path.

This curved path increases the path length of the electrons to the anode, which means they have a larger number of collisions with the argon atoms in the plasma. This, in turn, significantly improves the ionization probability because the collisions between the electrons and argon atoms result in the formation of more ions.

This additionally acting Lorenz force restricts the trajectories of the electrons. Therefore, the path of the electrons to the anode increases which significantly improves the ionization probability because of the larger number of collisions between argon atoms and electrons.

3. TiO_2: brief structure and properties

TiO_2 naturally exists in three polymorphs such as rutile (tetragonal), anatase (tetragonal), and brookite (rhombohedral) [33]. **Figure 6** shows the three polymorphic structures of TiO_2, which can be described based on their cell structure consisting of a TiO6 distorted octahedron. Each structure has a different degree of distortion of this octahedron, resulting in the characteristic differences observed between the polymorphs.

TiO_2 is an n-type semiconductor material that exhibits a bandgap ranging from 3.0 eV (for rutile) to 3.2 eV (for anatase), which corresponds to a light absorption edge in the UV range. One of the advantages of TiO_2 is that it is non-toxic, biocompatible, and has a high chemical stability. These properties, along with its exceptional electronic properties, make it an ideal material for use in photovoltaic applications, which convert solar energy into electricity. The discovery by Fujishima and Honda in 1972, which demonstrated the photocatalytic splitting of water in TiO_2 electrodes [35], has led to a growing interest in using TiO_2 as a photocatalyst for environmental purification [7], detoxification, and self-cleaning applications.

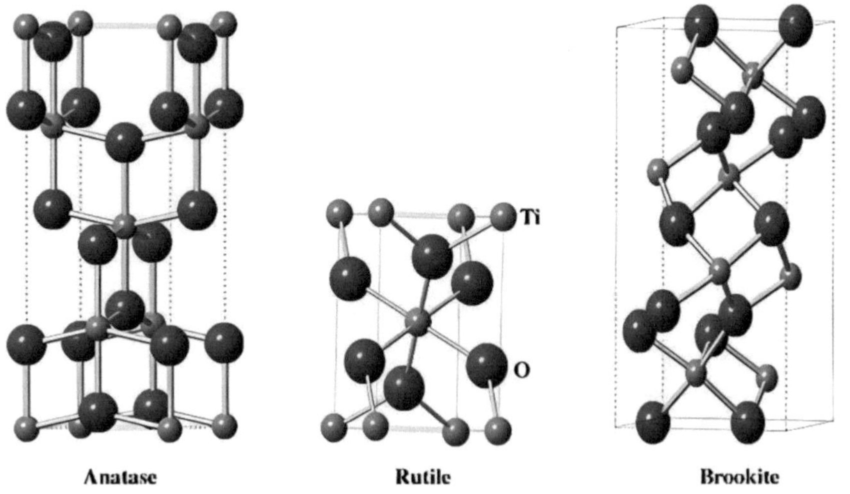

Figure 6.
TiO_2 main polymorphs—anatase, rutile, and brookite [34].

Anatase phase has proved to be a more efficient photocatalyst than rutile phase [36], that might be explained by the reduction ability of O_2 higher that rutile, despite the larger band gap of anatase [36]. Recently, amorphous TiO_2 has been considered very effective in antibacterial disinfection [22]. Is well known that TiO_2 photoactivity is hampered by the narrow range of UV wavelengths for photoactivation. Its energy gap is only sensitive to radiation in the ultraviolet (UV) region of the solar spectrum, which represents only 4% of the global solar radiation [37].

Almost 20 years ago it has been reported that the TiO_2 lattice doping with non-metallic atoms like N [38] can shift the absorption edge from UV to lower energies and thus increase visible absorption. Recently, a photonic band gap of 3.18 eV (390 nm) was measured for amorphous TiO_2 [22] which is slightly lower than the value reported to anatase (3.2 eV). TiO_2 absorbs photons and acquires enough energy ($h\nu$) to allow an electron in the valence band to jump to the conduction band.

This process (photocatalysis) gives rise to an electron (e^-)-hole (h^+) pair, in accordance with the reaction $TiO_2 + h\nu \leftrightarrow h^+ + e^-$, further responsible for the elimination of water toxic components by active species (•OH, •O_2, and H_2O_2) generated by redox reactions on the TiO_2 surface.

Rutile and anatase are stable phases at normal conditions an comprise identical TiO_6 octahedron building unit but with diversely sharing corners and edges giving rise to different configurations [22]. The TiO_6 in anatase are arranged in zigzag chains along {221}, sharing four edges, while in rutile, TiO_6 share two edges and link up in linear chains along {001} [39]. These structural differences give rise to different densities and electronic band structures between these two phases [22].

Moreover, the number of shared edges is related to the "energy of the structure" (and thus its stability). Rutile is more stable that anatase (metastable) being the number of shared edges per octahedron, respectively, two and four [22]. The distance Ti-Ti between the center of edge-sharing octahedra being smaller with the decrease of the number of shared edges which provided shorter Ti-Ti distances and a more closely packed crystal structure of rutile. Thus, there is a strong interaction in the Ti-Ti bond of rutile which has only two Ti atoms at the shortest distance. On the other hand, in anatase, the Ti-Ti interaction, instead, depends on four Ti atoms, which allows for a Ti-Ti distance greater than that of rutile. Therefore, rutile exhibits less blockage around each TiO_6 unit, leading to a more stable phase [40].

As in anatase there are four octahedrons, at a distance between them of 3.04 Å, while in rutile, despite its higher density, only two octahedrons are present at 2.96 Å [40], the distinct arrangement of TiO_6 octahedrons gives rise to different structures packaging that will condition the anatase-rutile transition.

In many synthetic routes, amorphous TiO_2 is often the first phase to form. The transformation of amorphous TiO_2 into anatase and/or rutile, usually occurs by effect of temperature, both in wet chemical methods, such as sol-gel [41] and also in DC reactive magnetron sputtering [42].

X-ray diffraction (XRD) and Raman spectroscopy are commonly used techniques for analyzing the crystallization process in materials science.

X-ray diffraction is a technique that involves shining X-rays onto a crystalline material and observing the resulting diffraction pattern. The diffraction pattern is characteristic of the crystal structure and provides information about the crystal lattice parameters, the orientation of the crystal grains, and the degree of crystallinity.

The XRD patterns of anatase (JCPDS card No. 96-900-9087) and rutile (JCPDS card No. 96-900-9084) phases shown in **Figure 7** provide important information about the crystal structures of these two polymorphs of TiO_2. In the XRD pattern of anatase, the main peak at 25.3° corresponds to the (101) plane of the crystal structure. This peak is relatively sharp and intense, indicating a high degree of crystallinity and a well-defined crystal structure. The presence of other peaks at lower angles also indicates the presence of other crystallographic planes in the anatase structure.

In the XRD pattern of rutile, the main peak at 27.4° corresponds to the (110) plane of the crystal structure. This peak is also relatively sharp and intense, indicating a

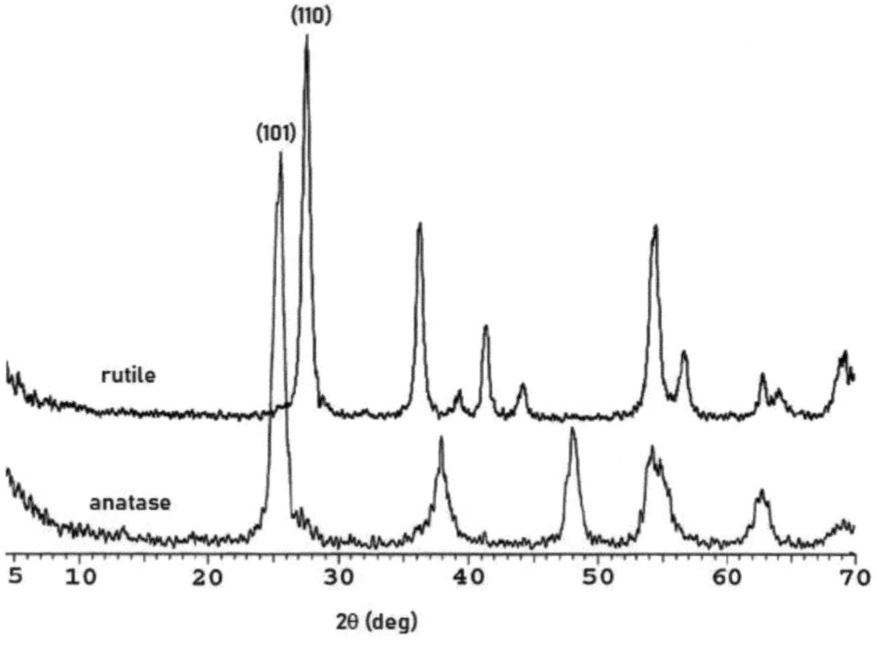

Figure 7.
XRD patterns for anatase and rutile TiO_2 phases. The insets are Miller indices of anatase and rutile phases. Adapted from [43].

well-defined crystal structure and a high degree of crystallinity. The presence of other peaks at higher angles also indicates the presence of other crystallographic planes in the rutile structure.

Raman spectroscopy, on the other hand, involves shining laser light onto a material and measuring the scattered light as a function of wavelength. The scattered light provides information about the vibrational modes of the atoms in the material, which are characteristic of the crystal structure.

By combining XRD and Raman spectroscopy, a more comprehensive understanding of the crystallization process in a material is obtained. XRD provides information about the long-range order and crystal structure, while Raman spectroscopy provides information about the short-range order and local structure.

Anatase has a tetragonal crystal structure and is characterized by Raman peaks at around 144, 399, and 519 cm^{-1}. The peak at 144 cm^{-1} is due to the symmetric stretching vibration of the Ti-O bond, while the peaks at 399 and 519 cm^{-1} are due to the bending modes of the Ti-O-Ti bond. The Raman spectrum of anatase is also characterized by a broad peak at around 639 cm^{-1}, which is due to the lattice vibrations of the TiO_6 octahedra. **Figure 8** shows the Raman spectrum of a pure anatase film.

Rutile, on the other hand, has a tetragonal crystal structure and is characterized by Raman peaks at around 143, 445, 610, and 880 cm^{-1}. The peak at 143 cm^{-1} is due to the symmetric stretching vibration of the Ti-O bond, while the peaks at 445, 610, and 880 cm^{-1} are due to the bending modes of the Ti-O-Ti bond. The Raman spectrum of rutile is also characterized by a sharp peak at around 237 cm^{-1}, which is due to the lattice vibrations of the TiO_6 octahedra.

When a material undergoes a phase transformation from the amorphous or liquid phase to a crystalline phase, the specific crystalline phase that forms depend on a number of factors, including the thermodynamic stability of the different phases and

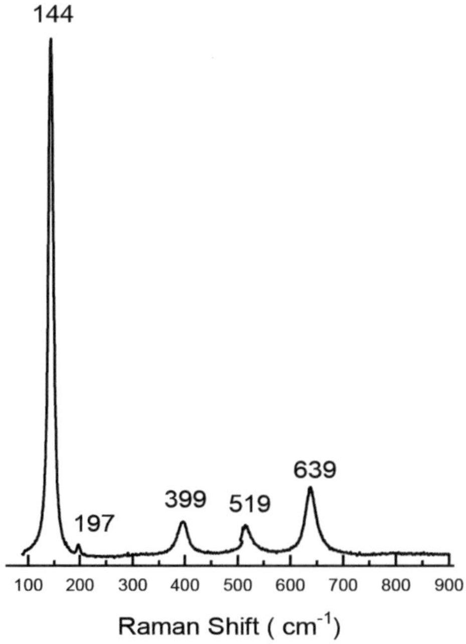

Figure 8.
Raman spectra of anatase TiO_2 phase.

the kinetics of nucleation and growth. In the case of TiO_2, the initial crystalline phase that forms is generally anatase, rather than rutile, because of its lower surface free energy compared to the rutile structure. Surface-free energy is a measure of the amount of energy required to create a unit area of a material's surface. Materials with lower surface free energy are typically more stable, because they have a lower tendency to form new surfaces or interfaces. In the case of TiO_2, the anatase structure has a lower surface free energy than the rutile structure, which means that it is more thermodynamically stable.

This difference in surface free energy is due to the different crystal structures of anatase and rutile. The anatase structure has a higher percentage of exposed (001) surfaces, which have a lower surface free energy compared to the (110) and (100) surfaces that are more prevalent in the rutile structure. As a result, the anatase structure is more stable and more likely to form during the crystallization process.

The surface roughness and microstructure can significantly influence the performance and hence the purpose of TiO_2 thin films. These characteristics depend on the deposition process, type of substrate, and chosen deposition parameters.

Liang et al. have produced TiO_2 films by the sol-gel method [44], highly compacts, continuous and smooth (**Figure 9**), exhibiting excellent self-cleaning properties. **Figure 10** shows SEM images of TiO_2 films (on glass substrates) prepared by reactive magnetron sputtering under different deposition conditions, namely plasma O_2 concentration (50% and 75%) and used power (500 and 1000 W) [18]. It can be seen clearly the differences in the morphology of the surface of TiO_2 coatings, as a function of different powers and concentrations of O_2. In general, the morphology is typically constituted by several agglomerates of nanoparticles (or grains) in the shape of a cauliflower but of different sizes, which are distributed over the surface of the substrate [18] in accordance with what was reported by Sério et al. [42]. There is a variation in the size of the agglomerates in the morphology of the films dependent on the $O_2/(Ar + O_2)$ ratio.

The value of the thickness (t_h) of the films allows estimating the deposition rate (v_d), in nanometers per minute.

$$v_d = \frac{t_h}{t}$$

where t is the deposition time. The thickness measurement is performed on SEM images in cross-section (as exemplified in the insert of **Figure 10a**). Regardless of the geometry, the surface is covered evenly.

Figure 9.
SEM image of the surface morphology of TiO_2 film deposited onto glass substrates by dipping-based sol-gel method. Adapted from [44].

Figure 10.
SEM images of surface morphology of TiO_2 films deposited onto glass substrates by reactive magnetron sputtering [18].

4. Characteristics of TiO_2 thin films deposited on natural fiber surfaces by DC reactive magnetron sputtering

The surface is the boundary that separates the material and the external environment. All alterations that occur in the sense of changing the surface structure will condition and tailor the material for certain applications. A wide range of surface modification processes, such as reactive magnetron sputtering, thermal oxidation, thermal evaporation, molecular-beam-epitaxy, chemical vapor deposition, sol-gel-assisted by dip-and/or-spin coating, spray pyrolysis, and electrodeposition have been used for used to produce thin film and tailor surfaces. However, in many of them, adhesion of the coating to the supports is still an important issue, which deserves special attention, especially in some types of substrates, such as natural fibers, which main composition is cellulose.

Cellulose is a polymeric chain with abundant hydroxyls (–OH) groups and other oxygen-containing functional groups –C=O, –C–O–C–, –CHO, and –COOH which makes the fiber surface potentially reactive [18]. These functional groups are available to bond to desired molecules and provide new properties and new applications for natural fibers [18]. The deposition of suitable coatings, such as TiO_2 films, allows the optimization of natural fibers by creation of new tailored properties of their surfaces, such as the photocatalytic ones, which are independent of that exhibited by the bulk fiber. Recently, nanostructured TiO_2 films successfully deposited on ginger lily fiber surfaces have been created by DC reactive magnetron sputtering [18].

The efficiency of the DC magnetron sputtering process to functionalize natural fibers (**Figure 11**) depends not only on the quantity cellulose reactive accessible

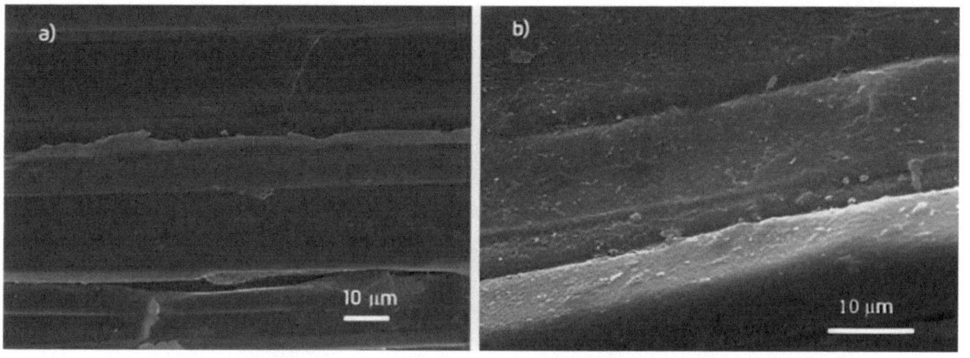

Figure 11.
SEM images of ginger lily fibers: (a) pristine; (b) after TiO_2 sputtered at 50% O_2–1000 W [45].

groups but also on sputtering conditions, such as the operating pressure, discharge power, O2 gas partial pressure, and deposition time.

The morphology of the films can be tailored by change of the partial pressure of the reactive gas (Ar/O_2) and the sputtering power [10]. Moreover, different film typology, namely dense and porous, can be obtained, as well as amorphous or crystalline (anatase and/or rutile) nanofilms. The influence of the O_2% in the discharge and the sputtering power on the amorphous/anatase phase transition, surface stoichiometry, and surface roughness of the films can be tailored.

4.1 Fibers preparation

Ginger lily fibers are obtained by mechanical extraction from the stems of the plant, after removing the leaves, as shown in **Figure 12a–c**. Before TiO_2 deposition by reactive magnetron sputtering (**Figure 12d**), fibers are cleaned successively in acetone, isopropanol, and deionized water to remove any organic contamination and further dried at low temperature (about 30°C).

4.2 Scanning electron microscopy (SEM) and energy dispersive X-ray analysis (EDX)

The EDX analysis (**Figure 13**) was used to determine the elemental composition of the TiO_2 films. The results show that as the power used during the reactive magnetron sputtering process increases, there is a corresponding increase in the intensity of the Ti peak. This finding supports the expected result that an increase in the number of Ti-O bonds contributes to the growth of the TiO_2 film. In other words, the higher the power used during the sputtering process, the greater the amount of titanium present in the film.

As the power and oxygen percentage during the reactive magnetron sputtering process are increased, the resulting TiO_2 films exhibit a nanostructured morphology in certain areas, similar to that seen in Zone 1 of Thornton et al.'s model [46]. This morphology is primarily due to the adatoms on the surface of the growing film having low mobility and the "shadow" effect. A nanostructured thin film exhibits nano-scale surface features, typically ranging in size from a few nanometers to several hundred nanometers. These features can include nanopores, nanocrystals, nanotubes, or other nano-items that are engineered into the film's surface by adjusting the deposition

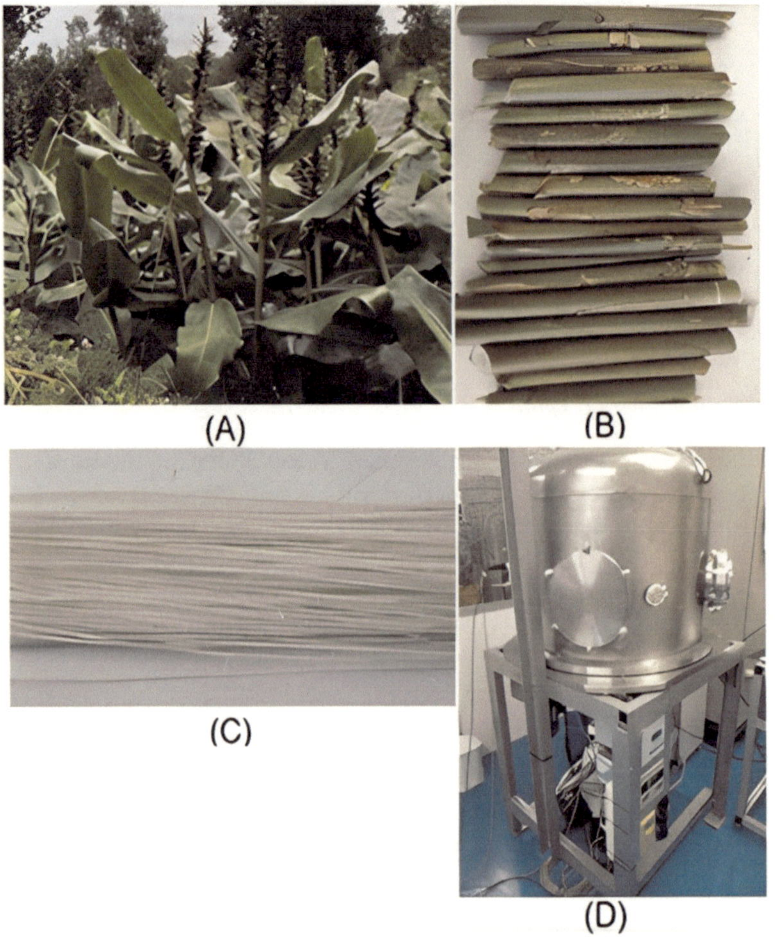

Figure 12.
Sequence of the ginger lily fiber preparation process for deposition of TiO_2 films by reactive magnetron sputtering. (a) Plant harvest; (b) stem preparation; (c) extraction of long fibers; (d) film deposition apparatus.

parameters, such as temperature, pressure, and substrate morphology. The structural and physical properties of the thin film can be controlled at the nanoscale. The nanostructured surface greatly increases the surface area of the coating and is well suited to photocatalytic applications due to its large surface area-to-volume ratio. The presence of pores, for example, increases the density of active sites with high accessibility of photons, but also facilitates diffusion and increases the adsorption capacity of pollutants. The size, shape, and distribution of the pores can be precisely controlled, allowing for the customization of the film's properties for specific applications.

In sputtering, the bombardment of the substrate surface with high-energy ions causes atoms to be ejected from the target and deposited onto the substrate surface. If the bombardment intensity is insufficient for film densification, the presence of pores can dominate the film's structure. When the deposited atoms do not have enough kinetic energy to overcome the surface diffusion and adhesion forces, they can accumulate on the substrate surface and form islands. These islands can coalesce and form a continuous film, but the presence of voids and pores between the islands can significantly affect the film's properties.

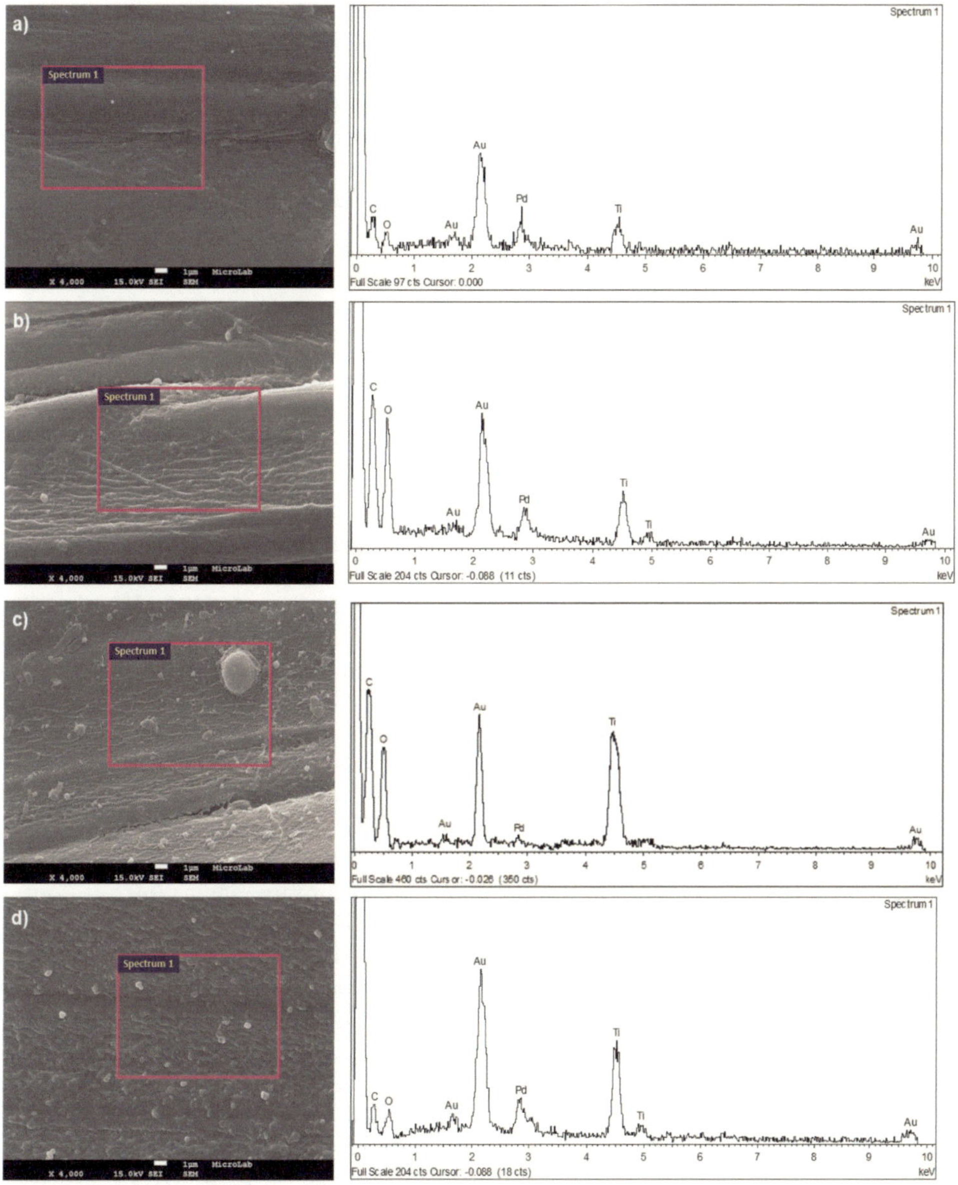

Figure 13.
SEM/EDX of the TiO_2 films deposited by DC reactive magnetron sputtering: (a) 50% O2–500 W, (b)75% O2–500 W, (c)50% O2–1000 W, (d)75% O2–1000 W.

The presence of pores in sputtered films can have both positive and negative effects on their properties. For example, in some applications, such as sensing or photocatalysis, the large surface area-to-volume ratio provided by the pores can enhance the film's activity. On the other hand, in other applications, such as barrier coatings or electronic devices, the presence of pores can reduce the film's performance and durability.

The greater availability of oxygen (75%) in the chamber during sputtering causes more ions to be generated, leading to an increase in the number of atoms bombarding the surface of the growing film. This results in a denser film which can lead to the formation of ridges and depressions on the film's surface.

Overall, the effect of the increased oxygen concentration in the chamber during sputtering is to create films with a high roughness topography, which can be advantageous or disadvantageous depending on the intended application. The ability to control the oxygen concentration and other process parameters during sputtering is therefore important for achieving the desired film properties.

4.3 X-ray photoelectron spectroscopy (XPS)

Figure 14a is shown the XPS survey spectra of the film 75% O2–1000 W. Carbon and oxygen lines dominate as expected because of the organic nature of the fiber. Intense Ti lines are also observed due to the TiO_2 film on the fiber surface. Typically, the fiber surface area can be divided into two kinds of regions: those covered with TiO_2 and those covered with organic material. These two regions are on different potentials, so that their reference binding energies are different. Nevertheless, the analysis can be performed by using the charge reference Ti 2p3/2 assumed to be at 458.5 eV, which is characteristic for TiO_2 phase [47] and the C 1 s line, with the smallest binding energy corresponding to adventitious carbon at 284.8 eV. It is believed that Ti at the surface of a "TiO_x material" is generally present as TiO_2. Since in this case of study, the measured Ti 2p lines clearly show only a single phase, as can be seen from **Figure 14b**, which confirms that only TiO_2 phase is present.

Deposition of TiO_2 increase the amount of oxygen at the fiber surface. This fact can be interpreted in three ways: (a) the reactive atmosphere during the deposition process contribute to significant surface oxidation of the surface; (b) the reactive atmosphere in the magnetron chamber etches (probably chemically) the surface and "opens" oxygen-rich phases laying below the carbon-rich surface layer; and (c) after the magnetron sputtering the samples are able to adsorb more water which is bound strongly so that it remains at the surface in vacuum [18].

During the deposition process, the increase in oxygen content mainly occurs for two reasons: (1) deposition of the TiO_2 film and (2) oxidation of the organic material. The latter occurs due to the presence of O–C–O and COOH groups in the fiber [18]. The XPS analysis in this study is related to the fitting of the C 1 s line, which have four contributions related to (a) C-C and C-H bonds, (b) C-OH and C-O-C bonds, (c) O-C-O bonds, and (d) COOH group [18]. From **Figure 14c**, C1 is attributed to the saturated C-C and C-H bonds. C2 at 287.0 eV is attributed to oxygen bound to two neighboring carbon atoms, forming a triangle. C3 at 288.7 eV can be attributed to carboxyl group (C=O)-OH, and C4 can be only attributed to –O–(C=O)–O– group [18].

4.4 Fourier transform infrared (FTIR) spectroscopy

The FTIR spectra observed in **Figure 15** show the presence of TiO_2 on the surface of the fibers. The peak observed between 800 and 450 cm^{-1}, at 670 cm^{-1}, is quite intense in the 75% O2–1000 W sample, being attributed to the Ti–O elongation, which is one of the characteristic peaks of the FTIR spectrum of TiO_2. Švagelj et al. [48] in a study of TiO_2 deposition on Al_2O_3 substrates, they reported the presence of the Ti–O elongation band, in the range of 640–700 cm^{-1}. This peak is associated with the

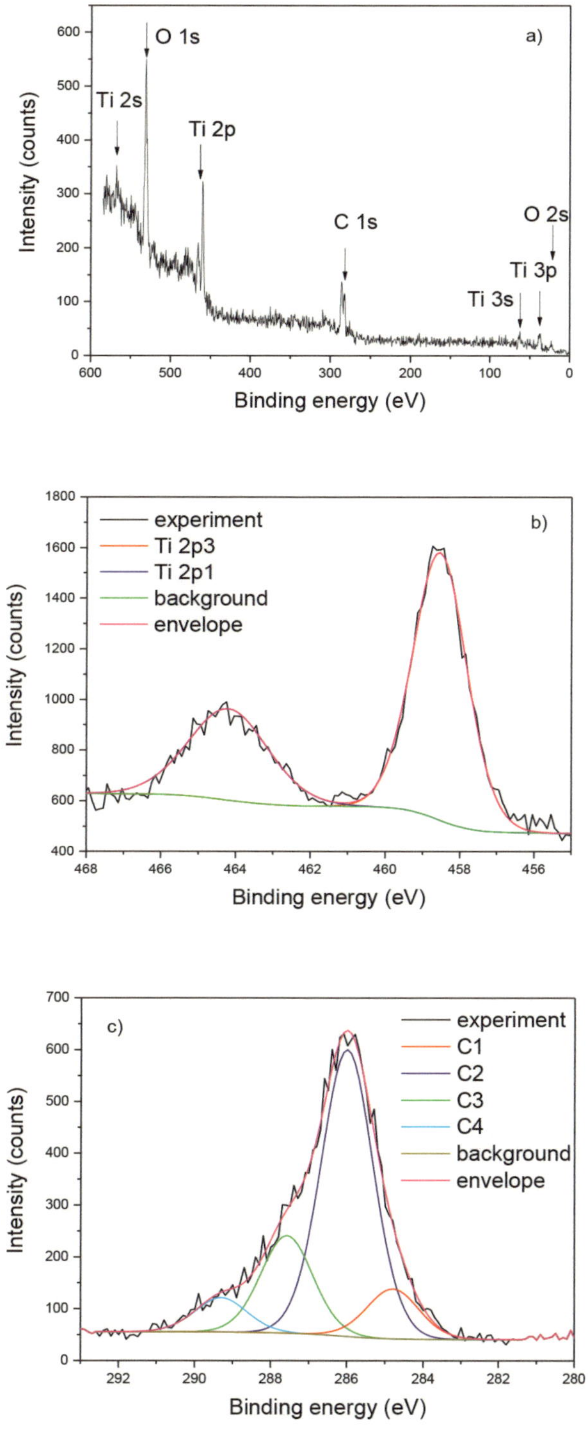

Figure 14.
(a) XPS survey spectrum from coated fiber 75% O2–1000 W, (b) high-resolution XPS spectrum of the line Ti 2p, and (c) high-resolution XPS spectrum of the line C 1 s taken from the pristine fiber. Adapted from [18].

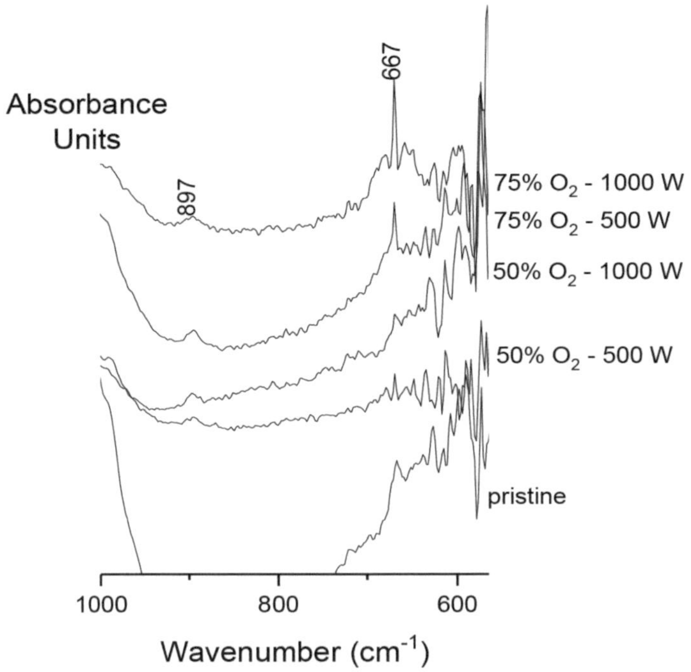

Figure 15.
FTIR spectra of the TiO_2 films deposited by DC reactive magnetron sputtering in the 900–500 cm^{-1}. Adapted from [18].

presence of O–Ti bonds in the TiO_2 film, which, in turn, bond to the surface of natural fibers, possibly by hydrogen bonding or van der Waals forces.

4.5 X-ray diffraction (XRD)

The structure of the deposited films is influenced by various deposition parameters, such as sputtering power, pressure, target-substrate distance, and the amount of reactive gases present in the deposition chamber.

The formation of a solid film during the sputtering process is affected by two factors: the heat generated by the substrate and the energy of the sputtered particles hitting the substrate. In situations where the substrate is not intentionally heated, it can still reach temperatures between 60 and 100°C due to the energy transfer from the sputtered particles. Normally, amorphous TiO_2 films require annealing at temperatures above 300°C to crystallize. However, Sério et al. [10] observed that crystallization occurred in as-sputtered TiO_2 thin films not because of the thermal energy, but rather due to the energy of the sputtered particles.

The sputtered particles could be from the target, such as atomic Ti, molecular TiO, molecular TiO_2, and TiO_2 clusters, as well as energetic electrons, negative ions (O^-), and neutrals reflected from the target (e.g., atoms of argon and oxygen) [10]. The films prepared with a sputtering power of 1000 W were found to be crystalline, likely due to the enhancement of plasma density in front of the substrate and an increase in the cluster growth rate with an augment in the sputtering power (**Figure 16**) [10].

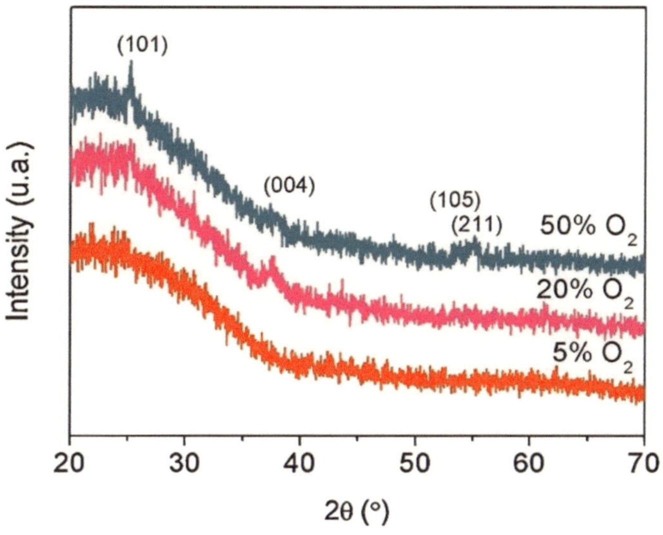

Figure 16.
XRD patterns of as-sputtered TiO_2 thin films deposited at 5% O2–500 W, 20% O2–1000 W, and 50% O2–1000 W [10].

5. Conclusion

Ginger lily fibers are a new sustainable resource for new product development. Its surface can be coated with a thin film of TiO_2 by reactive magnetron sputtering whose topography can be tailored depending on the power used and the percentage of oxygen in the chamber. Regarding the structure, the application of 500 W power allows obtaining a TiO_2 film with an amorphous structure, while the samples with twice the power showed polycrystalline structures. Anatase is the dominant phase in the films deposited at 1000 W. XPS and FTIR analyses revealed that ginger lily fibers can serve as a new sustainable resource for developing novel products, and the topography of the TiO_2-coated fibers can be tailored by adjusting the power and oxygen percentage during reactive magnetron sputtering.

Conflict of interest

The authors declare no conflict of interest.

Author details

Helena Cristina Vasconcelos[1,2*], Telmo Eleutério[1], Maria Gabriela Meirelles[1,3] and Susana Sério[4]

1 Faculty of Sciences and Technology, Azores University, Azores, Portugal

2 LIBPhys, NOVA School of Science and Technology, Monte da Caparica, Portugal

3 OKEANOS, Azores University, Horta, Portugal

4 Laboratory of Instrumentation, Biomedical Engineering and Radiation Physics (LIBPhys-UNL), Department of Physics, NOVA School of Science and Technology, NOVA University Lisbon, Almada, Portugal

*Address all correspondence to: helena.cs.vasconcelo@uac.pt

References

[1] Syduzzaman M, Al Faruque MA, Bilisik K, Naebe M. Plant-based natural fibre reinforced composites: A review on fabrication, properties and applications. Coatings 2020;**10**(10):973. https://doi.org/10.3390/coatings10100973

[2] Meyers MA, Chen PY, Lin YM, et al. Biological materials: Structure and mechanical properties. Progress in Materials Science. 2008;**56**:1-206

[3] Khan MZR, Srivasthava SK, Gupta MK. Tensile and flexural properties of natural fiber reinforced polymer composites: A review. Journal of Reinforced Plastics and Composites. 2018;**37**:1435-1455

[4] Pereira MJ, Eleutério T, Meirelles MG, Vasconcelos HC. *Hedychium gardnerianum* Sheph. ex Ker Gawl. from its discovery to its invasive status: A review. Botanical Studies. 2021;**62**(1):11. DOI: 10.1186/s40529-021-00318-5

[5] Vasconcelos HC, Eleutério T. Sustainable materials for advanced products. In: Leal Filho W, Azul AM, Doni F, Salvia AL, editors. Handbook of Sustainability Science in the Future. Cham: Springer; 2022. pp 1-17. DOI: 10.1007/978-3-030-68074-9_42-1

[6] Kumar A. Different Methods Used for the Synthesis of TiO_2 Based Nanomaterials: A Review. American Journal of Nano Research and Applications. 2018;**6**(1):1. DOI: 10.11648/j.nano.20180601.11

[7] Freire T, Fragoso AR, Matias M, Vaz Pinto J, Marques AC, Pimentel A, et al. Enhanced solar photocatalysis of TiO_2 nanoparticles and nanostructured thin films grown on paper. Nano Express. 2021;**2**:040002

[8] Shan AY, Ghazi TIM, Rashid SA. Immobilisation of titanium dioxide onto supporting materials in heterogeneous photocatalysis: A review. Applied Catalysis, A: General. 2010;**389**(1-2):1-8

[9] Bouarioua A, Zerdaoui M. Photocatalytic activities of TiO_2 layers immobilized on glass substrates by dip-coating technique toward the decolorization of methyl orange as a model organic pollutant. Journal of Environmental Chemical Engineering. 2017;**5**(2):1565-1574. DOI: 10.1016/j.jece.2017.02.025

[10] Sério S, Jorge MEM, Maneira MJP, Nunes Y. Influence of O2 partial pressure on the growth of nanostructured anatase phase TiO_2 thin films prepared by DC reactive magnetron sputtering. Materials Chemistry and Physics. 2011;**126**:73-81

[11] Cunha DL, Kuznetsov A, Achete CA, Machado AEDH, Marques M. Immobilized TiO_2 on glass spheres applied to heterogeneous photocatalysis: Photoactivity, leaching and regeneration process. PeerJ. 2018;**6**(6):e4464. DOI: 10.7717/peerj.4464 PMID: 29527416; PMCID: PMC5844248

[12] Horikoshi S, Watanabe N, Onishi H, Hikada H, Serpone N. Photodecomposition of a nonylphenol polyethoxylate surfactant in a cylindrical photoreactor with TiO_2 immobilized fiberglass cloth. Applied Catalysis. B, Environmental. 2002;**37**:117-129

[13] Saqib NU, Adnan R, Shah I, Arshad M, Inam M. Activated carbon, zeolite, and ceramics immobilized TiO_2 photocatalysts for the enhanced sequential uptake of dyes and Cd2+ ions. Journal of Dispersion Science and Technology. 2022:1-11. DOI: 10.1080/01932691.2022.2070497

[14] Jiao Y, Basiliko N, Kovala AT, Shepherd J, Shang H, Scott JA. TiO_2 based nanopowder coatings over stainless steel plates for UV-C photocatalytic degradation of methylene blue. Canadian Journal of Chemical Engineering . 2020;**98**:728-739. https://doi.org/10.1002/cjce.23653

[15] Amran SNBS, Wongso V, Halim NSA, Husni MK, Sambudi NS, Wirzal MDH. Immobilized carbon-doped TiO_2 in polyamide fibers for the degradation of methylene blue. Journal of Asian Ceramic Societies. 2019;**7**(3):321-330. DOI: 10.1080/21870764.2019.1636929

[16] Wang W, Yang R, Li T, Komarneni S, Liu B. Advances in recyclable and superior photocatalytic fibers: Material, construction, application and future perspective. Composites Part B: Engineering. 2021; **205**:108512. DOI: 10.1016/j.compositesb.2020.108512

[17] Morjène L, Schwarze M, Seffen M, Schomäcker R, Tasbihi M. Immobilization of TiO_2 semiconductor nanoparticles onto Posidonia Oceanica fibers for photocatalytic phenol degradation. Water. 2021;**13**:2948. https://doi.org/10.3390/w13212948

[18] Eleutério T, Sério S, Teodoro O. MND, Bundaleski N, Vasconcelos HC. XPS and FTIR studies of DC reactive magnetron sputtered TiO_2 thin films on natural based-cellulose fibers. Coatings. 2020;**10**:287. https://doi.org/10.3390/coatings10030287

[19] Li W, Zhang H, Song Y et al. Effect of the crystalline structure of cotton cellulose on the photocatalytic activities of cotton fibers immobilized with TiO_2 nanoparticles. Cellulose. 2022;**29**:644-6459 . https://doi.org/10.1007/s10570-022-04663-x

[20] Madsen B, Lilholt H. Physical and mechanical properties of unidirectional plant fibre composites: An evaluation of the influence of porosity. Composites Science and Technology. 2003;**63**(9): 1265-1272

[21] Tryba B. Immobilization of TiO_2 and Fe–C–TiO_2 photocatalysts on the cotton material for application in a flow photocatalytic reactor for decomposition of phenol in water. Journal of Hazardous Materials. 2008;**151**:623-627

[22] Gonçalves MC, Pereira JC, Matos JC, Vasconcelos HC. Photonic band gap and bactericide performance of amorphous sol-gel titania: An alternative to crystalline TiO_2. Molecules. 2018;**23**: 1677. https://doi.org/10.3390/molecules23071677

[23] Nisha T. Padmanabhan, Honey John, Titanium dioxide based self-cleaning smart surfaces: A short review, Journal of Environmental Chemical Engineering. 2020;**8**(5):104211. https://doi.org/10.1016/j.jece.2020.104211

[24] MironyukI F, SoltysL M, TatarchukT R, SavkaK O. Methods of titanium dioxide synthesis (review). Physics and Chemistry of Solid State. 2020;**21**(3): 462-477. https://doi.org/10.15330/pcss.21.3.462-477

[25] El-Roz M, Haidar Z, Lakiss L, Toufaily J, Thibault-Starzyk F. Immobilization of TiO_2 nanoparticles on natural *Luffa cylindrica* fibers for photocatalytic applications. RSC Advances, Royal Society of Chemistry. 2013;**3**(10):3438-3445

[26] Zhang Liuxue, Wang Xiulian, Liu Peng, Su Zhixing, Photocatalytic activity of anatase thin films coated cotton fibers prepared via a microwave assisted liquid phase deposition process, Surface and Coatings Technology. 2007;**201**(18):

7607-7614. https://doi.org/10.1016/j.surfcoat.2007.02.004

[27] Mattox DM. Physical Sputtering and Sputter Deposition (Sputtering). In: Mattox DM, editor. Handbook of Physical Vapor Deposition (PVD) Processing, Second Edition, Elsevier, 2010. pp. 236-286. ISBN 978-0-8155-2037-5. DOI:10.1016/b978-0-8155-2037-5.00007-1

[28] Shi F. Introductory Chapter: Basic Theory of Magnetron Sputtering. In: Shi F, editor. Magnetron Sputtering [Working Title] [Internet]. London: IntechOpen; 2018. Available from: https://www.intechopen.com/online-first/63559 DOI: 10.5772/intechopen.80550

[29] Abdelrahman MM. Journal of Physical Science and Application. 2015;**5**(2):128-142. DOI: 10.17265/2159-5348/2015.02.007

[30] Dane AE. Reactive DC Magnetron Sputtering of Ultrathin Superconducting Niobium Nitride Films [thesis]. S.M. in Electrical Engineering, Massachusetts Institute of Technology, Department of Electrical Engineering and Computer Science, 2015. Available at: https://dspace.mit.edu/bitstream/handle/1721.1/97257/910342499-MIT.pdf?sequence=1&isAllowed=y

[31] Online Electrical and Electronics Study. (n.d). Paschen Breakdown | Paschen's Law | Paschen curve. EEEGuide. Available from: https://www.eeeguide.com/paschen-breakdown/

[32] Mattox DM. Substrate ("Real") Surfaces and Surface Modification. In: Mattox DM. Handbook of Physical Vapor Deposition (PVD) Processing, Second Edition. Elsevier, 2010. pp. 25-72. ISBN 978-0-8155-2037-5. DOI:10.1016/b978-0-8155-2037-5.00002-2

[33] Thirugnanasambandan T. Review on titania nanopowder—processing and applications. Condensed Matter. 2017;**17**: 981

[34] Etacheri V, Di Valentin C, Schneider J, Bahnemann D, Pillai SC. Visible-light activation of TiO_2 photocatalysts: Advances in theory and experiments. Journal of Photochemistry and Photobiology C: Photochemistry Reviews. 2015;**25**:1-29. DOI: 10.1016/j.jphotochemrev.2015

[35] Fujishima A, Honda K. Electrochemical photolysis of water at a semiconductor electrode. Nature. 1972; **238**(5358):37-38. DOI: 10.1038/238037a0 PMID: 12635268

[36] Beltrán A, Gracia L, Andrés J. Density Functional Theory Study of the Brookite Surfaces and Phase Transitions between Natural Titania Polymorphs. The Journal of Physical Chemistry B. 2006;**110**(46):23417-23423. DOI: 10.1021/jp0643000

[37] Dette C, Pérez-Osorio MA, Kley CS, Punke P, Patrick CE, Jacobson P, et al. TiO_2 anatase with a bandgap in the visible region. Nano Letters. 2014; **14**(11):6533-6538. DOI: 10.1021/nl503131s

[38] Asahi R, Morikawa T, Ohwaki T, Aoki K, Taga Y. Visible-light photocatalysis in nitrogen-doped titanium oxides. Science. 2001;**293**: 269-271

[39] Gopal M, Chan WJM, DeJonghe LC. Room temperature synthesis of crystalline metal oxides. Journal of Materials Science. 1997;**32**:6001-6008. DOI: 10.1023/A:1018671212890

[40] Bellifa, A., Pirault-Roy, L., Kappenstein, C. et al. Study of effect of chromium on titanium dioxide phase

transformation. Bulletin of Materials Science 37, 669–677 (2014). https://doi.org/10.1007/s12034-014-0674-1

[41] Strohhöfer C, Fick J, Vasconcelos H, Almeida R. Active optical properties of Er-containing crystallites in sol–gel derived glass films. Journal of Non-Crystalline Solids. 1998;**226**(1-2): 182-191. DOI: 10.1016/s0022-3093(98)00365-2

[42] Sério S, Melo Jorge ME, Coutinho ML, Hoffmann SV, Limão-Vieira P, Nunes Y. Spectroscopic studies of anatase TiO_2 thin films prepared by DC reactive magnetron sputtering. Chemical Physics Letters. 2011;**508**(1–3): 71-75. DOI: 10.1016/j.cplett.2011.04.002

[43] Khataee AR, Aleboyeh H, Aleboyeh A. Crystallite phase controlled preparation, characterisation and photocatalytic properties of titanium dioxide nanoparticles. Journal of Experimental Nanoscience. 2009;**4**(2): 121-137. DOI: 10.1080/17458080902929945

[44] Liang Y, Sun S, Deng T, Ding H, Chen W, Chen Y. The Preparation of TiO_2 Film by the Sol-Gel Method and Evaluation of Its Self-Cleaning Property. Materials, MDPI, Basel. 2018 Mar 19;**11**(3):450. DOI: 10.3390/ma11030450

[45] Helena C. Vasconcelos, Telmo Eleutério. Recent developments in surface functionalization of natural fibers by DC reactive sputtering. Proceedings of the 5th International Conference on Natural Fibers - Materials of the Future. Ed. R. Fangueiro. Portugal: Sciencentris; 17–19 May 2021

[46] Thornton JA. Structure-zone models of thin films. In: Jacobson MR, editor. Modeling of Optical Thin Films. Proc. SPIE. Vol. 821. 1988. pp. 95-103

[47] Biesinger MC, Lau LWM, Gerson AR, Smart RSC. Resolving surface chemical states in XPS analysis of first row transition metals, oxides and hydroxides: Sc, Ti, V, Cu and Zn. Applied Surface Science. 2010;**257**(3): 887-898. DOI: 10.1016/j.apsusc.2010.07.086

[48] Švagelj Z, Mandi'c V, Curkovi'c L, Bioši'c M, Žmak I, Gaborardi M. Titania-coated alumina foam photocatalyst for memantine degradation derived by replica method and sol-gel reaction. Materials. 2020;**13**:227

Chapter 9

Effects of Gamma Radiation on the Structural, Optical, and Photocatalytic Properties of TiO_2 Thin Films and Nanostructures for Photovoltaic Applications

Aymen Bourezgui and Imen Kacem

Abstract

In this chapter, we delve into a comprehensive discussion of the complex impact of gamma rays on materials of titanium dioxide (TiO_2) and their practical use in photovoltaic contexts. Our goal is to gain a better grasp of the dynamic interplay between gamma irradiation and the performance of TiO_2 for better utilization in photocatalysis. We aim to explore how the employment of gamma-treated TiO_2 in photovoltaic applications can lead to amplified solar cell effectiveness and endurance. As we strive to enhance sustainable energy initiatives and extend the range of innovative prospects for TiO_2 materials, we also scrutinize the fundamental processes that drive these developments. Additionally, we contemplate prospective avenues for research such as identifying optimal gamma-ray parameters, assessing the durability of treated TiO_2, and studying the synergistic influence of combining gamma radiation with other treatments. Scientists and industrialists seeking to enrich the performance of TiO_2 materials in solar energy endeavors can benefit from this chapter as a valuable reference.

Keywords: gamma radiation, TiO_2 thin films, TiO_2 nanostructures, structural modifications, optical properties, photocatalytic activity, photovoltaic devices

1. Introduction

Different fields are interested in titanium dioxide (TiO_2) due to its exceptional traits that make it a versatile material and crucial area of research [1, 2]. With a high surface-to-volume ratio, TiO_2 thin films are able to enhance properties and therefore receive significant attention [3]. The fabrication of TiO_2 thin films has been accomplished through different synthesis techniques that include sol-gel, chemical vapor deposition, and physical vapor deposition [4–6]. Nevertheless, post-treatments like ionizing radiation [7] can enhance these thin films.

Ionizing radiation in the form of gamma rays has been used to modify the chemical and physical characteristics of diverse materials, such as thin films of TiO_2. Induction of defects and oxygen vacancies in TiO_2 through gamma radiation has been found to change its structural, optical, and photocatalytic properties [8]. Still, the methods responsible for these adjustments are currently under investigation.

This chapter seeks to offer a thorough examination of how gamma radiation exposure affects the physical properties of both TiO_2 thin films and nanostructures, emphasizing its applications in photovoltaic technology. An overview of how gamma radiation interacts with matter is provided in our discussion about the mechanism of TiO_2 materials under this effect. This is given in our first section. In this section, we delve into how gamma radiation specifically impacts both the structural and electronic traits of TiO_2. This includes examining how it creates defects and impurities within the material.

The third section thoroughly examines the optical effects of gamma radiation on titanium dioxide materials, which can lead to the creation of defects and alterations in electronic structure. These changes can result in improved visible light absorption and an overall enhancement of the material's photocatalytic properties.

In the fourth section, we investigate the impacts of gamma ray exposure on TiO_2 photocatalytic characteristics. Our goal is to comprehend the intricate correlation between gamma radiation and TiO_2 efficiency in various photocatalysis domains. Through exploring the underlying mechanisms, we can enhance TiO_2 applications, realize more effective photocatalysts, and accomplish eco-friendly solutions, ultimately resulting in increasing possibilities for innovation and utilization of TiO_2 materials.

In section five, we explore the prospects of utilizing gamma-treated TiO_2 materials for photovoltaic purposes. The study concentrates on comprehending the consequences of gamma radiation exposure on TiO_2 material's efficiency, stability, and endurance concerning solar cells. Through scrutinizing the fundamental mechanisms, our goal is to refine TiO_2 implementation, design more potent photovoltaic apparatuses, and promote sustainable energy alternatives.

In the sixth section, we outline the future research directions for TiO_2 materials in photovoltaic applications, which involve gamma rays treatment. Our primary focus is on finding the optimal gamma-ray treatment parameters and assessing the long-term stability of treated TiO_2. Additionally, we investigate the possible synergistic benefits of combining gamma radiation with other treatments like doping and annealing.

In summary, this chapter provides a thorough examination of the impact of gamma radiation on thin films and nanostructures of TiO_2, emphasizing its potential applications in photovoltaics. By exploring the underlying mechanisms that drive these effects, the latest advancements in this field, and the associated challenges and merits of gamma radiation treatment, this chapter aims to provide invaluable insights into utilizing this tool for optimizing the properties of TiO_2-based devices.

2. Gamma radiation-induced structural changes in TiO_2 materials

Significant alterations in the structural characteristics of TiO_2 materials can result from exposure to gamma radiation. The crystalline composition may undergo changes such as transitioning from the anatase stage to the rutile phase or the emergence of faults in the crystal lattice. For instance, a study noted that a dose of 10 kGy (1 Mra) of gamma irradiation could raise the anatase to rutile transformation temperature from $T_1 = 773$ K to $T_2 = 873$ K, indicating that the thermal stability of the TiO_2 phase can be influenced by gamma radiation [9]. Furthermore, the ability of TiO_2 nanoparticles to perform optimally in procedures such as gas sensing and photocatalysis can be impacted by exposure to gamma radiation because it lowers the surface area and porosity of the particles. By way of example, another study found that the specific surface area of TiO_2 nanoparticles declined from 233.2 to 107.9 m^2/g subsequent to being exposed to 300 kGy of gamma radiation [10]. The reduction in surface area was determined to be due to the particles agglomerating and their porous structure collapsing.

Significant impacts on the properties and performance of TiO_2 materials can result from structural changes induced by gamma radiation. The material's optoelectronics and photocatalytic characteristics may be altered due to changes occurring in both its crystal structure and morphology [11]. In addition to its effects on other properties, such as morphology and gamma radiation-induced structural changes in titanium dioxide (TiO_2), nanomaterials can reduce both surface area and porosity. Such alterations could potentially hinder optimal functioning for gas sensing or photocatalytic applications [10].

The investigation of gamma radiation effects on TiO_2 materials can involve various techniques such as X-ray diffraction, Raman spectroscopy, transmission electron microscopy, and positron annihilation spectroscopy [12]. A high anisotropy in stretched thermal and electrical conductivity of TiO_2 nanostructures was demonstrated by one study through DFT calculations [9]. Various studies have investigated the effects of gamma radiation on organic matter composed of one or more constituents and determined that the radiation caused significant alterations to their electronic and structural properties [13].

It is important to comprehend the alterations in the structures of TiO_2 nanomaterials and thin films caused by gamma radiation, as it can enhance their properties for numerous applications, such as photocatalysis, gas sensing, and electronics. For instance, research indicates that gamma radiation ameliorates the photocatalytic performance of TiO_2 nanoparticles in eliminating pollutants [14]. Furthermore, a separate examination employing ceramography notes the influence of gamma radiation on the electronic properties of TiO_2 thick film, displaying that the method has potential in electronic device production [15].

It can be concluded that TiO_2 material properties and performance can be greatly impacted by structural changes induced by gamma radiation. Therefore, it is crucial to utilize multiple techniques to study the impact of gamma radiation on TiO_2 materials in order to gain a better understanding of the underlying mechanism and enhance their effectiveness for diverse applications.

3. Gamma rays-induced optical effects in titanium dioxide materials

Gamma rays, as a form of high-energy electromagnetic radiation, can cause substantial changes in the optical properties of titanium dioxide (TiO_2) materials, which serve as a highly adaptable and commonly employed semiconductor [4]. Interaction between titanium dioxide (TiO_2) materials with gamma rays has been known to elevate photocatalytic properties, amend electronic structures, and lead to the formation of defects [5]. This section will analyze the different optical impacts caused by gamma rays in TiO_2 materials, cite examples, and debate prior experimental findings.

In their study, Bouregui et al. [16] explored the effect of gamma irradiation on TiO_2 thin films' optical properties. By examining how different doses of gamma radiation affected these properties, they found that the absorption edge shifted toward lower energies, indicating a decrease in bandgap energy. This change could be due to defects produced by gamma irradiation, which increase the number of defect states and enhance visible light absorption. Along with an increase in photocatalytic efficiency, Bouregui et al. [16] suggest that gamma irradiation may be a valuable technique for tailoring TiO_2 thin films' optical properties to meet specific application needs.

The TiO_2 lattice can experience defects and oxygen vacancies when gamma rays disrupt the Ti-O bonds [16]. This, in turn, leads to an increased degree of defect states, decreased amount of energy in the bandgap, and a subsequent reinforcement of visible light usage [4]. This process can significantly amplify the photocatalytic efficiency of the TiO_2 materials. Wang and colleagues [17] found that by diminishing the energy in the bandgap of gamma-irradiated TiO_2 from 3.2 to 2.9 eV, there was a higher degree of photocatalytic activity for the breakdown of organic pollutants.

Several other investigations have reported related findings, supporting the influence of gamma rays on the optical properties of TiO_2. In one study by Zuo et al. [18], gamma-irradiated TiO_2 nanoparticles were perceived to demonstrate amplified photocatalytic activity in the degradation of organic pollutant under visible light. The enhancement was credited to a reduction in the bandgap and an escalation in oxygen vacancies induced by gamma irradiation. Furthermore, another study by Kumar et al. [19] observed a decline in the bandgap energy and improved photocatalytic activity of TiO_2 nanoparticles following exposure to gamma rays, which they attributed to the creation of oxygen vacancies and other defects.

Furthermore, along with photocatalytic applications, gamma-irradiated TiO_2 materials seem to have the potential for solar energy conversion. TiO_2-based solar cells can benefit from the reduced bandgap energy and increased visible light absorption. A good example is the improvement in photovoltaic performance shown by gamma-irradiated TiO_2 nanotubes as reported by Choi et al. [20], which was due to the increase in visible light absorption and the gamma ray-induced modification of electronic structures.

In summary, TiO_2 materials can undergo significant alterations in their optical properties, which can improve visible light absorption and photocatalytic efficiency when exposed to gamma rays. Numerous studies have highlighted the potential of gamma-irradiated TiO_2 in areas like environmental remediation and solar energy conversion [16]. This is possible since the interaction between gamma rays and TiO_2 can produce defects, modify the electronic structures, and improve the photocatalytic properties. Subsequently, researchers should focus on refining the gamma irradiation protocol to achieve the optimal optical properties to enhance the applications of TiO_2 [14]. Hybrid materials are also an area of interest. Through the use of semiconductors or metal nanoparticles in conjunction with TiO_2, researchers can enhance the separation of photogenerated charge carriers, resulting in improved photocatalytic performance [21]. In hybrid systems, gamma-irradiated TiO_2 can be beneficial owing to the changes in optical properties resulting from gamma rays.

Research on gamma rays-generated optical outcomes in titanium dioxide materials has unraveled multiple remarkable phenomena that could be utilized to boost the efficiency of TiO_2 in diverse roles. Through comprehending the processes that underpin these outcomes and refining the gamma irradiation procedure, experts can forge fresh methods to engineer sublime TiO_2 materials boasting specific optical traits.

This research domain is ripe with prospects for novel inventions and the generation of eco-friendly and better-performing resolutions in environmental revitalization, photovoltaic transformation, and other domains.

4. The influence of gamma radiation on titanium dioxide photocatalysis

Titanium dioxide (TiO_2), a widely-employed photocatalyst, boasts reactive properties, commendable stability, cost-effectiveness, and biocompatibility. TiO_2 nanoparticles, of exceptional interest as photocatalysts, exhibit sterling photocatalytic activity since their nanometric size and high surface area promote such reactivity. Photocatalytic activity of TiO_2 materials has been linked to a number of variables, gamma radiation among them. The effect of gamma rays on the photocatalytic activity of TiO_2 nanoparticles has been investigated using a multitude of samples.

A research study by Bouregui et al. [16] examined the impact of gamma radiation on the photocatalytic performance of controlled atmosphere TiO_2 nanoparticles. The nanoparticles underwent gamma irradiation at varying doses for investigation into their optical, structural, and photocatalytic attributes. According to the results, gamma radiation exerted favorable effects on the optic and structural properties of the TiO_2 nanoparticles. Further, there was a rise in nanoparticle crystallinity while their bandgap energy decreased due to increased gamma radiation dosages. The photocatalytic activity of the nanoparticles also improved with gamma radiation dosages by enhancing their surface area and creating oxygen vacancies. Such vacancies served as recombination centers for electron-hole pairs, elevating the charge separation efficacy and photocatalytic activity of the nanoparticles.

The impact of gamma radiation on the photocatalytic features of TiO_2 thin films was investigated in other studies conducted by Semalti et al. [22] and Haldar et al. [23]. Semalti et al. found that photocatalytic activity was amplified by gamma radiation, particularly at higher doses, due to the accumulation of oxygen vacancies and the development of Ti^{3+} ions, which boosted carrier separation. In contrast, Haldar et al. found that TiO_2 nanoparticle photocatalytic activity decreased with gamma radiation exposure due to defects and impurities formed in the crystal lattice, leading to reduced efficiency of charge separation and decreased photocatalytic activity.

A report published by Sajjadi et al. [24] shared that TiO_2 nanoparticles' photocatalytic activity reduced after gamma irradiation. Their research showed that as gamma radiation dose increased, the degradation rate of methylene blue dye under UV light irradiation also reduced. This is due to the reduction of oxygen vacancies and hydroxyl radicals. Xue et al. [25], on the other hand, discovered a rise in photocatalytic activity of TiO_2 nanoparticles at higher doses of gamma radiation. However, they associated the decrease in activity at low doses to the formation of recombination centers for electron-hole pairs and oxygen vacancies, but the high concentration of Ti^{3+} ions enhanced the carriers' separation.

We may reconcile these two studies by considering that oxygen vacancies and Ti^{3+} ions may impact TiO_2 nanoparticle photocatalysis. Low gamma radiation may increase oxygen vacancies, limiting photocatalytic activity, but at higher doses Ti^{3+} ions may dominate oxygen vacancies, boosting photocatalytic activity.

Ti^{3+} ions may impact the photocatalytic activity of TiO_2 nanoparticles [26], which can also be influenced by nanoparticle size, shape, contaminants, dopants, and irradiation conditions. Gamma radiation is known to affect the electronic structure and surface area of TiO_2 crystals, potentially altering the generation and recombination of electron-hole pairs vital to photocatalytic activity. The degree and type of defects formed by gamma radiation are dependent on a range of variables such as radiation dose, nanoparticle type, and environmental factors.

TiO_2 materials remains an encouraging option for environmental use cases like purifying water and controlling air pollution despite contradictory findings. Using TiO_2-based photocatalysis in solar photocatalysis has proven to effectively decontaminate and disinfect water [27]. The utilization of sunlight activates the photocatalytic reaction in this technology, without requiring any external energy sources or chemicals. Moreover, nanotechnologies based on TiO_2 have been created to purify and recycle water [28]. Their performance under visible light irradiation [29] is enhanced by combining TiO_2 nanoparticles with other materials like graphene.

Photocatalytic performance of TiO_2 materials can be enhanced by altering its surface properties. TiO_2 nanoparticles that have high-energy {001} facets exhibit improved photocatalytic activity due to increased adsorption of reactant molecules and facilitated separation of charge carriers. This phenomenon has been previously established by researchers [30]. The photocatalytic activity of TiO_2 nanoparticles can be enhanced by modifying their surface with metals or nonmetals, which improves their electron transfer and surface plasmon resonance properties.

Machine learning methods were utilized to examine the photocatalytic mechanism of semiconducting materials like TiO_2. Photocatalytic processes involve various intricate steps such as adsorption-reaction mechanisms for reactants' generation-separation mechanisms for charged particles role played by defects impurities; all these procedures can be investigated with these techniques. To explore the photocatalytic mechanism of TiO_2 nanoparticles under UV and visible light irradiation, Wu et al. [31] used machine learning algorithms. This study revealed that nanoparticles' photocatalytic activity was most significantly influenced by its surface structure and defects. Additionally, other important factors include its adsorption energy of reactant molecules and its energy barrier for electron transfer.

To summarize, the influence of gamma radiation on the photocatalytic operation of TiO_2 nanoparticles is still being questioned. Although some studies observed an improvement in photocatalytic performance, others noted a decrease. The inconsistent findings could be ascribed to the different experimental parameters and TiO_2 nanoparticles employed in each analysis. Additional investigation is necessary to fully comprehend the fundamental mechanisms and optimize the conditions for employing gamma-irradiated TiO_2 nanoparticles in photocatalytic applications. TiO_2 remains a hopeful photocatalyst for environmental implementations, and its photocatalytic efficiency may be boosted by combining it with other substances or altering its surface attributes. Learning algorithms may provide useful understandings into the intricate techniques implicated in photocatalysis and lead the way in designing more dependable photocatalysts.

5. Potential uses of gamma-treated TiO_2 materials in photovoltaics

The employment of gamma-examined TiO_2 materials in photovoltaic devices has demonstrated various benefits over unexamined TiO_2 materials. Gamma radiation handling alters the surface qualities of TiO_2, ensuring in enhanced light absorption, charge separation, and charge transit characteristics [16]. This may ultimately lead to amplified potency in solar cells, as well as enhanced resilience and durability.

Dye-sensitized solar cells (DSSCs) could benefit greatly from using gamma-treated TiO_2 materials in photovoltaics. Their affordability, lightweight, and flexibility make DSSCs an attractive alternative to the conventional silicon-based solar cells [32]. Enhancement of the surface area, crystallinity, and charge transfer properties [33, 34] of TiO_2 materials is achieved through gamma radiation treatment, thereby improving their performance in DSSCs. Additionally, this therapy can be a hopeful method to boost the effectiveness of DSSCs. TiO_2 nanotubes treated with gamma radiation demonstrated higher DSSC efficiency compared with untreated ones, as discovered by Kim et al. [35], while Mahmoud et al. [36] found that using gamma radiation-treated TiO_2 nanoparticles also increased DSSC efficiency.

Perovskite solar cells have displayed potential in photovoltaic applications due to their proficient power conversion efficiency and economical manufacturing expenses. Gamma radiation-treated TiO_2 materials have been studied as electron transport layers in perovskite solar cells, with arousing outcomes. Liu et al. [37] noted a substantial progress in the conductivity and crystallinity of gamma radiation-treated TiO_2 films. This led to a better charge transfer performance and a consequential elevation in the efficiency of the perovskite solar cells.

The study of organic solar cells, which use organic materials rather than inorganic semiconductors, has revealed promising results for use in photovoltaics. Gamma-treated TiO_2 materials have been shown to improve power conversion efficiency by improving light absorption, charge separation, and mobility of charges properties [38].

Photovoltaic device performance may be enhanced by gamma radiation treatment. Likewise, it can contribute toward enhancing the stability and endurance of TiO_2 components that play a crucial role in their long-term deployment within solar panels. Gamma treatment applied on TiO_2 materials enhances their surface area and crystallinity, which leads to reduced recombination speed and increased lifespan of electrons. This eventually promotes greater stability [35, 36]. TiO_2 materials treated with gamma radiation have shown increased ability in resisting various forms of environmental stressors, among them being high temperature and humidity. Also, this can reinforce the resilience of solar panels [33].

In general, the utilization of gamma-treated TiO_2 substances in the photovoltaic industry presents a hopeful pathway for creating effective, enduring, and stable solar cells. Further examination is necessary to investigate the complete potential of gamma radiation therapy on TiO_2 materials and how it affects different photovoltaic devices. Solar energy's adoption as a renewable resource can be boosted with improvements in this field. Additionally, their involvement can lead to the progression of a more eco-conscious future.

6. Gamma rays treatment of titanium dioxide materials for photovoltaic applications: Future research directions

Further investigations are needed to find the optimum parameters for gamma-ray treatment of TiO_2 materials, which would result in desirable properties for photovoltaic applications [10]. The dose and the dose rate of gamma radiation are critical considerations in controlling the physical and chemical properties of the material. Gamma radiation can affect the crystal structure, morphology, and electrical features of TiO_2 materials in different ways, thus making it a promising research avenue to determine the optimal protocol for photovoltaic usage.

Combining gamma radiation with other treatments, such as doping and annealing, may further enhance the performances of titanium dioxide thin films and nanostructures used in photovoltaic applications. Doping with transition metal ions or non-metals can improve titanium dioxide material's photocatalytic activity and electronic properties [11], while annealing can enhance its conductivity and optical absorption properties by reducing defects and enhancing crystal quality [12]. Novel strategies for improving of titanium dioxide thin films and nanostructures properties for photovoltaic applications can be explored by mixing gamma radiation with such treatments.

It is essential to study the long-term stability of TiO_2 materials that have been exposed to gamma radiation [13], as gamma radiation has been shown to gradually diminish the properties of materials. Thus, it is crucial to identify the degradation processes and investigate the durability of TiO_2 materials after extended exposure to gamma rays. This knowledge is essential for ensuring the long-term performance and durability of photovoltaic applications that involve TiO_2 materials.

Future research should focus on enhancing the durability of irradiated TiO_2 materials, refining gamma radiation treatment settings, and exploring the use of gamma radiation in conjunction with other therapies. These strategies will help advance the comprehension of photovoltaics by utilizing more efficient and resilient TiO_2 materials.

7. Conclusion

Titanium dioxide thin films and nanostructures exposed to gamma radiation exhibit significant modifications in their structural, optical, and photocatalytic properties, rendering them promising candidates for various applications. Gamma irradiation of TiO_2 is being explored for its potential use as an electron transport layer in perovskite solar cells and as a photoanode in dye-sensitized solar cells. Experiments have confirmed long-term increases in power conversion efficiency and reliability, offering hope for practical usage of these materials in PV systems.

Future research avenues may include refining the gamma radiation treatment settings to achieve greater performance and investigating the compatibility of these materials with other photovoltaic systems such as organic photovoltaics and multi-junction solar cells. Comprehensive research must be conducted to ensure the safety of these materials in real-world applications.

In summary, gamma radiation treatment of TiO_2 is a promising technique for customizing the material's properties for use in photovoltaic cells. Gamma radiation-induced changes in the crystal structure, morphology, electronic properties, and photocatalytic activity of TiO_2 contribute to enhancements in photoanode and electron transport layer performance in solar cells. However, further studies are needed to determine the viability of gamma radiation-treated TiO_2 in practical applications.

Author details

Aymen Bourezgui[1,2*] and Imen Kacem[1,2]

1 Nanomaterials and Systems for Renewable Energy Laboratory, Research and Technology Center of Energy, Hammam Lif, Tunisia

2 National School of Advanced Sciences and Technologies of Borj Cedria (ENSTAB), University of Carthage, Tunis, Tunisia

*Address all correspondence to: bourezgui.aymen@gmail.com

References

[1] Pawar TJ, Contreras López D, Olivares Romero JL, Vallejo Montesinos J. Surface modification of titanium dioxide. Journal of Materials Science. 2023;**58**:1-44

[2] Kumarage GW, Hakkoum H, Comini E. Recent advancements in TiO_2 nanostructures: Sustainable synthesis and gas sensing. Nanomaterials. 2023;**13**(8):1424

[3] Roy P, Berger S, Schmuki P. TiO_2 nanotubes: Synthesis and applications. Angewandte Chemie International Edition. 2011;**50**(13):2904-2939

[4] Bourezgui A, Kacem I, Assaker IB, Gannouni M, Naceur JB, Karyaoui M, et al. Synthesis of porous TiO_2 thin films prepared with templating technique to improve the photoelectrochemical properties. Journal of Materials Science: Materials in Electronics. 2019;**30**(8):7757-7767

[5] Bourezgui A, Kacem I, Karyaoui M, Chtourou R. Optical and photocatalytic properties of porous TiO_2 thin films. Journal of Materials Science: Materials in Electronics. 2019;**30**(22):20036-20043

[6] Mattox DM. Handbook of Physical Vapor Deposition (PVD) Processing. 2nd ed. Norwich, NY: William Andrew (Elsevier); 2010

[7] Gopal NO, Narayanan V. Ionizing radiation: A versatile tool for advanced materials. Radiation Effects & Defects in Solids. 2014;**169**(1):1-26

[8] Wang J, Zhao H, He H. Gamma ray induced oxygen vacancy and its role in the photocatalytic activity of TiO_2 nanoparticles. RSC Advances. 2014;**4**(17):8532-8537

[9] Saleh H, Hiromitsu T. Effect of gamma irradiation on structural and optical properties of TiO_2 nanoparticles. Radiation Physics and Chemistry. 2017;**130**:181-186

[10] Arunachalam A, Dinesh R, Sanjeeviraja C. Influence of gamma radiation on the structural and optical properties of anatase TiO_2 nanoparticles. Radiation Physics and Chemistry. 2013;**91**:114-120

[11] Wu Y, Zhang X, Zhao H. A review of the effect of gamma radiation on the photocatalytic properties of TiO_2 nanostructures. RSC Advances. 2016;**6**(111):109333-109343

[12] Yu J, Wang W, Cheng B. Enhanced photocatalytic activity of TiO_2 powders (P25) via calcination treatment. Journal of Molecular Catalysis A: Chemical. 2008;**281**(1-2):29-35

[13] Biro LP, Kertesz K, Vertesy G, Lambin P. Effects of gamma radiation on the structural properties of single and multicomponent organic materials. Nuclear Instruments and Methods in Physics Research Section B: Beam Interactions with Materials and Atoms. 1997;**131**(1-4):303-311

[14] Gopal NO, Premkumar S, Nagarajan V. Enhancement of photocatalytic activity of TiO_2 nanoparticles by gamma irradiation. Radiation Physics and Chemistry. 2012;**81**(12):1906-1911

[15] Sharma KK, Reddy GK. Effect of gamma radiation on the electrical properties of TiO_2 thick film. Journal of Materials Science: Materials in Electronics. 2013;**24**(12):4931-4935

[16] Bourezgui A, Kacem I, Daoudi M, et al. Influence of gamma-irradiation on

structural, optical and photocatalytic performance of TiO_2 nanoparticles under controlled atmospheres. Journal of Electronic Materials. 2020;**49**:1904-1921. DOI: 10.1007/s11664-019-07887-z

[17] Wang X, Ma Y, Chen J, Chen L. Gamma-ray irradiation-induced modification of the structural and optical properties of TiO_2 thin films. Journal of Alloys and Compounds. 2011;**509**(5):L26-L29

[18] Zuo F, Wang L, Wu T, Zhang G, Borchardt D, Feng P. Self-doped Ti^{3+} enhanced photocatalyst for hydrogen production under visible light. Journal of the American Chemical Society. 2010;**132**(34):11856-11857

[19] Kumar S, Kumar V, Anand RS, Dhawan S, Chandra R. Structural, optical and gas sensing properties of TiO_2 nanoparticles synthesized by hydrothermal and sol-gel methods. Thin Solid Films. 2012;**520**(15):5043-5049

[20] Choi W, Termin A, Hoffmann MR. The role of metal ion dopants in quantum-sized TiO2: Correlation between photo reactivity and charge carrier recombination dynamics. Journal of Physical Chemistry. 1994;**98**(51):13669-13679

[21] Wei G, Ramana CV, Harris VG. Structural and electronic properties of titanium dioxide nanotubes: First-principles calculations. Physical Review B. 2004;**70**(23):235402

[22] Semalti A et al. Influence of gamma radiation on the photocatalytic properties of TiO_2 thin films. Journal of Materials Science. 2019;**54**(21):13792-13802

[23] Haldar A et al. Influence of gamma radiation on the photocatalytic activity of TiO_2 nanoparticles. Radiation Physics and Chemistry. 2021;**184**:article 109516

[24] Sajjadi S et al. Gamma irradiation effects on the photocatalytic activity of TiO_2 nanoparticles. Journal of Radioanalytical and Nuclear Chemistry. 2015;**303**(2):1091-1095

[25] Xue L et al. Effect of gamma radiation on the photocatalytic activity of TiO_2 nanoparticles. Journal of Hazardous Materials. 2011;**186**(1):104-110

[26] Chen X et al. Semiconductor-based photocatalytic hydrogen generation. Chemical Reviews. 2010;**110**(11):6503-6570

[27] Malato S et al. Decontamination and disinfection of water by solar photocatalysis: Recent overview and trends. Catalysis Today. 2009;**147**(1):1-59

[28] Zheng Y et al. TiO_2-based advanced oxidation nanotechnologies for water purification and reuse. Journal of Materials Chemistry A. 2018;**6**(1):21-42

[29] Zhang X et al. Enhanced photocatalytic performance of graphene-TiO_2 composite under visible light irradiation. Carbon. 2012;**50**(8):2772-2781

[30] Yu J et al. Synthesis and characterization of highly photocatalytic active TiO_2 film. Applied Surface Science. 2007;**253**(17):7121-7126

[31] Wu W et al. A review on photocatalytic mechanism of semiconducting materials with machine learning methods. Journal of Environmental Chemical Engineering. 2019;**7**(4) article 103213

[32] Prajapat K, Dhonde M, Sahu K, Bhojane P, Murty VVS, Shirage PM. The evolution of organic materials for efficient dye-sensitized solar cells. Journal of Photochemistry and

Photobiology C: Photochemistry Reviews. 2023;**4**:100586

[33] Zhang Y, Shi J, Zhang X, Wu L, Zhao H. Electrochemical synthesis of SnO2 nanorod arrays on FTO substrates and their photocatalytic properties. Journal of Nanoparticle Research. 2013;**15**(5):1-8

[34] Zhang X, Shi J, Zhang Y, Wu L, Zhao H. Facile synthesis of Ag_2S/TiO_2 nanotube arrays with enhanced visible-light-driven photocatalytic activity. Journal of Alloys and Compounds. 2013;**580**:46-51

[35] Kim H, Lee C, Lee J, Park H, Lee J, Lee S, et al. Gamma-ray effect on the efficiency of dye-sensitized solar cells using TiO_2 nanotubes. Journal of Alloys and Compounds. 2017;**725**:415-419

[36] Mahmoud WE, Adel M, El-Shishtawy RM, Hassan IM. Improvement of dye-sensitized solar cells efficiency using gamma irradiated TiO_2 nanoparticles. Journal of Materials Science: Materials in Electronics. 2018;**29**(2):1198-1203

[37] Liu Y, Chen M, Deng W. The effect of gamma radiation on TiO_2 films as electron transport layers in perovskite solar cells. Journal of Materials Science: Materials in Electronics. 2020;**31**(1):448-453

[38] Goh C, Lim T, Tijing L. Improved power conversion efficiency of organic solar cells using gamma-ray treated TiO_2 nanoparticles as an electron transport layer. Journal of Alloys and Compounds. 2017;**695**:1763-1770

Chapter 10

Plasmonic-TiO_2 Nanohybrid for Environmental and Energy Applications

Jaspal Singh and Ashwani Kumar Verma

Abstract

Engineering the plasmonic nanohybrid structures to provide the advancement in their optical and photocatalytic profiles is one of the important aspects for the development of several environmental and energy applications. Plasmonic nanohybrids, integration of semiconductors and noble nanoparticles provide efficient charge separation due to Schottky junction and plasmon nanoparticle induced electromagnetic field. Effective charge separation and electromagnetic features make plasmonic nanohybrids a promising candidate for SERS-based detection environmental detoxification and energy harvesting applications. In the present chapter, we will summarize and elaborate the different strategies and modification techniques to enhance photocatalytic-driven environmental and energy applications. Moreover, the current chapter also includes the detection of various harmful pollutant molecules and their decomposition under sunlight using several plasmonic nanohybrids. This chapter also reveals the origins of morphological, optical, and plasmonic variations on TiO_2 nanostructures for enhanced photocatalytic efficiency. We have also highlighted the probable mechanism due to the plasmonic nanoparticles' aspects over TiO_2 nanostructures and their future perspectives of advanced photocatalysis. This chapter provides the fundamental synthesis aspects of plasmonic nanohybrid and their possible usage in energy and environmental applications significantly. This chapter will provide a basic understanding for the readers to develop several plasmonic nanostructures for environmental applications.

Keywords: TiO_2, plasmonic nanohybrids, photocatalysis, energy, pollutants in water

1. Introduction

Nanostructured semiconductor-based photocatalysts have proven tremendous candidates after the outstanding water-splitting discovery by using TiO_2-based electrodes in 1972 [1]. Owing to the band structure properties, semiconductors have the ability to perform and enhance the redox reaction with different light exposure [2]. The nontoxicity, ease of availability, and cost-effectiveness properties of semiconductors make them fascinating for environmental and energy applications [3–5]. Several nanostructured semiconductors such as ZnO, CuO and TiO_2, WO_3, and V_2O_5 have been employed for different energy and environmental applications [6–10].

In comparison to other semiconductors, TiO_2 has been found more beneficial for photocatalytic and energy generation devices due to its high photostability, unique band gap, lower cost, and non-toxic nature [11]. TiO_2 is known to be an n-type semiconductor and is the most explored towards energy harvesting and energy generation applications owing to its fascinating electrical and optical behavior [12]. TiO_2 contains three crystalline phases with wide bandgap; anatase rutile (3 eV), brookite (3.1 eV), and (3.2 eV) [13]. Among these phases, the rutile phase is more stable as compared to the brookite and anatase phases. Brookite and anatase phases are stable only at low temperatures and can be transformed into rutile phases by using high-temperature thermal annealing. The anatase phase is found to be more efficient as a photocatalyst as compared to both other phases of TiO_2. Rutile TiO_2 is the most stable phase, while. As compared to the rutile and brookite phases, anatase TiO_2 is more suitable for energy-harvesting reaction processes [13]. The formation of the biphasic TiO_2 is also one of the effective approaches to improving the charge separation in TiO_2 without any external modification. Several research groups have demonstrated the improved photodecomposition ability of mixed-phase TiO_2 in comparison to single-phase TiO_2 [14–17]. Singh et al. [13] used the hydrothermal method and synthesized mixed-phasic TiO_2 nanoflowers and used them as photocatalysts for the water remediation applicators. They have shown that mixed-phase TiO_2 nanoflowers exhibited high photocatalytic activity owing to the creation of the heterojunction interfaces among the rutile and anatase phases of TiO_2.

As compared to the bulk, nanostructured TiO_2 contains superior photocatalytic efficiency due to their effective active sites and high surface-to-volume ratio, which provides a strong tendency for molecular interactions. TiO_2 with a wide band gap (3.2 eV) absorbs ultra-violet light, following the charge separation, yielding the photo-induced electrons in the conduction band and the complementary holes in the valance band. These photo-generated carriers are short-lived, so they quickly recombine and result in diminishing photocatalytic efficiency. In order to resolve these issues, several methods have been adopted by various research groups, such as doping with metals [18], non-metals [19], and, more recently, through surface modification by noble metal nanoparticles [20–22]. Metal nanoparticles functionalizing TiO_2 duplicate as an electron sink capturing electrons from TiO_2 and also help to furnish more charged carriers using its localized electric field or Surface Plasmon Resonance (SPR) [20, 21]. With the attachment of plasmonic metal nanoparticles (MNP), the reduction in the recombination rate takes place by the migration of electrons from the conduction band of TiO_2 to MNP. In addition, the electromagnetic field generated by the plasmonic nanoparticles attached over the TiO_2 surface under electromagnetic radiation also helps to reduce the recombination rate. Owing to the SPR effect of plasmonic nanoparticles, the plasmonic nanoparticles modified TiO_2 enable visible light adsorption. Thus plasmonic nanoparticles functionalized TiO_2 nanostructures are expected to exhibit superior energy harvesting efficiency as compared to bare nanostructured TiO_2 [23]. **Figure 1** reveals the working mechanism of the plasmonic nanoparticles functionalized TiO_2 nanostructures. Bare TiO_2 nanostructures contain a high probability of charge recombination. With the modification with plasmonic nanoparticles, efficient charge transfer occurs; consequently, a reduction in the recombination rate takes place. A high density of electrons in the conduction band produces superoxide radicals by reacting with the surface oxygen, while holes transform the water molecule into hydroxyl radicals [23]. These two unsaturated radicals effectively control the different energy generation and environmental applications, such as H_2 production and water purification, respectively. Plasmonic nanoparticles functionalized TiO_2 nanostructures

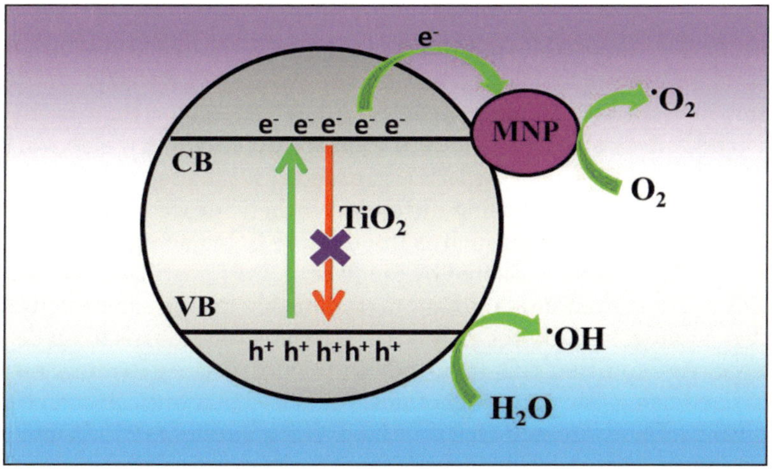

Figure 1.
Scheme reveals the efficient charge transfer mechanism in TiO_2 using plasmonic noble metal nanoparticles.

significantly improve its efficiency in various applications such as sensors [24], solar cells [25], photocatalytic activity [26], energy storage [27], and energy production [28]. Various physical and chemical techniques such as sol-gel [29], impregnation method [30], sputtering [31] pulse laser deposition method [32], and photo-deposition method [33] have been embraced by different research groups for the fabrication of plasmonic nanocomposites such as Ag-TiO_2 and Au-TiO_2.

2. Environmental applications of plasmonic-TiO_2 nanohybrid

In this section, plasmonic-TiO_2 nanostructure-driven environmental applications have been explained in detail. Environmental applications include the SERS-based detection of pollutant molecules solution and the breakdown of pollutant molecules in water under light illumination. This section also explains the underline mechanism responsible for the outstanding performance of plasmonic-TiO_2 nanohybrid for photocatalytic water purification and SERS-based detection process. This section will provide an in-depth understanding for the readers and encourage them to develop several plasmonic nanostructures for environmental applications.

2.1 Photodecomposition of organic pollutants in water

Plasmonic-TiO_2 nanohybrids have been proven the outstanding photocatalyst which can effectively decompose industrial pollutants [20–22]. Plasmonic-TiO_2 nanohybrids photocatalyst with efficient charge separation properties reveals the light-harvesting ability from UV to visible regions, which make them fascinating for photodecomposition reactions. Various research groups studied the modification of plasmonic nanoparticles over different morphologies of TiO_2 for photocatalytic applications [34–36]. The attachment of plasmonic nanoparticles (Ag, Au) with TiO_2 nanostructures is an effective way to improve their photodecomposition behavior. The existence of plasmonic nanoparticles on nanostructured TiO_2 forms the Schottky junctions interface and captures the photo-generated electrons, and decreases the rate of recombination in TiO_2. Moreover, due to the presence of plasmonic nanoparticles over TiO_2, their light-harvesting ability significantly extended [20, 21, 37]. Previous reports indicate the remarkable improvement in TiO_2 nanostructures with the modification of plasmonic nanoparticles such as Ag and Au [20–23, 34–37].

Uniformly interspersed Ag nanoparticles on 1-D TiO_2 nanostructures have been a favorite choice for their efficient SPR excitation ability, feasibility, electron mobility, and cost-effectiveness. Ag-dispersed TiO_2 nanorods [38], nanobelts [39], and nanowires [34] have been synthesized using different synthetic procedures for effective decomposition of a range of dyes such as Rhodamine B, Malachite Green, MB, R6G, etc. A contemporary study conducted by Singh et al. [40] postulates electron scavenging and LSPR of Ag as predominant factors responsible for the exceedingly high efficiency of their synthesized photocatalyst nanocomposites towards toxin degradation. Three different Ag-modified TiO_2 nanorods with increasing Ag loading, respectively, were synthesized using one pot wet chemical process. Their optical, structural, and morphological profiles were well characterized. Using sunlight (950 W/cm^2) as their illumination source, they degraded model organic dyes methyl orange and methylene blue. In another report, Bian et al. [41] informed the synthesis of Au-attached TiO_2 nanohybrids using the impregnation reaction and demonstrated photodegradation of chlorophenol, MB, and RhB 4 pollutant solution under Xe lamp exposure. In their photocatalysis studies, 10 μM of chlorophenol, MB, and RhB were decomposed in 4 hours, 20 minutes, and 2 minutes, respectively. Li et al. [42] reported the formation of Au-functioned TiO_2 nanostructures by a wet chemical process. Pyrolysis method-driven TiO_2 nanostructures were attached with plasmonic Au nanoparticles with various concentrations (2, 5, and 10 wt%). The prepared Au-TiO_2 nanohybrids (100 mg) exhibited photodegradation by diminished RhB (20 mg/L) molecules solution in 180 minutes under visible light exposure. Singh et al. [43] reported the Ag-TiO_2 thin films formation by combining the sol-gel method with the spin-coating method. **Figure 2(a-d)** show the morphological changes in modified Ag-TiO_2 thin films in comparison to the bare Ag-TiO_2 and. They tuned the photocatalytic profiles

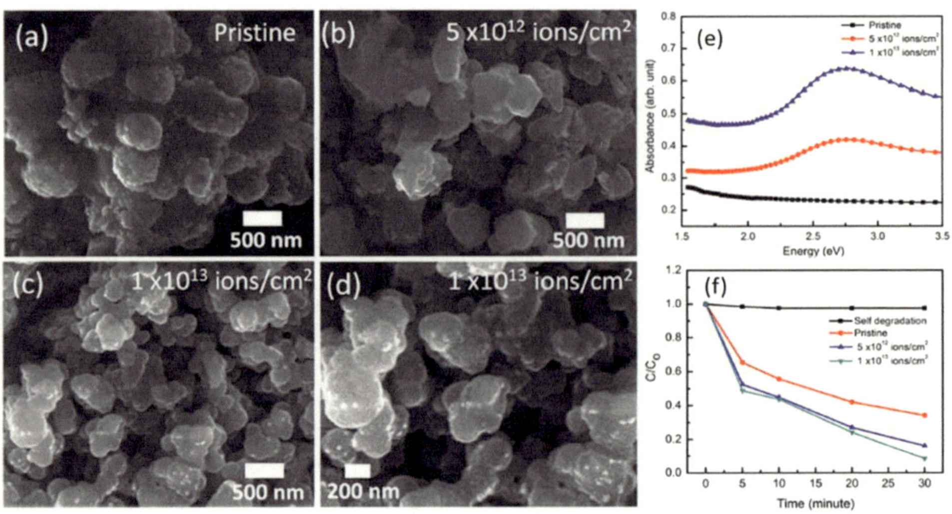

Figure 2.
SEM images of (a) bare Ag-TiO_2 thin films and (b-d) ion beam modified Ag-TiO_2 thin films, (e) UV-Visible spectra of bare and modified Ag-TiO_2 thin films, (f) photocatalytic activity of bare and modified Ag-TiO_2 thin films under sunlight. Reprinted from the reference [43].

of Ag-TiO_2 thin films by using the ion-beam irradiation technique. They modulate the optical properties effectively, and consequently, the light-harvesting behavior of the Ag-TiO_2 was largely affected. They have observed a significant shift in Ag SPR peak from 456 to 446 nm under different ion fluence values (**Figure 2(e)**). In their photocatalytic activity, they have demonstrated the breakdown of 5 μM MB pollutant molecule solution. They have shown that the most efficient Ag-TiO_2 thin film decomposed the MB solution in 30 minutes of sunlight exposure (**Figure 2(f)**).

Wang et al. [44] successfully prepared Au-TiO_2 nanohybrids with the joint effort of the sol-gel method with the solvothermal method. In their photocatalytic studies, 500 mg/L of the fabricated photocatalyst was employed for the decomposition of 50 mg/L MB solution under the exposure of a Hg lamp (500 W). In their studies, they have highlighted that Au-modified porous TiO_2 nanostructures exhibited superior photodegradation nature as compared to bare porous TiO_2. The superior photodecomposition nature of the porous Au-TiO_2 nanostructures could be assigned to the effective charge separation and improved light harvesting properties owing to the SPR effect in plasmonic nanoparticles. Recently, Singh et al. [20] reported that Au nanoparticles encapsulated TiO_2 nanospheres by using the chemical precipitation method. They have engineered the size of Au nanoparticles by varying the concentration of Au precursors during TiO_2 growth. **Figure 3(a-e)** shows the morphology and elemental mapping of Au-encapsulated TiO_2 spheres. With the variations in the size of Au nanoparticles, their optical properties improved significantly. Modified Au-TiO_2 nanohybrids absorb the entire visible region (500–800 nm) and majorly contributed to increased photocatalytic activity (**Figure 3(f)**). The quenching in the rate of recombination with the variations in the size of Au nanoparticles is further confirmed by the reduction in the intensity of photoluminescence spectroscopy.

They have highlighted that the outstanding photocatalytic behavior of the Au-TiO_2 nanohybrids can be assigned to the formation of the Schottky junction interface among the Au nanoparticles and TiO_2. Owing to the creation of the Schottky junction, the density of electrons in the conduction band significantly improves the photodecomposition reaction and enhances the degradation rate of the pollutants solution. In their photocatalytic studies, 10 μM of solution three industrial well-known pollutant molecules solution (MB, MG, and MO) were decomposed in 30, 20, and 30 minutes of sunlight exposure, respectively, by using 5 mg of the photocatalyst. Singh et al. [45] prepared the Au-TiO_2 thin films by combining the spin coating method with the thermal evaporation method and employed them for the decomposition of a 5 μM solution of MB under natural sunlight exposure. They have demonstrated that 30 minutes of sunlight exposure can decompose a significant amount of pollutants.

Enhanced photocatalytic activity of Au-TiO_2 nanohybrids thin film can chiefly correspond to different electronic transfers occurring at the composite surface (**Figure 4**). When Au-TiO_2 photocatalyst with adsorbed MB molecules is illuminated by a light source of appropriate wavelength, two major electron pathways get activated. Primarily, upon excitation, electrons from the valence band of TiO_2 gain energy and subsequently get transferred to its conduction band, leaving a hole behind and thus generating a pair of photo-induced charge carriers. The presence of Au nanoparticles in the vicinity helps the composite system by taking up the photo-induced electrons from the conduction band of TiO_2, thereby suppressing recombination and felicitating catalytic reactions at the surface by acting as an electron sink. Secondly, Au nanoparticles upon light interaction, exhibit surface plasmon resonance (SPR), driving the generation of increased electron-hole pairs in TiO_2 on account of intense electric fields localized around these plasmonic nanostructures close to the

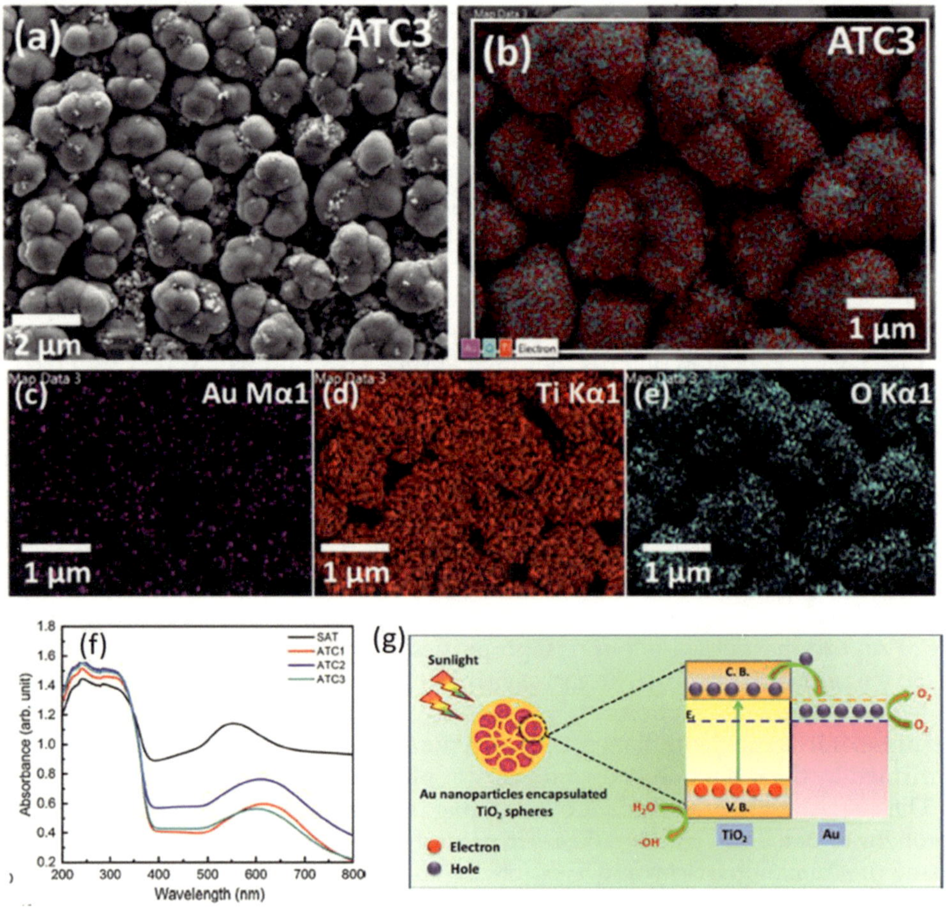

Figure 3.
FESEM images of (a) Au-TiO_2 nanohybrids revealing the Au encapsulated TiO_2 spheres, (b) combined mapped FESEM image of Au-TiO_2 nanohybrids, elemental mapping showing the individual distribution of (c) Au, (d) Ti and (e) O, (f) UV-DRS curves for synthesized Au-TiO_2 nanohybrids with tuned optical properties, (g) Scheme for enhanced photodecomposition of Au-TiO_2 nanohybrids. Reprinted from the reference [20].

TiO_2 surface. This phenomenon is referred to as Plasmon Resonance Energy Transfer (PRET) [45]. These plasmonic nanoparticles have been reported to serve as "hot spots" allowing increased electron-hole generation and improved photocatalytic reactions [45]. Though Au-TiO_2 electron exchange is an important step towards improved catalytic behavior of the nanocomposite system, it is not the key rate-determining step for the reaction. As a consequence of Schottky junction formation, electronic flow is maintained from TiO_2 to Au until equilibrium is attained. [45] Apart from Au, the adsorbed pollutant molecules also contribute to increased photocatalytic activity by absorbing light leading to electron transfer into the TiO_2 conduction band. The photo-generated electrons produced through different pathways react with oxygen molecules at the surface, converting them into superoxide radicals ($•O_2^-$), while the holes in the valance band react with water molecules generating hydroxyl radicals (•OH). These unsaturated highly active radicals transform the pollutant molecules into their degradation products [20, 21].

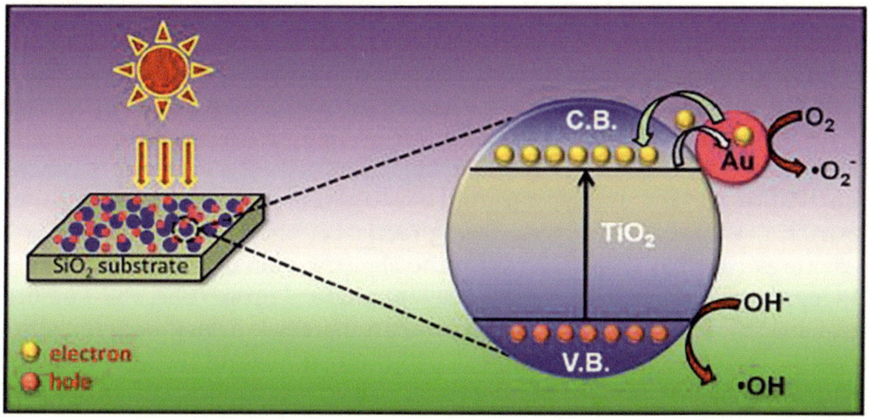

Figure 4.
Schematic picture revealing the efficient charge transfer among Au and TiO_2 for enhanced photodecomposition. Reprinted from the reference [45].

From the above discussion and excellent studies, it can be concluded that the tremendous properties of plasmonic-TiO_2 nanohybrids have significantly improved the photodecomposition process by providing additional light adsorption and reduction in the recombination rate.

2.2 Detection of pollutants in water

The contamination of pollutant molecules in the form of organic dyes, pesticides, pharmaceuticals, and biomolecules requires proper monitoring due to the serious impacts on human health, aquatic life, and the terrestrial environment [46]. As a fingerprint recognition technique, Raman spectroscopy reveals deep chemical and structural insights such as chemical bonding and molecular interactions and thus enables the identification of unknown molecular species [47]. The Raman spectral features consist of bands corresponding to vibrational or rotational transitions; however, its direct applicability is hindered due to inherently weak Raman scattering. Surface-enhanced Raman spectroscopy (SERS) has proven its uniqueness as a reliable spectroscopic technique for the direct identification of bio-molecules and chemicals owing to its selectivity and sensitivity for the sensing of ultralow concentrations of analyte molecules [48]. Nowadays, As a sensitive, specific, non-destructive, and label-free identification technique, SERS has shown enormous potential in several fields inducing medical sciences, life sciences, and analytical chemistry for the targeted identification of chemical and biological hazardous trace species. Primarily, the overall signal enhancement through SERS is governed by electromagnetic as well as chemical enhancement mechanisms. The modified SERS spectral features indicate the adsorption or chemical interaction of analytes with suitable SERS substrates and, thus, imply the involvement of charge transfer-initiated chemical enhancement along with the predominated electromagnetic enhancement. The formation of the charge-transfer complex is critically dependent on the atomic-scale substrate's surface properties along with the surface density of the adsorbates; therefore, the chemical state and topography of the surface strongly determine the overall SERS enhancement [49, 50].

Initially, a wide range of isotropic and structurally modified anisotropic noble metallic nanostructures such as nanorods, nanotriangles, nanowires, pyramids, nanostars (NSs), nanocubes, nanodendrites (NDs), nanoflowers (NFs), and core-shell NPs in the form of nano-crystalline aggregates and roughened surfaces were utilized for the ultra-sensitive sensing of different analytes. Along with efficient signal enhancement capabilities, the stability, reproducibility, cost-effectiveness, biocompatibility, and spectral background of SERS substrates are the vital parameters restricting the reliable and long-term practical applicability of plasmonic nanostructures-based SERS platforms for the identification of analyte molecules [51]. Owing to their high surface area, excellent physical properties, and flexible and low-cost fabrication peculiarities, semiconductor materials have shown the potential to address the issues with noble plasmonic SERS substrates and recently emerged as promising platforms for SERS-based molecular detection applications utilizing the chemical enhancement and the associated charge-transfer between the analyte and semiconductor's surface. The signal enhancement produced by semiconductor SERS (SC-SERS) substrates relies on the chemical enhancement mechanism and depicts lesser enhancement than traditional plasmonic materials [52]. However, semiconductor SERS substrates recently gained significant research attention as the prospective potential candidates with improved SERS performance. The high sensitivity and resolution of the SERS technique enable the study of the chemical and physical nature of the interaction among molecule and semiconductor nanostructures, as well as the properties of semiconductor nanomaterials. Direct proximity among the adsorbate molecule and SERS substrate initiates the charge transfer (CT) between them and contributes to some distinct spectral features [53, 54]. The varied adsorption direction of the adsorbate molecules on the SERS substrate could produce distinctly resolved spectral features, which in turn provide significant insights into charge-transfer-induced chemical enhancement mechanisms.

Interestingly, the synthesis simplicity, high stability, chemical inertness, non-toxicity, and corresponding bio-compatibility of nano-structures titanium (IV) oxide (TiO_2) make him a potential candidate for stable SERS signal requirements along with the excellent SERS response. A rich variety of novel TiO_2 NPs-based SERS substrates have been fabricated through varying the nanostructure's morphology, including spherical nanoparticles (NPs), nanowires [NWs], and nanorods [NRs] employing the various fabrication strategies such as sol-gel, hydrothermal, and solvothermal methods [55]. Especially, the three-dimensional (3D) architectures such as nanoflowers [NFs] and porous nanomaterials provide the supply larger surface area for the adsorption of analyte molecules and further support the chemical enhancement in SERS. The limited Raman intensity enhancements from semiconductor TiO_2 substrates can be improved by optimizing the micro/nano-structured morphology or more efficiently through incorporating plasmonic nanomaterials. Musumeci and coworkers reported the strong SERS signal enhancements of adsorbed enediol molecules over the pure TiO_2 NPs surface through the essential formation of the charge-transfer (CT) complex of TiO_2 with the molecule. The enhancement was attributed to the increased surface atoms and the consecutive adsorbed analytes molecules as well as borrowing the intensity from the allowed continuum state transitions (**Figure 5**) [56].

Anatase and rutile phases of TiO_2 NPs-based SERS substrates were prepared by Yang and coworkers utilizing the sol-hydrothermal method and explored for the TiO_2 phase-dependent study towards the sensitive identification of 4-mercaptobenzoic acid probe molecules significantly contributed by the TiO_2-to-molecule charge transfer mechanism [57].

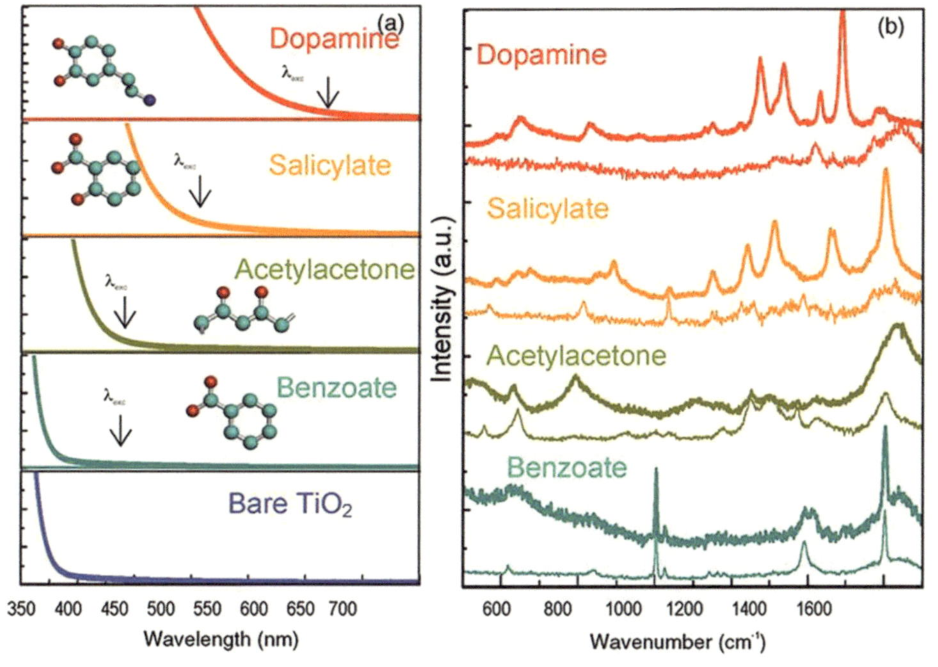

Figure 5.
(a) Absorption, and (b) normal Raman (0.1 M ligands) and SERS spectra of 5 nm pure TiO_2 NPs (1×10^{-4} M NP colloidal solution) after the modification with different (4×10^{-2} M) ligands. In figure λ_{exc} represents the laser excitation wavelength. Reprinted from the reference [56].

As shown in **Figure 6**, the charge-transfer initiated SERS activity of analyte molecule adsorbed TiO_2 NPs is dominated by the three different kinds of transition (i) the excitation of electrons from the valence band (VB) to the conduction band (CB) of TiO_2 and their subsequent transition to LUMO of the adsorbed molecules with excitation larger than the TiO_2 band gap energy (~3.2ev), (ii) the excitation of electrons from HOMO to LUMO of adsorbed molecules followed by the injection into TiO_2 CB, and (iii) the excitation of electrons from TiO_2 VB to surface state energy levels and their transfer to the LUMO of the adsorbed molecules [58]. The abundance availability of TiO_2 surface defects or surface oxygen vacancy defects can bind the electrons and result in the formation of surface state energy levels. Lin et al. reported

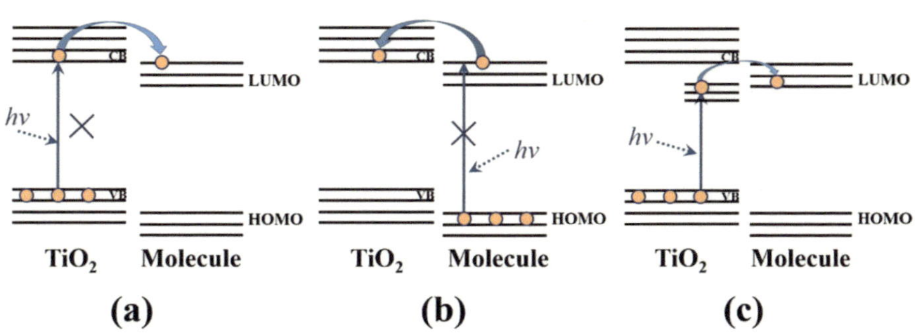

Figure 6.
Different modes representing the charge transfer between the TiO_2 and adsorbed analyte molecule. Reprinted from the reference [58].

the ultrahigh SERS activity with EF ~10^5 for the 4-NBT probe molecule utilizing photo-induced charge transfer employing crystal-amorphous core-shell black TiO_2 NPs by a controllable solid-state synthesis method [59]. Electrochemical anodization of titanium substrates was employed for the fabrication of TiO_2 Nanotube Arrays for the improved detection of cytochrome C with EFs ~10^4 and LOD down to the ~10^{-7} M concentration [59]. Notably, the crystalline femtosecond laser generated TiO_2 nanofibers in the form of three-dimensional chained nanoparticles provided the resultant EFs of ~1.3×10^6 of the CV dye molecules, which is comparable enhancement produced by noble metal-based substrates. This significant contribution to SERS is assigned to the fabrication of 3D nanonetworks, including the combined contribution of nanogap, nanocluster, and plasmonic hybridization [60].

Nowadays, the noble metal-semiconductor nanocomposites have attracted significant research attention as noble SERS substrates utilizing the synergistic interaction between the noble metal and semiconductor surfaces. The relatively poor SERS activity of pure TiO_2 can be significantly improved by incorporating plasmonic materials with TiO_2 substrates through the composite formation. Notably, noble metal NPs can be simply incorporated on the TiO_2 surface through the photocatalytic process and therefore offers the ease of fabrication of noble metal-TiO_2 hybrid nanocomposites [61]. The SERS activity of semiconductor substrates is dominantly assigned to metal-molecule charge-transfer contribution and the correspondingly involved chemical enhancement mechanism results in the weak SERS activity of pure TiO_2 NPs-based substrates. The considerably increased photophysical characteristics can be achieved through the attachment of metal nanoparticles with TiO_2, through moving the titania's Fermi level to more negative potentials and inhibiting the electron-hole pair recombination [62].

Initially, some studies reported the fabrication and SERS performance of hybrid Ag/TiO_2 and Au/TiO_2 composites [55]. The excitation of SPR of AgNPs produces enhanced electromagnetic fields at the AgNPs surface and, in turn, the generation of hot electrons at the metal-semiconductor interface. The correspondingly formed Schottky junction promotes the transfer of generated hot-electrons towards the conduction band of TiO_2 from the metal-semiconductor interface. Thus, the SERS activity of Ag/TiO_2 hybrid substrates is mainly governed by the three contributions, namely, the plasmonic NPs, the transition of electrons between the highest occupied molecular orbital (HOMO) energy level, and the lowest unoccupied molecular orbital (LUMO) energy level of analyte molecule and initiation of charge transfer between the Fermi level of Ag nanoparticle and HOMO level of the analyte molecule, as explained by Zhao et al. for adsorbed R6G molecules over surface of Ag/TiO_2 nanowire substrates [63].

Fu at el. employed the modified photocatalytic method for the AgNPs deposited TiO_2 films for the detection of rhodamine 6G dye molecules [64]. However, the fast photocatalytic process results in the inhomogeneous synthesis of AgNPs under UV exposure and so restricts the applicability of the photocatalytic method for reproducibly uniform deposition of AgNPs over TiO_2 surface. Especially, Song and coworkers demonstrated the applicability of metal composite nanofibers by detecting 4-MPy molecules adsorbed over the AgNPs-coated TiO_2 nanofibers by an enhancement factor of ~10^5 [65]. In another report, Wang and coworkers fabricated the in-situ AgNPs-deposited well-ordered TiO_2 nanosheets utilizing hydrothermal synthesis exhibiting excellently uniform and renewable SERs activity with EF ~10^8 for 4-mercaptobenzoic acid [66]. Yang et al. utilized the wet chemical approach for the preparation of Ag/ TiO_2 core-shell NPs and reported higher sensitivity as compared to Ag cores through synergistic SERS enhancement. The prepared ultra-thin coated Ag/TiO2 core-shell NPs were employed for the 4-MBA and tetramethyl thiuram disulfide (TMTD) insecticide with significantly lowered detection limits of 10^{-11} M and 10^{-10} M concentrations, respectively [67]. Notably, Lu et al. synthesized the hierarchical micro/nanostructured TiO_2/Ag architectures utilizing a combined approach of femtosecond laser structuring and hydrothermal treatment. The prepared SERS substrates demonstrated four-fold enhancement compared to the hydrothermally fabricated bare Ti surface along with LODs down to ~10^{-14} M R6G concentration [68]. In addition, the applicability of Ag/TiO_2 hybrid SERS substrates was also explored towards the detection of various biomolecules such as antibiotics, circulating tumor cells (CTC), uracil DNA glycolase (UDG), and bacteria E. coli and S. aureus [69–72]. Recently, photo-catalytically synthesized TiO_2@Ag nanostructured SERS bioprobe substrates were developed by Xu et al. targeted towards the ultralow (~10^{-14} M) concentration detection of R6G molecules and further utilized to specifically recognize CTC with LOD down to 1 cell per mL (**Figure 7**) [73].

Therefore, the recent SERS studies unveiled the amplified SERS performance of hybrid Ag/TiO_2 substrates comparable to noble plasmonic NPs based substrates with detectability down to picomolar and femtomolar concentration range.

Furthermore, TiO_2 substrates coated with Au nanohybrid structures were widely utilized to obtain higher sensitivity, stability, and superior compatibility for detecting biological molecules. The incorporation of Au with TiO_2 substrates improves the biocompatibility of the substrates along with the promoted stability of the SERS signal due to the non-oxidizing properties of Au. For instance, Jiang et al. proposed Au-deposited TiO_2 (Au-TiO_2) nanocomposites through photocatalytic reduction of $HAuCl_4$ on TiO_2 NPs [74]. The prepared Au-TiO_2 nanocomposites produced recyclable and sensitive SERS detection of 4-MBA molecules with the lowest detectable concentration of ~10^{-8} M, which benefitted from the synergistic contribution of metals and semiconductors. The employment of nanometer-sized inter-particle gaps with densely packed AuNPs deposited on vertically aligned large-area TiO_2 nanosheets (NSs) reported superior sensitivity along with the ultralow detectability of ~10^{-14} M crystal violet concentration (**Figure 8**) [75].

Interestingly, Singh et al. employed the combined thermal deposition and spin coating procedure with the improved optical response and targeted towards efficiently sensitive and simultaneous identification of two organic pollutants rhodamine

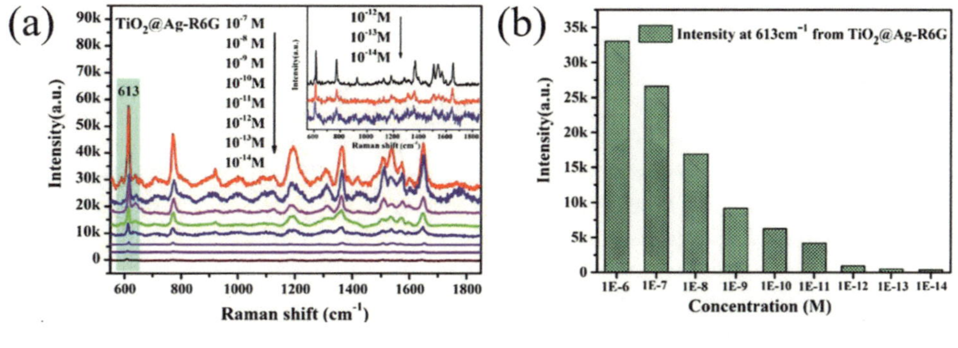

Figure 7.
(a) SERS signal of different concentration of R6G molecules adsorbed over TiO_2@Ag nanostructure. The inset depicts the SERS spectra at low R6G concentrations (10^{-12}, 10^{-13}, and 10^{-14} M), and (b) SERS intensity at ~613 cm^{-1} R6G peak adsorbed over TiO_2@Ag–R6G SERS system at different R6G concentrations. Reprinted from the reference [73].

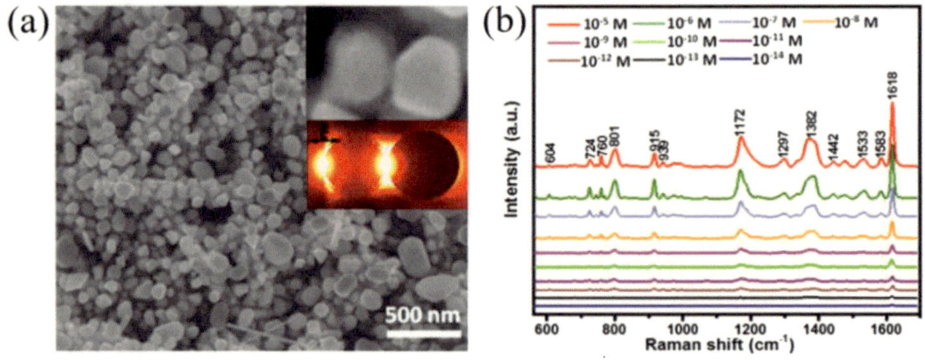

Figure 8.
(a) SEM image depicts the TiO_2 NS/Au NPs@SiO_2/Au NPs sample, and the inset shows a typical SEM image, and (b) SERS spectra of CV molecules with TiO_2 NS/Au NPs@SiO_2/Au NPs-based substrates. Reprinted from the reference [75].

6G (R6G) and methylene blue (MB), exhibiting the Raman intensity enhancement factors of the order of ~10^7 along with excellent multiplexed detection capabilities [45]. Photo-reduced growth of AuNPs over TiO_2 (TiO_2/AuNWAs) nanowire arrays used as recyclable and sensitive SERS platform for detecting ~10^{-9} M concentration of R6G molecules. In addition, the comparative SERS study of bare TiO_2, AuNPs, and TiO_2/Au NWAs reveal remarkably improved SERS performance of TiO_2/Au NWAs substrates [76]. Recently, AuNPs decorated ordered porous TiO_2 thin films based SERS sensors were optimized by Liang and coworkers for detection of rhodamine 6G probe molecules [77]. The improved bio-compatibility of Au/TiO_2 nanostructures offers an additional degree of freedom for detecting various biomolecules with high sensitivity. The highly sensitive detection of adenine biomolecules was performed by Jiang and coworkers utilizing Au/TiO_2/Au nanosheet substrates through sputtered deposition of AuNPs with a detection limit of ~10^{-7} M concentration [78].

Apart from monometallic NPs deposited TiO_2 SERS substrates, bimetallic Au/Ag nanoparticles (NPs) decorated well-aligned TiO_2 nanorod arrays (NRAs) with strong absorption from 400 to 1300 nm spectral region generated the superior NIR-SERS activity for antibiotic chloramphenicol and ciprofloxacin in real-world water samples with nano-molar detection limits [79]. Additionally, Borges et al. demonstrated the incorporation of bimetallic Ag-Au NPs with TiO_2 and the strong LSPR band and correspondingly improved SERS activity [80]. Along with noble (Ag and Au) metals, some recent SERS studies also focused on the implementation of poor plasmonic materials for the development of hybrid TiO_2 substrates. To prove the applicability of non-plasmonic materials, the uniformly deposited PtNPs onto the vertically aligned TiO_2 nanorod arrays (Pt@TiO_2 NTAs) for the detection of R6G through effective prevention of electron-hole pairs recombination [81]. In another study by Jiang et al., the SERS activity of Fe_2O_3@TiO_2 hybrid substrates was investigated employing exosomal miRNAs via integrated miRNA-triggered hot-spot SERS and Fe_3O_4@TiO_2-based exosome accumulation [82].

In summary, the excellent physical, chemical, and biological properties of metal-oxide TiO_2 semiconductor enables broad SERS detection applications through facile incorporation of plasmonic materials utilizing synergistic interaction between TiO_2 and metallic NPs. The SERS substrate design, material composition, and synthesis strategy strongly govern the stability, reproducibility, and compatibility with biomolecules and subsequently offers a new perspective of SERS substrate development benefitting from the charge-transfer induced chemical enhancement SERS mechanism.

3. Energy applications of plasmonic-TiO_2 nanohybrid

In this section plasmonic-TiO_2 nanostructure-driven energy production applications have been elucidated in detail. This section includes the applicability of plasmonic-TiO_2 nanohybrids for photovoltaics applications. This section explains the basic mechanism responsible for the tremendous usage of plasmonic-TiO_2 nanohybrid for photovoltaic applications. This section will expect to enhance knowledge and understanding of plasmonic nanostructures-based solar cells.

As we have discussed in the previous sections, the plasmonic nanoparticles can significantly control the charge kinetics in TiO_2 nanostructures and can additionally provide visible light absorption ability. By using these outstanding properties, different research groups reported the improved performance of plasmonic nanohybrid-based dye-sensitized solar cells (DSSC) [83–85]. After the discovery of TiO_2-based DSSC in 1991 by O'Regan and Grätzel the development of DSSC has been made using other semiconductors [86]. DSSC known to be the third generation of photovoltaics has similar working as we observed in the photosynthesis process. Photoanode in plasmonic nanohybrids based DSSC formed by the layer of plasmonic-TiO_2 nanohybrid which is further covered by the light-sensitive commercially available dye molecules such as black dye, which help to harvest solar light. Firstly, a light-sensitive dye molecule starts the conversion of light energy into electric energy by using a plasmonic-TiO_2 photoanode, electrolyte, and counter electrode. The basic process in Plasmonic-TiO_2 based DSSCs has been depicted in **Figure 9** below.

Dissanayake et al. [87] used Ag and Au-decorated TiO_2 nanostructures via chemical reduction methods. The fabricated Ag-TiO_2 and Au-TiO_2 nanohybrids were employed to design the photoanodes of DSSCs. In their study, they highlighted that the obtained PCE values for pure TiO_2, Ag-TiO_2, and Au-TiO_2-based photoanodes were 5.12%, 6.51%, and 6.23%, respectively. In addition, they have concluded that

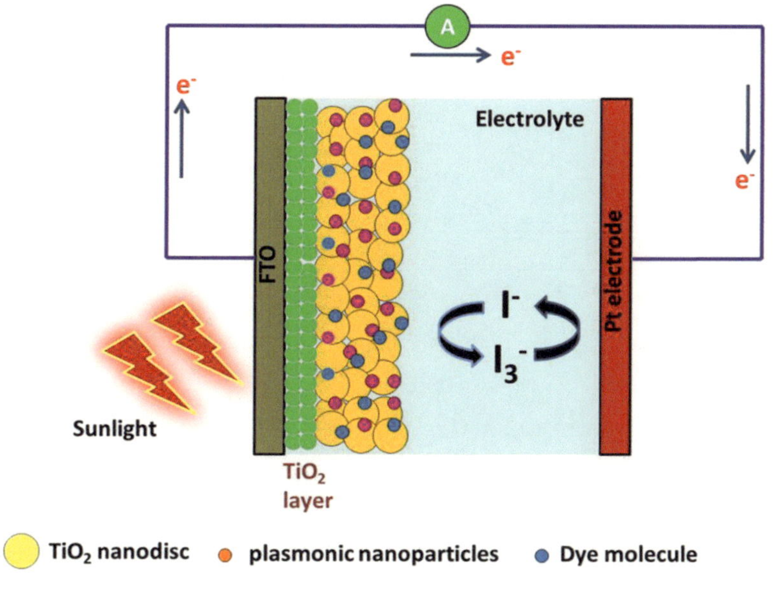

Figure 9.
Schematic diagram revealing the basic process of plasmonic-TiO_2 based photoanodes of DSSC.

Ag-functionalized TiO_2 plasmonic nanohybrids exhibited higher PCE values compared to pure TiO_2 and Au-functionalized TiO_2 nanohybrids.

Vaghasiya et al. [88] fabricated Ag-TiO_2 hybrid thin films and used them as photoanodes in DSSCs. The Ag-functioned TiO_2 nanohybrids contained thin films were found to result in superior PCE than bare TiO_2-based photoanodes. The computed values of J_{sc}, V_{oc}, FF, and PCE for Ag-TiO_2 thin films were found to be 5.7 mA/cm^2, 0.621 V, 54%, and 1.9%, respectively. The estimated J_{sc}, V_{oc}, FF, and PCE for pure TiO_2 thin films were 4 mA/cm^2, 0.507 V, 46%, and 1.1%, respectively.

Guo et al. [89] prepared core-shell of Ag@TiO_2 nanostructures and employed them to design DSSCs. They have modulated the Ag content for enhanced photocurrent density. In their photovoltaic studies, they have highlighted that the optimum concentration of Ag (0.15 wt %) provides the maximum photo conversion efficiency (PCE = 5.33%) in comparison to bare TiO_2-based DSSC (PCE = 3.96%).

Lim et al. [90] successfully prepared Ag-functionalized N-TiO_2 nanohybrids by using thermal annealing combined with the chemical method. The obtained nanohybrids were embraced to fabricate the photoanode of DSSC. They varied the Ag loading by varying the concentration from 2.5 to 20 wt% over the surface of N-TiO_2 and obtained a significant value of PCE of 8.15%. **Figure 10(a-b)** showing the surface morphology of the Ag-TiO_2 nanohybrids while their elemental mapping is presented in **Figure 10(c-d)**. The improvement in the charge separation due to the attachment of Ag nanoparticles over TiO_2 is assured by PL spectroscopy (**Figure 10(e)**). They have highlighted that the PCE value is higher for the N-TiO_2-based photoanodes as compared to the PCE of photoanodes based upon pure TiO_2 (2.19%), N-TiO_2 (2.93%) and Ag-TiO_2 (4.86%) (**Figure 10(f)**). In their study, they showed that increments in the Ag content significantly improve the PCE of DSSC.

In another study, Wang et al. [91] prepared Ag nanoparticles functionalized TiO_2 nanostructures via a hydrothermal process. They have used Ag nanoparticles to functionalize feather-like TiO_2 nanostructures. They have varied the Ag content over the

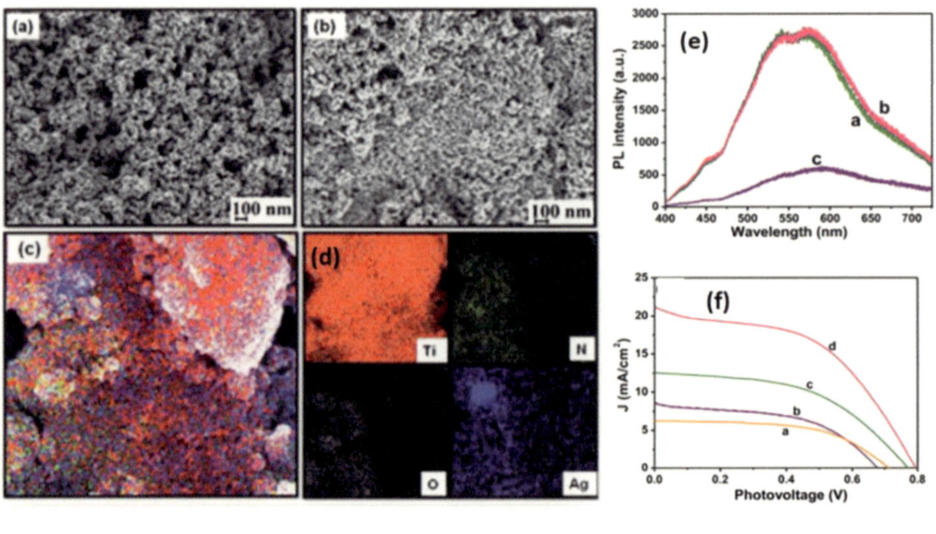

Figure 10.
(a-b) SEM images of Ag functionalized N-TiO_2 nanostructures, (c-d) Elemental mapping of Ag functionalized N-TiO_2 nanostructures showing the existence of Ti, O, N, and Ag atoms, (e) Photoluminescence spectra of N-TiO_2 and Ag decorated N-TiO_2, (f) Current density characteristic of fabricated DSSC using TiO_2, N-TiO_2, and Ag decorated N-TiO_2 nanostructures. Reprinted from the reference [90].

surface of TiO_2 varying from 1 to 7.5 wt %. They concluded that the optimum amount of Ag content on TiO_2 can remarkably enhance the PCE value of Ag-TiO_2-based photoanode. With the modulation of Ag loading on the TiO_2 nanostructures, the obtained photo conversion efficiency increased from 6.19% to 6.74%. Bhullar et al. [92] studied the effect of Ag implantation in TiO_2 thin films for DSSCs application. They have doped the Ag ions by varying the fluence value from 10^{13} to 10^{16} per cm^2. They have highlighted that due to the incorporation of Ag ion, the optical absorption is significantly improved and also controls the recombination rate, which significantly contributed to enhanced the PCE of DSSCs. They demonstrated that at a fluence value of 10^{14}, the Ag-doped photoanode significantly showed 21.82% better efficiency as compared to the bare TiO_2-based photoanode of DSSCs.

Muduli et al. [93] informed the formation of Au-TiO_2 nanohybrid by using the hydrothermal process. The prepared Au nanoparticles functionalized TiO_2 nanostructures were embraced for the fabrication of photoanodes of DSSCs. They observed the superior PCE for Au-TiO_2-based photoanodes as compared to the pristine TiO_2-based photoanode. They have concluded that due to the Schottky junction creation among Au and TiO_2 nanostructures, the quenching in the recombination rate takes place, and consequently, the enhancement in the PCE of Au-TiO_2-based photoanodes takes place as compared to bare TiO_2. Alamu et al. [94] prepared Ag-decorated TiO_2 nanohybrids and applied them for the fabrication of photoanodes of DSSCs. They have used natural plant extract of *Azadirachta indica* and *Lawsonia inermis* with the commercially available dye N719. They have revealed that the modification of Ag nanoparticles over TiO_2 particles enables effective band gap narrowing and improvement in the optical absorption in the visible region. In their study, they concluded that Ag-TiO_2-based photoanode with the usage of natural dyes exhibited tremendous photoconversion efficiency. The combined contribution of plasmonic nanoparticles with natural dye sensitizer effectively enhances the PCE for bare TiO_2-based photoanode as compared to bare TiO_2 nanoparticles based photoanodes.

Ran et al. [95] decorated the TiO_2 nanostructures with Ag, Au nanoparticles and Ag, Au nanowires for application in DSSCs. They have highlighted that the improved PCE could correspond to the high mobility of plasmonic nanostructures and the SPR effect of Ag and Au nanostructures. Plasmonic nanoparticles functionalized TiO_2-based photoanodes attained a higher PCE value as compared to the pristine TiO_2-based photoanodes. Under sun simulator exposure, a PCE of 5.74% was found in the Ag nanowires decorated TiO_2. The obtained efficiency of Ag-TiO_2-based photoanode attained a 25.3% improvement in comparison to photoanodes prepared by using pure TiO_2 film (4.58%). Improved electron mobility properties of Ag nanowires with enhanced optical absorption majorly contributed and enhanced the PCE of photoanodes. In addition, Ag nanowire also provides an enhancement in the light scattering, which is also favorable to improve the PCE of DSSCs.

Nbelayim et al. [96] reported the preparation of Ag@TiO_2 core-shell with variations in the Ag content from 0.1 to 1% by wet chemical synthesis process. The synthesized Ag@TiO_2 with varied Ag loading was used to fabricate the photoanode of DSSCs. In their study, they optimized the Ag doping concentration for the enhanced PCE of the DSSCs and concluded that 1% of Ag doping showed the maximum PCE. They have concluded that 1% Ag doping provides the optimum band alignment for injection from the Ag nanoparticles to TiO_2. In addition, the optimum concentration of Ag nanoparticles can effectively control the recombination rate in TiO_2, which majorly contributed to enhancing the PCE of the fabricated DSSCs. The PCE value of the Ag@TiO_2 and Ag-doped TiO_2 was found to be significantly enhanced as compared to the pristine sample. Song et al. [97] informed the fabrication of different morphology of Ag nanostructures (spherical and multi-shaped) and Au nanoparticles and explored their effect on the PCE of the DSSCs. In order to fabricate photo anodes of DSSCs, individual Ag, Au nanoparticles, multi-shaped Ag, Au, and their mixture were employed. Each plasmonic nanostructure and its combination were mixed with the mesoporous TiO_2 and were employed to fabricate photo anodes of DSSCs. Among all samples, a photoanode containing a multishape of Ag and Au with TiO_2 was found to be the most efficient for the photovoltaics performance. The enhanced PCE performance of multishaped Ag and Au nanostructures can be assigned to the wider optical absorption as compared to the spherical Ag and Au nanoparticles.

Villanueva-Cab et al. [98] prepared Au-decorated TiO_2-based DSSCs and adjusted the concentration of Au nanoparticles on TiO_2 for efficiently enhanced PCE. In their study, they have highlighted several parameters, such as charge collection efficiency, light absorption efficiency, and charge injection efficiency of dye majorly influenced the PCE of DSSC. Moreover, they showed that the existence of various Au content on the TiO_2 surface majorly influences the collection efficiency and consequently, PCE of DSSC varied.

We have successfully elaborated the importance of various parameters which majorly affect the photo-conversion efficiency of plasmonic nanostructures-based DSSCs. Plasmonic nanohybrids with their fascinating optical and electronic properties largely increase the PCE of the DSSCs. Plasmonic nanoparticles with TiO_2 tune the band alignment and significantly control the charge injection properties. Apart from this, plasmonic nanostructures over the TiO_2 surface also quench the rate of recombination and increase the lifetime of the charge carriers. Moreover, plasmonic nanoparticles also enhance the charge conduction properties of photoanodes which is also beneficial for the improvement in the PCE value of DSSCs. Thus this chapter explains the preparation and characterization of plasmonic-TiO_2 nanohybrids and their usage for water purification, SERS-based detection, and photovoltaic applications. This chapter provides a wide overview related to the preparation and employment of plasmonic nanostructures for different environmental and energy applications. This chapter also includes the importance of specific parameters of plasmonic-TiO_2 nanohybrids, which largely influence their performance for different energy and environmental applications and thus provide a better understanding to the readers.

4. Conclusion

In brief, plasmonic nanohybrids have the tremendous ability to show remarkable performance for environmental sensing and energy harvesting applications. Noble metal nanoparticles functionalized TiO_2 nanostructures-based studies provided significant evidence for their capability to address the environmental detoxification and energy crises problems. This chapter provides the recent trends, development, and applicability of plasmonic-TiO_2 nanohybrids for energy and environmental applications. This chapter includes fundamental aspects of the plasmonic-TiO_2 nanohybrid designing and brief details for the mechanism for SERS-based detection, environmental remediation, and solar cell applications. The fascinating properties of plasmonic nanohybrids include tunable enhanced optical absorption and efficient charge separation properties which open up several new opportunities for the researcher to engineer them for targeted new applications. In this chapter, we have presented the roadmap of each application by using the plasmonic-TiO_2 nanohybrids which can provide a better understanding for the readers to develop the plasmonic nanohybrids for particular applications. Certainly, the research field dealing with plasmonic nanohybrids is growing rapidly and in the future, properties, and applications will be explored for the plasmonic-TiO_2 nanohybrids.

Author details

Jaspal Singh[1*] and Ashwani Kumar Verma[2]

1 Sungkyunkwan University, Suwon, South Korea

2 Indian Institute of Technology Delhi, New Delhi, India

*Address all correspondence to: jaspal2125@gmail.com; jaspal.2125@skku.edu

References

[1] Singh J, Khan SA, Shah J, Kotnala RK, Mohapatra S. Nanostructured TiO_2 thin films prepared by RF magnetron sputtering for photocatalytic applications. Applied Surface Science. 2017;**422**:953-961

[2] Fujishima A, Zhang X, Tryk DA. TiO_2 photocatalysis and related surface phenomena. Surface Science Reports. 2008;**63**:515-582

[3] Singh J, Kumar S, Manna AK, Soni RK. Fabrication of ZnO–TiO_2 nanohybrids for rapid sunlight driven photodegradation of textile dyes and antibiotic residue molecules. Optical Materials. 2020;**107**:110138

[4] Singh J, Soni RK. Controlled synthesis of CuO decorated defect enriched ZnO nanoflakes for improved sunlight-induced photocatalytic degradation of organic pollutants. Applied Surface Science. 2020;**521**:146420

[5] Singh J, Palsaniya S, Soni RK. Mesoporous dark brown TiO_2 spheres for pollutant removal and energy storage applications. Applied Surface Science. 2020;**527**:146796

[6] Singh J, Juneja S, Palsaniya S, Manna AK, Soni RK, Bhattacharya J. Evidence of oxygen defects mediated enhanced photocatalytic and antibacterial performance of ZnO nanorods. Colloids and Surfaces B: Biointerfaces. 2019;**184**:110541

[7] Singh J, Manna AK, Soni RK. Sunlight driven photocatalysis and non-enzymatic glucose sensing performance of cubic structured CuO thin films. Applied Surface Science. 2020;**530**:147258

[8] Singh J, Juneja S, Soni RK, Bhattacharya J. Sunlight mediated enhanced photocatalytic activity of TiO_2 nanoparticles functionalized CuO-Cu_2O nanorods for removal of methylene blue and oxytetracycline hydrochloride. Journal of Colloid and Interface Science. 2021;**590**:60-71

[9] Hoffmann MR, Martin ST, Choi W, Bahnemann DW. Environmental applications of semiconductor photocatalysis. Chemical Reviews. 1995;**95**:69-96

[10] Hu X, Li G, Yu JC. Design, fabrication, and modification of nanostructured semiconductor materials for environmental and energy applications. Langmuir. 2010;**26**:3031-3039

[11] Singh J, Soni RK. Two-dimensional MoS_2 nanosheet-modified oxygen defect-rich TiO_2 nanoparticles for light emission and photocatalytic applications. New Journal of Chemistry. 2020;**44**:14936-14946

[12] Guo Q, Ma Z, Zhou C, Ren Z, Yang X. Single molecule photocatalysis on TiO_2 surfaces: Focus review. Chemical Reviews. 2019;**119**:11020-11041

[13] Singh J, Soni RK. Fabrication of hydroxyl group-enriched mixed-phase TiO_2 nanoflowers consisting of nanoflakes for efficient photocatalytic activity. Journal of Materials Science: Materials in Electronics. 2020;**31**:12546-12560

[14] Likodimos V, Chrysi A, Calamiotou M, Fernández-Rodríguez C, Doña-Rodríguez JM, Dionysiou DD, et al. Microstructure and charge trapping assessment in highly reactive mixed phase TiO_2 photocatalysts. Applied Catalysis B: Environmental. 2016;**192**:242-252

[15] Garcia JC, Nolan M, Deskins NA. The nature of interfaces and charge trapping sites in photocatalytic mixed-phase TiO_2 from first principles modeling. The Journal of Chemical Physics. 2015;**142**:024708

[16] Hou H, Shang M, Wang L, Li W, Tang B, Yang W. Efficient photocatalytic activities of TiO_2 hollow fibers with mixed phases and mesoporous walls. Scientific Reports. 2015;**5**:1-9

[17] Luo Z, Poyraz AS, Kuo CH, Miao R, Meng Y, Chen SY, et al. Crystalline mixed phase (anatase/rutile) mesoporous titanium dioxides for visible light photocatalytic activity. Chemistry of Materials. 2015;**27**:6-17

[18] Basavarajappa PS, Patil SB, Ganganagappa N, Reddy KR, Raghu AV, Reddy CV. Recent progress in metal-doped TiO_2, non-metal doped/codoped TiO_2 and TiO_2 nanostructured hybrids for enhanced photocatalysis. International Journal of Hydrogen Energy. 2020;**45**:7764-7778

[19] Ghumro SS, Lal B, Pirzada T. Visible-light-driven carbon-doped TiO_2-based nanocatalysts for enhanced activity toward microbes and removal of dye. ACS Omega. 2022;**7**:4333-4341

[20] Singh J, Soni RK. Tunable optical properties of Au nanoparticles encapsulated TiO_2 spheres and their improved sunlight mediated photocatalytic activity. Colloids and Surfaces A: Physicochemical and Engineering Aspects. 2021;**612**:126011

[21] Singh J, Tripathi N, Mohapatra S. Synthesis of Ag–TiO_2 hybrid nanoparticles with enhanced photocatalytic activity by a facile wet chemical method. Nano-Structures and Nano-Objects. 2019;**18**:100266

[22] Shuang S, Lv R, Xie Z, Zhang Z. Surface plasmon enhanced photocatalysis of Au/Pt-decorated TiO_2 nanopillar arrays. Scientific Reports. 2016;**6**:1-8

[23] Singh J, Soni RK. Efficient charge separation in Ag nanoparticles functionalized ZnO nanoflakes/CuO nanoflowers hybrids for improved photocatalytic and SERS activity. Colloids and Surfaces A: Physicochemical and Engineering Aspects. 2021;**626**:127005

[24] Yang R, Jiang G, Liu J, Wang Y, Jian N, He L, et al. Plasmonic TiO_2@Au NPs//CdS QDs photocurrent-direction switching system for ultrasensitive and selective photoelectrochemical biosensing with cathodic background signal. Analytica Chimica Acta. 2021;**1153**:338283

[25] Kaur N, Singh DP, Mahajan A. Plasmonic engineering of TiO_2 photoanodes for dye-sensitized solar cells: A review. Journal of Electronic Materials. 2022;**31**:1-9

[26] Truppi A, Petronella F, Placido T, Margiotta V, Lasorella G, Giotta L, et al. Gram-scale synthesis of UV–vis light active plasmonic photocatalytic nanocomposite based on TiO_2/Au nanorods for degradation of pollutants in water. Applied Catalysis B: Environmental. 2019;**243**:604-613

[27] Rajangam K, Amuthameena S, Thangavel S, Sanjanadevi VS, Balraj B. Synthesis and characterisation of Ag incorporated TiO_2 nanomaterials for supercapacitor applications. Journal of Molecular Structure. 2020;**1219**:128661

[28] Wu B, Liu D, Mubeen S, Chuong TT, Moskovits M, Stucky GD. Anisotropic growth of TiO_2 onto gold nanorods for plasmon-enhanced hydrogen production from water reduction. Journal of the American Chemical Society. 2016;**138**:1114-1117

[29] Zhao B, Chen YW. Ag/TiO_2 sol prepared by a sol-gel method and its photocatalytic activity. Journal of Physics and Chemistry of Solids. 2011;**72**:1312-1318

[30] Rodríguez-Martínez C, García-Domínguez ÁE, Guerrero-R obles F, Saavedra-Díaz RO, Torres-Torres G, Felipe C, et al. Synthesis of supported metal nanoparticles (Au/TiO_2) by the suspension impregnation method. Journal of Composites Science. 2020;**4**:89-16

[31] Yu HL, Wu QX, Wang J, Liu LQ, Zheng B, Zhang C, et al. Simple fabrication of the Ag-Ag_2O-TiO_2 photocatalyst thin films on polyester fabrics by magnetron sputtering and its photocatalytic activity. Applied Surface Science. 2020;**503**:144075

[32] Olvera-Rodríguez I, Hernández R, Medel A, Guzmán C, Escobar-Alarcón L, Brillas E, et al. TiO_2/Au/TiO_2 multilayer thin-film photoanodes synthesized by pulsed laser deposition for photoelectrochemical degradation of organic pollutants. Separation and Purification Technology. 2019;**224**:189-198

[33] Bhardwaj S, Pal B. Photodeposition of Ag and Cu binary co-catalyst onto TiO_2 for improved optical and photocatalytic degradation properties. Advanced Powder Technology. 2018;**29**:2119-2128

[34] Yao YC, Dai XR, Hu XY, Huang SZ, Jin Z. Synthesis of Ag-decorated porous TiO_2 nanowires through a sunlight induced reduction method and its enhanced photocatalytic activity. Applied Surface Science. 2016;**387**:469-476

[35] Li D, Zhang X, Zhang W. Designing a new reaction system by stacking use of Ti mesh supported Ag/N-TiO_2 nano-sheets for enhanced photocatalytic degradation of bisphenol A. Chemical Engineering Journal. 2021;**405**:126867

[36] Wang J, Liu W, Li H, Wang H, Wang Z, Zhou W, et al. Preparation of cellulose fiber–TiO_2 nanobelt–silver nanoparticle hierarchically structured hybrid paper and its photocatalytic and antibacterial properties. Chemical Engineering Journal. 2013;**228**:272-280

[37] Chen Y, Huang W, He D, Situ Y, Huang H. Construction of heterostructured g-C_3N_4/Ag/TiO_2 microspheres with enhanced photocatalysis performance under visible-light irradiation. ACS Applied Materials & Interfaces. 2014;**6**:14405-14414

[38] Paul KK, Giri PK. Role of surface plasmons and hot electrons on the multi-step photocatalytic decay by defect enriched Ag@ TiO_2 nanorods under visible light. Journal of Physical Chemistry C. 2017;**121**:20016-20030

[39] Liang R, Fong LCMLC, Arlos MJ, Leeuwen JV, Shahnam E, Peng P, et al. Photocatalytic degradation using one-dimensional TiO_2 and Ag-TiO_2 nanobelts under UV-LED controlled periodic illumination. Journal of Environmental Chemical Engineering. 2017;**5**:4365-4373

[40] Singh J, Satpati B, Mohapatra S. Structural, Optical and plasmonic properties of Ag-TiO_2 hybrid plasmonic nanostructures with enhanced photocatalytic activity. Plasmonics. 2017;**12**:877-888

[41] Bian Z, Tachikawa T, Zhang P, Fujitsuka M, Majima T. Au/TiO_2 superstructure-based plasmonic photocatalysts exhibiting efficient charge separation and unprecedented activity. Journal of the American Chemical Society. 2014;**136**:458-465

[42] Li B, Hao Y, Shao X, Tang H, Wang T, Zhu J, et al. Synthesis of hierarchically

porous metal oxides and Au/TiO_2 nanohybrids for photodegradation of organic dye and catalytic reduction of 4-nitrophenol. Journal of Catalysis. 2015;**329**:368-378

[43] Singh J, Sahu K, Mohapatra S. Ion beam engineering of morphological, structural, optical and photocatalytic properties of Ag-TiO_2-PVA nanocomposite thin film. Ceramics International. 2019;**45**:7976-7983

[44] Wang X, Dornom T, Blackford M, Caruso RA. Solvothermal synthesis and photocatalytic application of porous Au/TiO_2 nanocomposites. Journal of Materials Chemistry. 2012;**22**:11701-11710

[45] Singh J, Manna AK, Soni RK. Bifunctional Au–TiO_2 thin films with enhanced photocatalytic activity and SERS based multiplexed detection of organic pollutant. Journal of Materials Science: Materials in Electronics. 2019;**30**:16478-16493

[46] Kumar A, Choudhary P, Kumar A, Camargo PH, Krishnan V. Recent advances in plasmonic photocatalysis based on TiO_2 and noble metal nanoparticles for energy conversion, environmental remediation, and organic synthesis. Small. 2022;**18**:2101638

[47] Zulkifli SN, Rahim HA, Lau W-J. Detection of contaminants in water supply: A review on state-of-the-art monitoring technologies and their applications. Sensors and Actuators B: Chemical. 2018;**255**:2657-2689

[48] Raman CV, Krishnan KS. A new type of secondary radiation. Nature. 1928;**121**:501-502

[49] Langer J, de Aberasturi DJ, Aizpurua J, Alvarez-Puebla RA, Baumberg JJ AB, Bazan GC, et al. Present and future of surface-enhanced Raman scattering. ACS Nano. 2020;**14**:28-117

[50] Perez-Jimenez AI, Lyu D, Lu Z, Liu G, Ren B. Surface-enhanced Raman spectroscopy: benefits, trade-offs and future developments. Chemical Science. 2020;**11**:4563-4577

[51] Han XX, Rodriguez RS, Haynes CL, Ozaki Y, Zhao B. Surface-enhanced Raman spectroscopy. Nature Reviews Methods Primers. 2021;**1**:87

[52] Lopez-Lorente AI. Recent developments on gold nanostructures for surface enhanced Raman spectroscopy: Particle shape, substrates and analytical applications. A Review. Analytica Chimica Acta. 2021;**1168**:338474-338421

[53] Prakash J, Sun S, Swart HC, Gupta RK. Noble metals-TiO_2 nanocomposites: From fundamental mechanisms to photocatalysis, surface enhanced Raman scattering and antibacterial applications. Applied Materials Today. 2018;**11**:82-135

[54] Ji W, Xue X, Ruan W, Wang C, Ji N, Chen L, et al. Scanned chemical enhancement of surface-enhanced Raman scattering using a charge-transfer complex. Chemical Communications. 2011;**47**:2426-2428

[55] Xia L, Chen M, Zhao X, Zhang Z, Xia J, Xu H, et al. Visualized method of chemical enhancement mechanism on SERS and TERS. Journal of Raman Specroscopy. 2014;**45**:533-540

[56] Adesoye S, Dellinger K. ZnO and TiO_2 nanostructures for surface-enhanced Raman scattering-based bio-sensing: A review. Sensing and Bio-Sensing Research. 2202;**37**:100499-100412

[57] Musumeci A, Gosztola D, Schiller T, Dimitrijevic NM, Mujica V,

Martin D, et al. SERS of Semiconducting nanoparticles (TiO_2 hybrid composites). Journal of the American Chemical Society. 2009;**131**:6040-6041

[58] Yang L, Gong M, Jiang X, Yin D, Qin X, Zhao B, et al. Investigation on SERS of different phase structure TiO_2 nanoparticles. Journal of Raman Specroscopy. 2015;**46**:287-292

[59] Yang L, Jiang X, Ruan W, Zhao B, Xu W, Lombardi JR. Observation of enhanced Raman scattering for molecules adsorbed on TiO_2 nanoparticles: Charge-transfer contribution. Journal of Physical Chemistry C. 2008;**112**:20095-20098

[60] Lin J, Ren W, Li A, Yao C, Chen T, Ma X, et al. Crystal-amorphous core-shell structure synergistically enabling TiO_2 nanoparticles' remarkable SERS sensitivity for cancer cell imaging. ACS Applied Materials & Interfaces. 2020;**12**:4204-4211

[61] Dong Y, Wu N, Ji X, Laaksonen A, Lu X, Zhang S. Excellent trace detection of proteins on TiO_2 nanotube substrates through novel topography optimization. Journal of Physical Chemistry C. 2020;**124**:27790-27800

[62] Maznichenko D, Venkatakrishnan K, Tan B. Stimulating multiple SERS mechanisms by a nanofibrous three-dimensional network structure of titanium dioxide (TiO_2). Journal of Physical Chemistry C. 2013;**117**:578-583

[63] Cong T, Zhang Y, Huang H, Li C, Fan Z, Pan L. Ag nanoparticles synthesized on black-titanium dioxide by photocatalytic method as reusable substrates of surface enhanced Raman spectroscopy. Chem. 2022;**10**:441-411

[64] Song W, Wang Y, Zhao B. Surface-enhanced Raman scattering of 4-mercaptopyridine on the surface of TiO_2 nanofibers coated with Ag nanoparticles. Journal of Physical Chemistry C. 2007;**111**:12786-12791

[65] Zhao X, Zhang W, Peng C, Liang Y, Wang W. Sensitive surface-enhanced Raman scattering of TiO_2/Ag nanowires induced by photogenerated charge transfer. Journal of Colloid and Interface Science. 2017;**507**:370-377

[66] Fu X, Pan L, Li S, Wang Q, Qin J, Huang Y. Controlled preparation of Ag nanoparticle films by a modified photocatalytic method on TiO_2 films with Ag seeds for surface-enhanced Raman scattering. Applied Surface Science. 2016;**363**:412-420

[67] Wang Y, Li M, Wang D, Han C, Li J, Wu C, et al. Fabrication of highly uniform Ag nanoparticle-TiO_2 nanosheets array hybrid as reusable SERS substrates. Colloids and Interface Science Communications. 2020;**39**:100324-100328

[68] Yang J, Song G, Zhou L, Wang X, You L, Li J. Highly sensitively detecting tetramethylthiuram disulfide based on synergistic contribution of metal and semiconductor in stable Ag/TiO_2 core-shell SERS substrates. Applied Surface Science. 2021;**539**:147744-147749

[69] Lu J, Yang J, Singh SC, Zhan Z, Yu Z, Xin W, et al. Hierarchical micro/nanostructured TiO_2/Ag substrates based on femtosecond laser structuring: A facile route for enhanced SERS performance and location predictability. Applied Surface Science. 2019;**478**:737-743

[70] Jing M, Zhang H, Li M, Mao Z, Shi X. Silver nanoparticle-decorated TiO_2 nanotube array for solid-phase microextraction and SERS detection of antibiotic residue in milk. Spectrochimica Acta. Part A, Molecular

and Biomolecular Spectroscopy. 2021;**255**:119652-119658

[71] Xu Y, Zhang D, Lin J, Wu X, Xu X, Akakuru OU, et al. Ultrahigh SERS activity of the TiO_2@Ag nanostructure leveraged for accurately detecting CTCs in peripheral blood. Biomaterials Science. 2022;**10**:1812-1820

[72] Huang S, Wu C, Wang Y, Yang X, Yuan R, Chai Y. Ag/TiO_2 nanocomposites as a novel SERS substrate for construction of sensitive biosensor. Sensors and Actuators B: Chemical. 2021;**339**:129843-129848

[73] Yang Y, Zhang Z, Hea Y, Wang Z, Zhao Y, Sun L. Fabrication of Ag@TiO_2 electrospinning nanofibrous felts as SERS substrate for direct and sensitive bacterial detection. Sensors and Actuators B: Chemical. 2018;**273**:600-609

[74] Jiang X, Sun X, Yin D, Li X, Yang M, Han X, et al. Recyclable Au-TiO_2 nanocomposite SERS-active substrates contributed by synergistic charge-transfer effect. Physical Chemistry Chemical Physics. 2017;**19**:11212-11210

[75] Wang X, Zhu X, Shi H, Chen Y, Chen Z, Zeng Y, et al. Three-dimensional-stacked gold nanoparticles with sub-5 nm gaps on vertically aligned TiO_2 nanosheets for surface-enhanced Raman scattering detection down to 10 fm scale. ACS Applied Materials & Interfaces. 2018;**10**:35607-35614

[76] Zhao X, Wang W, Liang Y, Fu J, Zhu M, Shi H, et al. Visible-light-driven charge transfer to significantly improve surface-enhanced Raman scattering (SERS) activity of self-cleaning TiO_2/Au nanowire arrays as highly sensitive and recyclable SERS sensor. Sensors and Actuators B: Chemical. 2019;**279**:313-319

[77] Liang S, Guan T, Yin S, Krois E, Chen W, Everett CR, et al. Template-induced growth of sputter-deposited gold nanoparticles on ordered porous TiO_2 thin films for surface-enhanced Raman scattering sensors. ACS Applied Nano Materials. 2002;**5**:7492-7501

[78] Jiang L, Liang X, You T, Yin P, Wang H, Guo L, et al. A sensitive SERS substrate based on Au/TiO_2/Au nanosheets. Spectrochimica Acta, Part A: Molecular and Biomolecular Spectroscopy. 2015;**142**:50-54

[79] Jiao A, Cui Q, Li S, Li H, Xu L, Tian Y, et al. Aligned TiO_2 nanorod arrays decorated with closely interconnected Au/Ag nanoparticles: Near-infrared SERS active sensor for monitoring of antibiotic molecules in water. Sensors and Actuators B: Chemical. 2022;**350**:130848-130849

[80] Borges J, Ferreira CG, Fernandes JPC, Rodrigues MS, Proença M, Apreutesei M, et al. Thin films of Ag-Au nanoparticles dispersed in TiO_2: influence of composition and microstructure on the LSPR and SERS responses. Journal of Physics D: Applied Physics. 2018;**51**:205102-205114

[81] Cai J, Huang J, Ge M, Iocozzia J, Lin Z, Zhang K-Q, et al. Immobilization of Pt nanoparticles via rapid and reusable electropolymerization of dopamine on TiO_2 nanotube arrays for reversible SERS substrates and nonenzymatic glucose sensors. Small. 2017;**13**:1604240-1604212

[82] Jiang S, Li Q, Wang C, Pang Y, Sun Z, Xiao R. In situ exosomal microrna determination by target-triggered SERS and Fe_3O_4@TiO_2-based exosome accumulation. ACS Sensors. 2021;**6**:852-862

[83] Pugazhendhi K, Praveen B, Sharmila DJ, Mary JS, Kumar PN, Bharathilenin V, et al. Plasmonic TiO_2/Al@ ZnO nanocomposite-based novel

dye-sensitized solar cell with 11.4% power conversion efficiency. Solar Energy. 2021;**215**:443-450

[84] Lim SP. Plasmonic nanocomposite as photoanode. Interfacial Engineering in Functional Materials for Dye-sensitized Solar Cells. 2019:193-211

[85] Li YY, Wang JG, Liu XR, Shen C, Xie K, Wei B. Au/TiO_2 hollow spheres with synergistic effect of plasmonic enhancement and light scattering for improved dye-sensitized solar cells. ACS Applied Materials & Interfaces. 2017;**9**:31691-31698

[86] O'regan B, Grätzel M. A low-cost, high-efficiency solar cell based on dye-sensitized colloidal TiO_2 films. Nature. 1991;**353**:737-740

[87] Dissanayake MA, Kumari JM, Senadeera GK, Thotawatthage CA. Efficiency enhancement in plasmonic dye-sensitized solar cells with TiO_2 photoanodes incorporating gold and silver nanoparticles. Journal of Applied Electrochemistry. 2016;**46**:47-58

[88] Vaghasiya JV, Sonigara KK, Fadadu KB, Soni SS. Hybrid AgNP–TiO_2 thin film based photoanode for dye sensitized solar cell. Perspectives on Science. 2016;**8**:46-49

[89] Guo K, Li M, Fang X, Liu X, Sebo B, Zhu Y, et al. Preparation and enhanced properties of dye-sensitized solar cells by surface plasmon resonance of Ag nanoparticles in nanocomposite photoanode. Journal of Power Sources. 2013;**230**:155-160

[90] Lim SP, Pandikumar A, Huang NM, Lim HN. Facile synthesis of Au@ TiO_2 nanocomposite and its application as a photoanode in dye-sensitized solar cells. RSC Advances. 2015;**5**:44398-44407

[91] Wang Y, Zhai J, Song Y. Feather-like Ag@ TiO_2 nanostructures as plasmonic antenna to enhance optoelectronic performance. Physical Chemistry Chemical Physics. 2015;**17**:5051-5056

[92] Bhullar V, Devi D, Singh F, Chopra S, Debnath AK, Aswal DK, et al. Ag implanted TiO_2 nanoparticle/nanofibers composites for dye sensitized solar cells applications. Solar Energy. 2022;**241**:109-119

[93] Muduli S, Game O, Dhas V, Vijayamohanan K, Bogle KA, Valanoor N, et al. TiO_2–Au plasmonic nanocomposite for enhanced dye-sensitized solar cell (DSSC) performance. Solar Energy. 2012;**86**:1428-1434

[94] Alamu GA, Adedokun O, Bello IT, Sanusi YK. Plasmonic enhancement of visible light absorption in Ag-TiO_2 based dye-sensitized solar cells. Chemical Physics Impact. 2021;**3**:100037-100011

[95] Ran H, Fan J, Zhang X, Mao J, Shao G. Enhanced performances of dye-sensitized solar cells based on Au-TiO_2 and Ag-TiO_2 plasmonic hybrid nanocomposites. Applied Surface Science. 2018;**430**:415-423

[96] Nbelayim P, Kawamura G, Kian Tan W, Muto H, Matsuda A. Systematic characterization of the effect of Ag@ TiO_2 nanoparticles on the performance of plasmonic dye-sensitized solar cells. Scientific Reports. 2017;**7**:1-12

[97] Song DH, Kim HS, Suh JS, Jun BH, Rho WY. Multi-shaped Ag nanoparticles in the plasmonic layer of dye-sensitized solar cells for increased power conversion efficiency. Nanomaterials. 2017;**7**:136-111

[98] Villanueva-Cab J, Montaño-Priede JL, Pal U. Effects of plasmonic nanoparticle incorporation on electrodynamics and photovoltaic performance of dye sensitized solar cells. Journal of Physical Chemistry C. 2016;**120**:10129-10136

Section 3

Medicinal Application of TiO_2

Chapter 11

Nano Titania Applications in Cancer Theranostics

Rida e Maria Qazi, Zahra Sajid, Chunqiu Zhao, Fawad Ur Rehman and Afsar Ali Mian

Abstract

Titanium is one of the most abundantly utilized nanomaterials for human consumption. Biomedical applications of nano titania include sunscreens, drug delivery, prosthetic implants, bioimaging probes, and antimicrobial and antirheumatic agents for various treatment of diseases, including autoimmune disease, neurogenerative diseases, cardiovascular, musculoskeletal, and cancer. Its applications as a drug delivery vehicle and photosensitizer in cancer therapy and diagnosis are highly appreciated, especially for skin and natural cavities applications. The reactive oxygen species (i.e., H_2O_2, $OH^{\cdot}$, OH_2, $^1O_2$, etc.) generation properties of nano titania after activation with light or ultrasound make it ideal for apoptosis induction in neoplastic cells. In addition, the singlet oxygen ($^1O_2$) generating properties make it suitable for bioimaging deep-seated and superficial tumors after activation. Nano titania is highly biocompatible with negligible adverse effects. In this chapter, we will focus on the anticancer effects of nano titania on various types of cancers by employing it as a drug delivery vehicle and sensitizer for external source-activated modalities viz. photodynamic and sonodynamic therapy.

Keywords: nano titania, anticancer effects, theranostics, photodynamic therapy, sonodynamic therapy

1. Introduction

Nanotechnology has opened a new avenue to investigate and explore the potentials of materials at the nanoscale with known functionality at the macroscale. The biomedical applications of nanoscale materials are supported by the evidence that most of the cellular organelles, cell membranes, protein ligands, and DNA sizes are ranged from 2 to 20 nm [1]. The interaction of materials with cellular organelles at the nanoscale can significantly enhance their desired biomedical application with enormous traceability. Nanotechnology is applicable in various areas of the healthcare system due to the distinguished biological and physicochemical properties of nanomaterials. Various nanostructures with distinct characteristics have been utilized in drug delivery, diagnostic probes, prosthetic implants, and biotechnological applications. Out of many, titanium dioxide (TiO_2) has been extensively utilized [2].

TiO_2 are metallic oxide nanoparticles, widely used, and are of great interest in modern therapeutics. They are semiconductive, highly stable, and possess anticorrosive and antibacterial characteristics. Titanium is the second most abundantly consumed metal, with daily 1–2 mg/kg consumption for children and 0.2–0.7 mg/kg for adults in the USA [3]. It is well distributed on the earth's crust and abundantly found in T, $TiCl_4$, and TiO_2. The anatase is the most reactive crystalline form of TiO_2 compared to brookite, rutile, and TiO_2-B1 as various polymorphs [4]. Titanium is well recognized for its exceptional characteristics, such as low weight, good mechanical strength, high wear resistance, and biocompatibility [5, 6]. They are less toxic than other nanomaterials and relatively economical to fabricate [7, 8]. Anatase and rutile exist in a tetragonal structure, whereas brookite is rhombohedral [9]. Moreover, an amorphous form of TiO_2 can also be found [10].

Their white appearance is attributed to their high refractive index and is used in skin care products as a white pigment. They possess catalytic activity upon exposure to UV light and can be utilized for water treatment to remove the chemicals from them [8]. In addition, TiO_2 has also been used as an additive in food products [11–14]. TiO_2 is one of the most produced nanoparticles due to its wide range of applications [15]. TiO_2 has been employed in biomedical applications such as molecular imaging, drug delivery system, and therapeutic approaches alongside conventional therapies or substitutes [16, 17]. Akira Fujishima was the first to discover its anticancer effect against human cervical cancer cells (HeLa). Photoactivation with UV light could generate hydroxyl ($OH^{\cdot}$), per hydroxyl ($H_2O^{\cdot}$), and singlet oxygen ($^1O_2$) as Reactive Oxygen Species (ROS) [18]. These ROS then interfere with cellular signal pathways and induce apoptosis by damaging the mitochondria. Different biomedical applications of nano titania are shown in **Figure 1**. This chapter focuses on combining various applications of titanium NPs in biomedicine, especially in various cancer

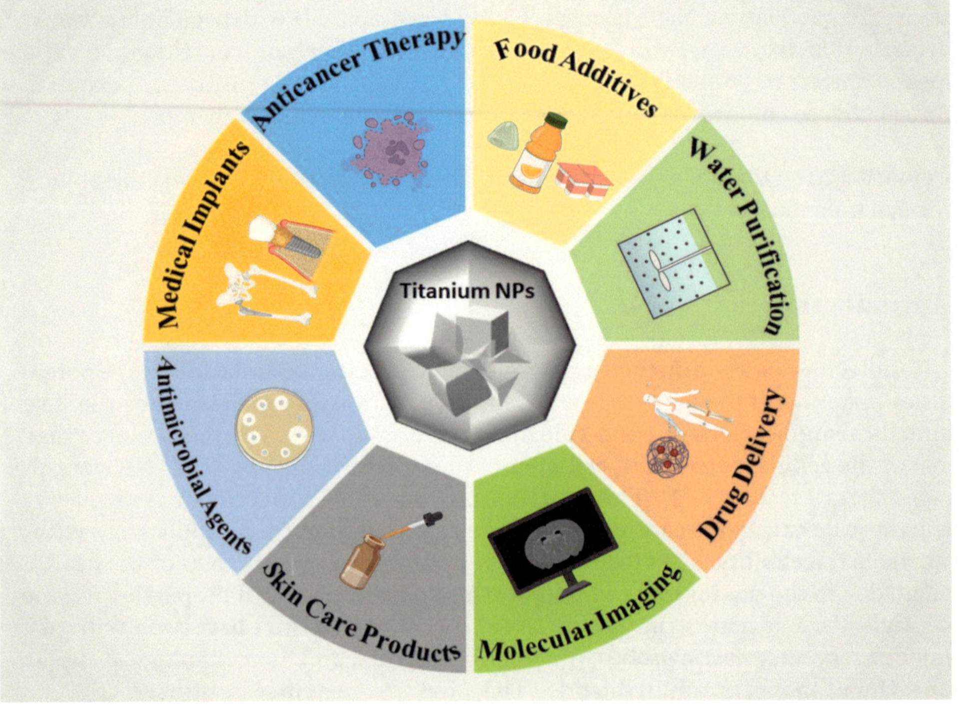

Figure 1.
Various biomedical applications of titanium-based nanoparticles (developed by using BioRender).

therapeutics and diagnostic purposes. We will also spotlight its applications in the specialized modalities viz. photodynamic and sonodynamic therapy as photosensitizers. In targeted cancer therapies, the use of nano titania as a delivery vehicle is highly favorable and this will be the main focus of this chapter.

2. Antimicrobial activity of Titania nanoparticles

Antimicrobial activity is one of the major applications of biomedical science. Pathogenic microbial species such as *Escherichia coli*, *Klebsiella pneumoniae*, *Staphylococcus epidermidis*, *Staphylococcus aureus*, and *Proteus vulgaris* are known to affect humans by causing various infections. Prescribed antibiotics occasionally cannot kill or cause growth inhibition of these pathogenic bacteria, and they often develop multidrug resistance. Therefore, there is an urgent need to develop novel and nano-based therapy to eradicate bacterial infections. Mahendran et al. synthesized biomolecule-coated TiO_2 nanocatalysts by using rhizome extracts. They observed that nano titania catalysts showed robust antimicrobial activity. This potential antimicrobial activity was produced against *P. aeruginosa* and *S. epidermidis*. They further observed the resistance against nano titania catalysts in gram-positive than gram-negative bacteria [19].

Fungal diseases cause deterioration in mangoes post-harvesting, affecting their quality and shelf-life. In the last few years, edible coatings have been investigated to preserve fruits and vegetables. Nano titanium dioxide is an immensely active nanomaterial with antibacterial, anti-ultraviolet, super lipophilic, and non-toxic characteristics. Chitosan is a good food preservative, antioxidant, and antibacterial agent for coating fruits and vegetables. Xing et al. used Chitosan (CTS) and TiO_2 composite coating and analyzed its antifungal properties against Colletotrichum gloeosporioides (MA), Cladosporium oxysporum (ME), and Penicillium steckii (MF). They found that CTS/TiO_2 composite exhibited a better antifungal effect than chitosan coating alone. CTS/TiO_2 coating killed the molds, induced leakage of intracellular proteins and nucleic acid, disrupted the cell membrane integrity, retard the mycelial growth, and increased the conductivity value of fungal suspensions [20]. Maneerat and Hayata used TiO_2 coating films and examined the antifungal effect. They showed a significant reduction in the penicillium rot development in apples and lemons [21].

3. Sonodynamic therapy

Sonodynamic Therapy (SDT) has recently gained much attention as a new anticancer treatment strategy that is relatively cheap, minimally invasive, and possesses deep penetration power. In this therapy, ultrasound waves activate the sonosensitizers (sound-sensitive agents), killing tumor cells by producing ROS [22]. The use of ultrasound offers some advantages in comparison to the use of light in cancer treatment which includes sonoporation (cell permeabilization mediated by ultrasound waves) and deeper penetration (depending on the frequency of ultrasound) which could be up to 15 cm in soft tissues [23–25]. Sonosensitizers refer to the use of chemical compounds that could increase the cytotoxicity of ultrasound. Nano-sonosensitizers are considered potent sonosensitizers, as compared to conventional organic sonosensitizing agents, owing to their high bioavailability achieved by improved pharmacokinetics, pharmacodynamics, and biodistribution properties. Generally, nano-sonosensitizers can be categorized into two main types: (1) nanoparticles which include TiO_2, and (2) nanoparticles assisted sonosensitizers, consisting of nanoparticles loaded with organic molecules with controlled release at the target site [26]. Among many nanoparticles, the use of TiO_2 NPs is preferred because of their inert behavior in the biological system, easy fabrication, and cost-effectiveness. TiO_2 is a semiconductor with a large energy band gap, allowing for electron transitions from the valence to the conduction band when exposed to UV light. This facilitates the generation of free radicals, including the enormously reactive singlet oxygen. However, UV radiations are not clinically ideal due to low penetration power. Using ultrasound can overcome this due to its greater in vivo penetration ability with low frequency [27]. Various studies have reported the use of TiO_2 NPs as anticancer agents in vitro and in vivo systems, especially when combined with ultrasound irradiation.

TiO_2 NPs, in association with high-intensity ultrasound waves, were used for sonodynamic therapy of squamous cell carcinoma cells (HSC-2). The authors reported that the toxicity of TiO_2 with ultrasound was much higher than that of TiO_2 or ultrasound alone, which increased with the increase in intensity and exposure time [28]. SDT with TiO_2 NPs was evaluated for the treatment of melanoma. C32 (melanoma cell line) was treated with ultrasound waves of 1 MHz frequency. The apoptotic effect was more significantly observed in the TiO_2-based SDT than in either treatment alone. In addition, the apoptotic percentage of cells was increased by 2.73 times than untreated cells [29]. Aksel et al. reported that TiO_2 NPs mediated sonodynamic, photodynamic, and Sono-Photodynamic (SPDT) Therapy for prostate cancer. SDPT combines sonodynamic therapy and photodynamic therapy along with TiO_2 NPs as sensitizers. The results showed a reduction in cancer cell viability after TiO_2-mediated sono-photodynamic therapy. The production of singlet oxygen affects the intrinsic pathway, which might be responsible for producing antiapoptotic effects [30].

4. Photodynamic therapy

Photodynamic Therapy (PDT) is an emerging non-invasive therapy that received clinical approval. This therapy is preferred over conventional anticancer treatments due to its high efficacy, specificity, and subtle side effects [1, 31]. This therapeutic strategy utilizes photosensitizers (chemicals, drugs) with light in the presence of molecular oxygen to stimulate the generation of ROS, thereby inducing tumor cell death. However, the combination of PDT and drug is expected to produce a more significant effect as an anticancer treatment since PDT alone is relatively inefficient in eradicating cancer [32–35]. The photosensitizer should ideally enter the target cells/tissues without affecting the neighboring healthy tissues (**Figure 2**).

Moreover, the treatment can be confined to an elevated concentration of photosensitizers. This promising strategy can be applied to inhibit microbial growth and treat cancer and infectious diseases [35]. The effectiveness of PDT relies on the type of photosensitizers used. Several materials, including inorganic [33], organic, and porphyrin-based materials [34], have been used as photosensitizers in PDT. However, several drawbacks have been associated with these materials, such as inadequate dispersion in water and photostability. In addition, these materials cannot absorb light of longer wavelength, i.e., greater than 700 nm, which results in improper light penetration and subsequent reduction in cell killing effect. This causes unwanted toxicity and damage to cancer and normal cells or tissues.

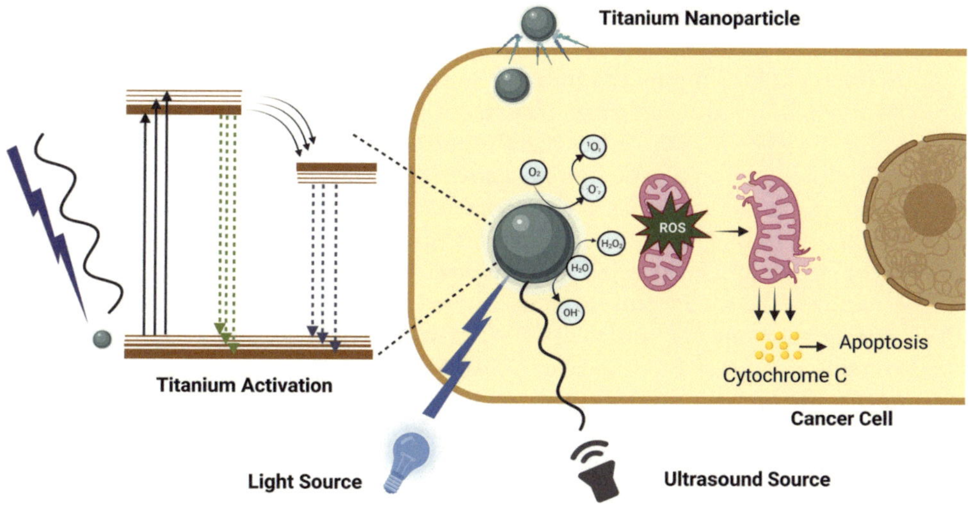

Figure 2.
TiO_2 NPs-based photodynamic or sonodynamic therapy of cancer cells. The ROS generated after photo-sonoactivation results in mitochondria damage leading to cytochrome c release to induce apoptosis in cancer cells (developed by using BioRender).

Metal oxide nanoparticles have been widely studied as photosensitizing agents in PDT due to the drawbacks associated with porphyrin-based photosensitizing agents. TiO_2 NPs gained immense interest due to their distinct characteristics, enabling them to effectively kill tumor cells upon optical irradiation. Irradiation of TiO_2 NPs, with an energy greater than or equal to the bandgap, causes the redox reaction on the surface of these NPs, which leads to the generation of reactive oxygen species, including superoxide anions, hydrogen peroxide, and hydroxyl radicals [36, 37]. TiO_2 is more stable than other conventional photosensitizers because they are nanosized particle with anti-photodegradable stability. TiO_2 NPs have been used as photosensitizers in several types of tumor cell lines, which include HepG2 (hepatocellular carcinoma cells) [38], HeLa (cervical cancer cells) [39], MDA-MB-468 and MCF7 (breast cancer cells) [40], leukemia cells (K592) [41], and lung cancer cells (NSCLC) [42].

TiO_2 NPs are considered marvelous photosensitizers; however, their possible toxicity impedes their applicability in PDT [8, 43]. TiO_2 can be excited in its pristine form by short-wavelength ultraviolet irradiation. Lagopati et al. conducted a study in which they used TiO_2 as photosensitizers against breast cancer cells (MCF7 and MDA-MB-468). TiO_2 nanostructures were prepared by using the sol-gel technique. The results showed significant effects of the applied modification against MDA-MB-468 cells [44]. Modifying TiO_2 NPs with Quantum Dots (QDs) have received significant attention since they allow TiO_2 to absorb light of much longer wavelengths and, thereby, deeper tissue penetration. In PDT, QDs usually possess dual-function properties and act as energy transducers and carriers for photosensitizers. Ramachandran et al. synthesized TiO_2 NPs by microwave-assisted synthesis and TiO_2 conjugated with N-doped graphene QDs (N-GQDs/TiO_2) by two-pot hydrothermal method. N-GQDs/TiO_2 nanocomposites generated ROS, particularly singlet oxygen, upon activation with the light of the near-infrared region. This induced cell death in MDA-MB-231 cells more significantly than in the HS27 cell line (human foreskin fibroblasts) [45].

5. Drug delivery vehicle

Nano titania holds a higher reputation among various nanodrug delivery materials due to its amenability to a vast array of surface functionalization for targeting tissues, easy forming composites with other metals, porous texture, and highly biocompatible nature [46]. Its excretion also occurs via a standard excretory route, i.e., the hepato-urinary system. Nano titania has been reported to carry not only anticancer drugs but also other types of drugs, such as dexamethasone [47], DNA fragments [48], norfloxacin [49], ciprofloxacin [50], and aspirin [51], etc.

TiO_2 nanowhiskers were employed in cancer therapeutics to deliver Temozolomide (TMZ) to Glioblastoma Multiforme (GBM) orthoptic models. These TiO_2 nanowhiskers traversing the Blood-Brain Barrier (BBB) were accelerated by ultrasonication. Additionally, the ultrasound could also assist in releasing TMZ from TiO_2 and generate ROS to induce apoptosis [52]. Likewise, Kim et al. have also reported ultrasound-driven doxorubicin delivery to cancer cells by TiO2 nanoparticles [53]. Among other anticancer drugs, 5 fluorouracil drug delivery to cancer cells by ZnO-doped TiO_2 was performed by Faria et al. The ZnO doping could shift their absorption from UV (TiO_2 only) to red (TiO_2-ZnO), making it a perfect candidate for photodynamic therapy [54]. Liposome-covered TiO_2 nanotubes have also delivered the 5 fluorouracil to HeLa cells [55]. Doxorubicin's successful loading on TiO_2 nanotubes and efficient delivery to cancer cells is another example of TiO_2 employment as a drug delivery vehicle. The drug release was lower pH dependent [56]. Similarly, paclitaxel delivery via Polyethylene Glycol (PEG) and folic acid surface decorated TiO_2 nanoparticles was reported by Venkatasubbu et al. [57].

Not only in cancer theranostics but $TiO_{2's}$ role as a vehicle in other diseases, including rheumatoid arthritis, has also been explored. The porphyrin derivative, i.e., Tetra Sulphonatophenyl Porphyrin (TSPP), was loaded on TiO_2 nanowhiskers by an adsorption process assisted by its porous nature [58–60]. The TiO_2 could deliver the TSPP to inflamed tissue and release it upon photoactivation with 532 nm light.

6. Anticancer effects

Cancer remains a critical global threat due to severe complications such as unbearable physical pain, severe cytotoxicity, side effects, and compromised therapeutic efficacy of conventional therapeutic strategies, including surgical interventions, chemo- and radiotherapy [61–73]. Various studies are aimed at investigating the new therapeutic approaches, including Photodynamic Therapy (PTD), Chemodynamic Therapy (CDT), Sonodynamic Therapy (SDT), Photothermal Therapy (PTT), Starvation Therapy (ST), and Immunotherapy (IMT) having lower side effects and high-level efficiency [26, 74–79]. New therapeutic approaches have been effectively applied as a substitute to conventional therapies and merged with imaging techniques for diagnosis, which is quite optimistic for the diagnosis and treatment of cancer [80, 81]. Cancer theranostics, a combination of diagnostics and treatment, has recently gained much interest [82]. Several therapeutic strategies can be integrated with various imaging techniques to synthesize multifunctional tumor-targeted nanoprobes, having a significant therapeutic effect and improving tumor identification [83].

In recent years, a newly established field of nanomedicine has been instigated to offer various solutions. Nanomedicine is the implementation of nanomaterials, possessing particle sizes ranging from 1 to 100 nm, to diagnose, observe, prevent,

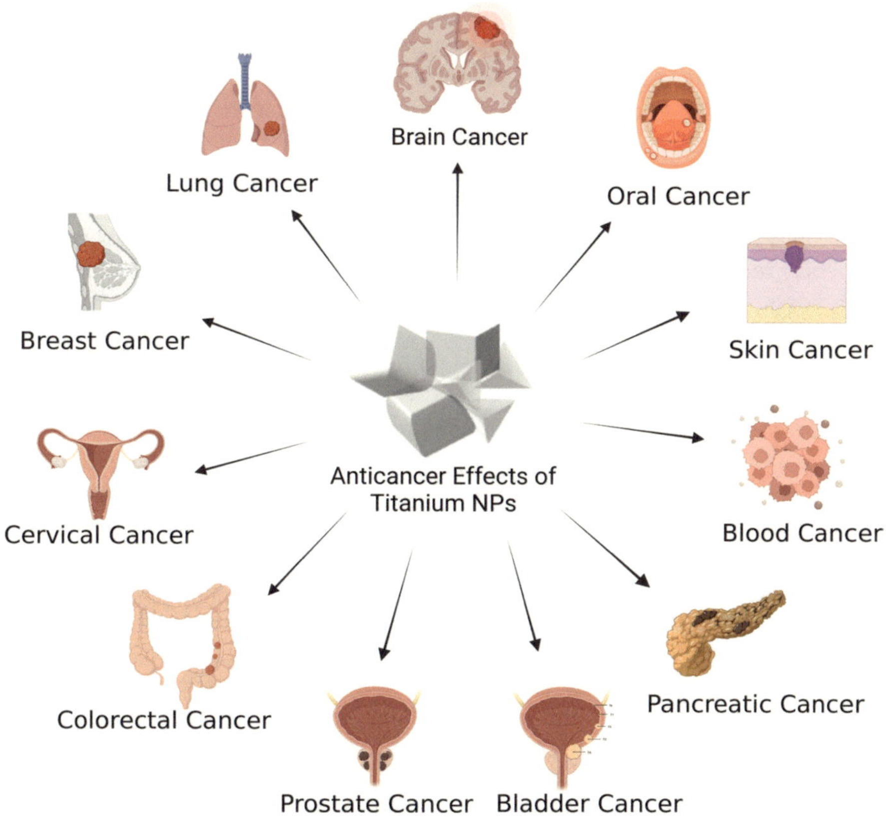

Figure 3.
Different types of cancer that can be treated with nano titania (developed by using BioRender).

and treat disease [84]. Nanoparticles (NPs) have been extensively used as anticancer therapeutic agents, particularly in cargo delivery, i.e., genes, chemotherapeutic drugs, or contrast agents [70, 85–87], or alone, using their inherent toxicity, e.g., associated with the release of reactive oxygen/nitrogen species [88, 89]. Additionally, nanoparticles can be coated with a chemical or biological material to facilitate their stealth characteristics and minimize their tendency to aggregate in biological fluids. Moreover, they can be coupled with selected ligands to enhance their targeted cell delivery [90]. NPs can impulsively accumulate in the tumors because of the Enhanced Permeability and Retention (EPR) effect. They can easily pass through the tumor vasculature due to large pores, and inadequate lymphatic drainage allows their retention, expediting their therapeutic efficacy without being associated with the targeted ligands [91]. Nano titania-based anticancer therapy is well-known (**Figure 3**). Below are various types of cancers treated with nano titania.

6.1 Breast cancer

Breast cancer is the primary cause of mortality in women ranging from 35 to 55 years of age in industrialized countries. The prevalence of breast cancer is relatively high because the breast is among the most vulnerable organ to malignancy (after the liver, lungs, and stomach) [92, 93]. Conventional treatment modalities include surgery, chemo-, radio- and hormonal therapy, or a combination of these therapeutic options [94–96]. The complete removal of the tumor is challenging due to limited access to the region for surgery, side effects associated with conventional therapy, and the development of drug resistance. Hence, the five-year survival rate is limited to 20% [97]. Recently, pembrolizumab and atezolizumab, immunotherapeutic drugs, have received FDA approval. However, only triple-negative breast cancer patients can use these therapeutic drugs [98]. Therefore, designing a targeted drug delivery technique for anticancer therapy with minimal cytotoxicity in normal tissues is persistently required [99]. In this context, nanoparticles seemed to be a promising approach possessing low cytotoxicity, target specificity, mature drug distribution in the tumor, and fast elimination of the drug from the body [99–102].

TiO_2 nanoparticles are among the prominent nanoparticles with both in vitro and in vivo applications. TiO_2 nanoparticles exhibit distinct morphology and surface chemistry, adequate biocompatibility, employ intrinsic biological activity, reduced side effects, and insignificant eco-toxicity [103]. Previously, it was reported that TiO_2 induces ROS generation by interfering with the EGFR signaling cascade, leading to apoptosis induction in tumor cells compared to nearby physiological cells [104]. However, there is little information about the therapeutic role of TiO_2 in breast cancer compared to conventional therapeutic drugs, i.e., doxorubicin is lacking. Doxorubicin is among the most effective therapeutic drugs in ovarian and breast cancer [105]. However, its clinical application is restricted due to adverse effects, of which cardiotoxicity is the most significant [106]. Iqbal et al. synthesized TiO_2 NPs from leaf extract of Zanthoxylum armatum and evaluated their safety and anticancer activity. They demonstrated that TiO_2 NPs and doxorubicin were equally effective against breast cancer in vivo and ex vivo. TiO_2 NPs exhibited anticancer activity by inducing ROS-dependent cell death in 4 T1 breast cancer cells. In vivo analysis in 4 T1 breast cancer cells containing BALB/c mice revealed that TiO_2 NPs exerted doxorubicin comparable to anticancer activity and without any cardiotoxicity and body weight alteration as compared to doxorubicin [107].

Kim et al. analyzed the possible cytotoxicity in breast cancer cells. They used two cell lines, Hs578T and MDA-MB-231, which overexpress Epidermal Growth Factor Receptor (EGFR). EGFR is a transmembrane protein activated by binding growth factors and transmitting cellular signals inducing cell survival and propagation. They tried to elucidate the effect of alterations in extracellular signaling receptors mediated by TiO_2 nanoparticles rather than focusing on the toxicity induced by TiO_2-mediated ROS generation. They showed that the cytotoxicity caused by TiO_2 nanoparticles in breast tumor cells is due to the interference in the EGFR-regulated signaling pathway, which reduced cell adhesion, survival, and propagation, thus inducing apoptosis [104]. Mahendran et al. used *Gloriosa superba* rhizome extract to synthesize crystalline TiO_2 nanocatalysts. These TiO_2 nanocatalysts caused exorbitant mitochondrial depolarization and DNA damage when treated with MCF-7 cells, primarily due to the persistent release of TiO_2 nanoparticles and the generation of free radicals [19].

6.2 Pancreatic cancer

Pancreatic cancer is the third major contributor of deaths caused by cancers in the United States [108], with a five-year survival rate of about 10% only [109, 110]. Only about 15–20% of cancer patients can avail the surgical treatment due to delayed diagnosis [111], and even after tumor resection, the five-year survival rate remains about 20% only [112–114]. Immune Checkpoint Blockade (ICB) therapeutic approaches have been developed which are based on the applicability of monoclonal antibodies against PD-L1 (programmed cell death ligand 1) and CTLA-4 (cytotoxic T-lymphocyte antigen 4), able to support tumor eradication and protection from recurrence and metastasis [115–118]. However, these approaches failed to exhibit significant results in patients diagnosed with pancreatic cancer [119–121]. Hence, the combination of ICB and therapeutic approaches, able to enhance T-cell infiltration and activation in the tumor, can be promising for treating and preventing tumor relapse and metastasis [122–124].

Ultrasound exposure represents a non-invasive, inexpensive, and well-portable therapeutic tool [125–127] and is well-studied in the perspective of cancer treatment, in addition to its general utilization in imaging systems [126, 128–130]. Ultrasound-activated sonodynamic therapy (SDT) can cause tumor cell death by inducing high levels of ROS generation, causing apoptotic or necrotic immunogenic cell death [131, 132]. Titanium diselenide ($TiSe_2$) is a 2D transition metal dichalcogenide extensively used in photodynamic therapy due to its good photoresponsivity [133]. Chen et al. synthesized TiSe2 nanosheets and evaluated the combination of $TiSe_2$-mediated sonodynamic therapy with PD-1 blockage for pancreatic cancer treatment in vitro using Pac02 cells and in vivo model of pancreatic cancer. They reported the generation of ROS by $TiSe_2$ nanosheets upon exposure to non-invasive US irradiation and induction of immunogenic death of malignant cells, thereby promoting the maturation of dendritic cells and infiltration of activated T cells within the tumor. Besides inhibiting primary pancreatic tumor growth, this combinatorial therapeutic approach also inhibited the growth of distant tumors and lung metastasis [134].

6.3 Lung cancer

The limited therapeutic efficiency of Non-Small Cell Lung Carcinoma (NSCLC) is due to the resistance to chemotherapeutic drugs. The median survival rate is about 6 months only. Nanoparticles are progressively emerging as a new tool against drug resistance because of their limited toxicity and ability to act on numerous targets in cancer cells due to their distinct physicochemical features [135]. Two-dimensional (2D) titanium carbide (Ti_2C) possesses ultra-high surface area and enhanced cell membrane penetration ability as compared to other conventional nanoparticles [136]. It also contains many reactive groups that can be utilized as potent protein interaction sites affecting their structure and function. The chemo drug resistance reversal ability of Ti_2C was evaluated by Zhu et al. by using the characteristics of 2D Ti_2C on the NSCLC cell line. The cells were treated with cisplatin, the standard drug for treating end-stage NSCLC, with and without Ti_2C. They found that Ti_2C reversed the resistance of NSCLC to cisplatin by reducing the antioxidant reserves in the cells and decreasing the expression of primary drug resistance genes. They also reported drug resistance reversal in the NSCLC model in vivo [135]. Balachandran et al. synthesized TiO_2 nanoparticles using a novel wet chemical technique using titanium tetra isopropoxide precursor, characterized by SEM, TEM, XRD, and UV–visible spectroscopic analysis. The synthesized nanoparticles exhibited good photocatalytic activity and were evaluated for anticancer effect in A549 (lung cancer) cells. The cells were treated with TiO_2 and exposed to UV light. After 4 hours, TiO_2 caused approximately 85% of cell decomposition [137].

6.4 Colorectal cancer

Colorectal Cancer (CRC) is among the most common malignancy in humans. Its prevalence is increasing despite several advances in therapeutic and diagnostic interventions. CRC is caused due to gradual transformation of epithelial cells found in the intestinal lumen to tumor cells. Cancer treatment aims to utilize an anticancer agent that can induce apoptosis. These days, nanoparticles (NPs) are considered novel anticancer agents. Nanosized titanium dioxide nanoparticles (TiO_2 NPs) with about <100 nm diameter possessing whiteness and opacity are publicly accepted. The biological properties of TiO_2 NPs depend on their size, physicochemical properties, and surface area since particles with a large surface area are more chemically reactive [138]. Wei et al. reported the green synthesis of TiO_2 from the extract of *Calendula officinalis* and evaluated its effects on colorectal carcinoma cell lines WiDr, LS123, DLD-1, and SW1417 [SW-1417]. TiO_2 reduced the viability of all colorectal carcinoma cells in a dose-dependent manner [139]. Vigneshwaran et al. synthesized TiO_2 nanoparticles from Lactobacillus and evaluated its cytotoxic effects on the HT-29 cell line. They reported ROS generation in HT-29 cells by the treatment with TiO_2 NPs and the induction of apoptosis by intrinsic pathway [140].

6.5 Cervical cancer

Cervical cancer is the malignancy of the uterine cervix. It is ranked fourth in commonly occurring cancer in women globally and second in the low and medium Human Development Index (HDI) [141]. The key risk factors include late menopause, increasing age, obesity, elevated estrogen levels, breast cancer, no childbirth, diabetes mellitus, and tamoxifen use. Some gene mutations can also cause cervical cancer [142]. The treatment strategies for cervical cancer include radiotherapy, immunotherapy, and chemotherapy [143]. Due to the severe adverse effects of chemotherapeutic drugs, research interest has been transferred to metallic nanoparticles [144–146].

Titanium nanoparticles can be used with other nanoparticles, such as zinc and silver, to evaluate their anticancer effects on cervical cancer cell lines [147]. Ag/AgBr/TiO_2 nanoparticles effectively eliminated xenograft tumors due to their photocatalytic activity [148]. Thermodynamic therapeutic potential, bioimaging, and doxorubicin delivery to cervical cancer cells by hybridized TiO_2 and zinc phthalocyanine nanoparticles were also studied [149]. Yurt et al. synthesized zinc phthalocyanine and hybridized it with TiO_2 to evaluate their photodynamic therapeutic effect and nuclear imaging potential. Intracellular localization of ZnPc and ZnPc/TiO_2 in cervical adenocarcinoma (HeLa) and breast cancer cells was observed. High uptake of ZnPc/ZnPc-TiO_2 by the cervical and breast cancer cells suggested their use as cancer theranostic agents [150]. TiO_2 has also been reported to enhance caspase-3 activity and prevent the growth of HeLa cells [151].

6.6 Brain cancer

The brain is probably the most mature organ of the human body, so its protection is a crucial issue [152]. Despite several advancements in developing therapeutic and diagnostic procedures, brain cancer is a great challenge to treat, and a successful therapeutic strategy still cannot be established. The major hurdles to establishing a successful treatment strategy for brain tumors include tumor recurrence, acquired resistance to chemotherapeutic agents, and complex central nervous system structure [153]. Glioblastoma is the most common and dangerous tumor in adults. Despite the availability of various treatments, such as chemotherapy, radiotherapy, and surgical resection, the prognosis is still inferior. Following the diagnosis, the life expectancy of glioblastoma patients is just 12–15 months, and the five-year survival rate is approximately 5% [154].

The blood-brain barrier (BBB) is a highly selective interface responsible for maintaining homeostasis, protecting from harmful agents, and providing all necessary molecules to the brain [155]. Brain disorders and tumors require the drug to cross the BBB to exert its therapeutic effect. Several lipophilic therapeutic agents can pass through the BBB, but due to its selective permeability, several other medications fail to cross it [156, 157]. Various pharmacological agents are considered potentially harmful external agents by the BBB. Thus they are removed by the efflux system, degraded by the enzymes, or hindered from crossing the BBB [158]. Only molecules smaller than 400 Daltons or less than nine hydrogen bonds are BBB permeable. Therefore, several nanomedicine-based approaches have been suggested to facilitate drug delivery across the BBB in the recent past [159, 160].

Nanoparticles have gained much interest in this regard [161–163]. It has been reported that engineered nanomaterials can cause neurotoxicity [164]. TiO_2-NPs can induce neurotoxicity due to their ability to cross BBB [165–167]. They are potential candidates for treating glioblastoma multiforme (GBM) and other tumor types. Gene and protein expression analysis revealed the reduction of antitumor drug resistance and metastasis by inhibiting angiogenesis. These characteristics would make TiO_2 promising therapeutic agents against cancer, particularly if other chemotherapeutic agents can be combined. Fuster et al. evaluated the anticancer effects of TiO_2 NPs and ZnO-NP on the T98G glioblastoma cell line and reported that TiO_2 is a more effective anticancer agent than ZnO. They demonstrated that TiO_2 exposure disrupted the BBB and induced neuroinflammation and suggested the necessity of risk assessment regarding the TiO_2 toxicity in the central nervous system [168]. Using ultrasound-sensitive piezoelectric nanoparticles, Marino et al. delivered electric stimulations to distant glioblastoma cells. Barium titanate NPs were functionalized with antibodies against transferrin receptors to target BBB and glioblastoma cells. The distant ultrasound-mediated piezo-stimulation caused a significant reduction in the proliferation of glioblastoma cells in vitro and greatly enhanced the chemotherapeutic sensitivity when combined with temozolomide [169].

6.7 Prostate cancer

Cancer is the major cause of global mortality after cardiopulmonary arrest [170]. Prostate cancer is the fifth most common cancer worldwide and ranked second in men among common cancer types [171]. The onset of cancer can be characterized by delayed progression, tumor markers, detectable preneoplastic abrasion, and high prevalence [172]. Surgery is a successful option in some cases. However, after a few years, tumor recurrence can shorten chemotherapy as a valuable therapeutic option for prostate cancer. However, associated side effects such as toxicity, fatigue, difficulty breathing, low white blood cell count, and blood clotting hamper their efficacy for tumor eradication [173]. Recently, targeted drug delivery and stimulus-responsive release have minimized toxicity and improved drug delivery and accumulation at the target site [174, 175].

Different inorganic nanoparticles such as TiO_2, graphene oxide, iron oxide, and porous silica have been used for drug delivery and anticancer therapeutic agents [173]. TiO_2 NPs are considered potent drug carriers and photosensitizers due to their low cost, toxicity, and non-photobleaching characteristics [176, 177]. ROS generation by ultrasound-activated TiO_2 NPs has been reported by various studies [29, 178, 179]. However, in comparison to light, ultrasound scattering in the tissue is weaker, making it penetrate deeply without losing energy [33]. Previous studies revealed that combining TiO_2 with rare earth or noble metals can increase ROS quantum yield [29, 180]. Ayca et al. synthesized TiO_2 and ZnO NPs. They showed the potent inhibition of the growth of prostate cancer cells (DU-145) by TiO_2 and ZnO_2 nanocomposites [173]. Ultrasound-activated multifunctional system based on TiO_2:Gd@DOX/FA for MRI-guided therapy for prostate cancer was developed by Yuan et al. [181]. This system acts as a sonosensitizer for sonodynamic therapy and drug nanocarriers for pH-responsive drug release. Gd doping to TiO_2 improved their sonodynamic ability and their performance in MRI. In vitro and in vivo anticancer treatment proved the efficacy of TiO_2:Gd/DOX/FA in inhibiting cancer by ultrasound-responsive chemo-sonodynamic therapy without damaging other organs and as MRI agents. Aksel et al. showed the formation of apoptotic bodies in the PC3 prostate cancer cell line by TiO_2 NPs-mediated photo-sonodynamic therapy [30].

6.8 Bladder cancer

Urothelial bladder cancer is among the most widespread malignancies [182]. It is categorized into two subgroups, i.e., Muscle-Invasive Bladder Cancer (MIBC) and Non-Muscle-Invasive Bladder Cancer (NMIBC). Most bladder cancers are NMIBC at diagnosis. Frequent tumor relapse is found in about 50–70% of NMBIC [183], and 10–15% tend to progress into MIBC [3, 184]. Chemotherapy or Bacillus Calmette-Guérin (BCG) and post-transurethral resection are the therapeutic interventions used [185]. Other therapeutic options are under investigation, including photodynamic therapy, radiotherapy, immunotherapy, gene therapy, and nanodrug delivery system using nanoparticles [186]. Among many therapeutic options, a photodynamic theory is less invasive than any surgical intervention [187]. Under physiological conditions, TiO_2 NPs possess promising photodynamic characteristics and are suitable materials for cancer treatment. Studies reported the development of $Ti(OH)_4$ in which peroxide was coated on TiO_2 nanoparticles [188, 189]. $Ti(OH)_4$ could absorb visible light and showed equivalent photocatalytic activity upon exposure to UV radiations with 90% greater photocatalytic efficiency than TiO_2 NPs. Moreover, $Ti(OH)_4$ can generate hydroxyl radicals when it comes in contact with water, even after numerous photodegradation cycles [188]. In another study, a bladder cancer cell line, MB49, was treated with various concentrations of $Ti(OH)_{4,}$ and the results demonstrated that photo exposure of $Ti(OH)_4$ stimulated ROS generation and induced dose-dependent necrosis in cancer cells [190]. Black TiO_2 NPs were used as photosensitizers triggered by near-infrared light with maximum 808 nm absorbance by T24 cells (bladder cancer cells). The cells were incubated with TiO_2 NPs and irradiated at 808 nm. The results showed concentration-dependent enhanced antitumor activity by the black TiO_2 NPs. Hence, black TiO_2 was proven a potent anticancer agent, promising photosensitizer, and maximally active at near-infrared and visible light [191].

6.9 Skin cancer

Skin cancer is the most common human malignancy due to the uncontrolled growth of tumor cells associated with the dermis and epidermis. Patients need recurrent treatment due to the aggravated and repetitive growth of tumor cells and, therefore, suffer from treatment-associated side effects and toxicity. Though the topical chemotherapeutic option is associated with less severe side effects, it is impeded due to the rapid liquifying characteristic of the polymers used in the therapy and tormenting-sized microneedles [192, 193].

Melanoma is a type of skin cancer that appears in melanocytes (skin cells) [194]. Melanocytes are the producers of melanin, which gives color to the skin [4, 195]. Ultraviolet radiations are the leading cause of melanoma, adversely affecting DNA repair, skin cell growth [196], immunosurveillance, and apoptosis. These adverse reactions allow the activation of oncogene or deactivation of tumor suppressor genes and subsequent tumor development [197]. Clinically, nanoparticles are shown to have the ability of tumor reduction and lessen the side effects [198–200]. Conventional anticancer therapies, including chemotherapy, radiotherapy, and surgery, are associated with the risk of harming adjacent healthy cells. This problem can be overcome using chemotherapeutic agents conjugated nanoparticles that can precisely target tumor cells [201, 202]. TiO_2 NPs possess unique characteristics and have been applied in various fields [203]. They also have immunomodulatory effects [204].

Titanium dioxide nanotubes (TNT) offer a larger surface for carrying molecules and have distinct physicochemical properties. They are potent anticancer agents. They have been conjugated with quercetin to evaluate their effect against melanoma. Quercetin is a flavonoid found in fruits and leafy vegetables and possesses antioxidant, antiviral, and anticancer effects. The in vitro anticancer effect of quercetin-conjugated TNT (TNT-Qu) was evaluated on melanoma cells (B16F10). The results showed inhibitory effects of TNT-Qu on the migration of B16F10 cells, enhanced DNA fragmentation, and cell cycle arrest in the cells. Moreover, TNT-Qu was more cytotoxic to the B16F10 cells than quercetin or TNT alone [205]. The anticancer effect of TNT-Qu was also evaluated on the B16F10 mouse melanoma model and two-stage chemical carcinogenesis in vivo model. The study's results demonstrated enhanced antitumor effects of TNT-Qu than either of the two alone by the topical application of TNT-Qu. TNT-Qu treatment inhibited tumor growth and increased the survival time of the two-stage chemical carcinogenesis mice models [206]. TiO_2 exhibits full-size dependent immunomodulatory effects in the nanorod form [207]. TiO_2 NPs were hydrothermally converted to nanorods that greatly enhanced the loading efficiency of resveratrol, which would be a great anticancer agent for skin cancer [208]. Polyvinyl Alcohol (PVA) is biocompatible, hydrophilic, and biodegradable [209]. PVA nanofibers are a dressing material for wound healing [210, 211]. Conjugating a polymeric form of PVA with a pharmaceutical agent improves EPR and facilitates the slow and sustained release of the incorporated drugs [212]. Ekambaram et al. reported the anticancer effect of the green synthesized TiO_2 nanorods loaded with resveratrol-incorporated nanofibers against skin cancer cells (A431). They found inhibition in cancer cell growth by activating caspase enzymes [213].

6.10 Hematological malignancies

Hematological malignancies originate from the bone marrow or blood and result from the acquisition of genetic abnormalities that lead to unrestrained proliferation, resistance to cell death, and evasion of the immune system [214]. The occurrences of hematological malignancies, including leukemia, multiple myeloma, lymphoma, myelodysplastic syndromes, and myeloproliferative neoplasm, continuously increase despite recent advances which increased the five-year rate in many types of hematological malignancies [215]. Photodynamic therapy (PDT) has advantages over conventional anticancer therapy, including no risk of drug resistance and controllable ROS generation by controlled dosimetry [216–218]. TiO_2 NPs have been used in many cancer types [40, 42, 219–221], but the biggest hurdle is the high energy band gap of TiO_2 (anatase, 3.2 EV) which needs the excitation by detrimental UV radiations. Doping of TiO_2 with metal/non-metals resolves this issue by making TiO_2 able to activate by absorbing light of longer wavelengths [222–224]. N-TiO_2 exhibits anticancer activity and higher capability of ROS production in comparison to TiO_2 NPs [39, 225, 226]. N-TiO_2 was used as a photosensitizer in PDT for leukemia cells. Upon activation with visible light, N-TiO_2 photosensitizers induced ROS-mediated autophagy in leukemia cells (K562), which increased with the increasing doses of light and photosensitizer. In addition, low doses of PDT also showed enhanced ROS and autophagy in normal peripheral lymphocytes. However, the typical human cell model showed no cytotoxic or inhibitory effects [41].

Acute lymphoblastic leukemia occurs due to the abnormal growth of white blood cells in the bone marrow [227, 228]. It is the most common cancer in children 2–5 years of age [229]. The treatment advancements show 90% effectiveness in curing the disease, but relapse and drug resistance remain the most significant clinical challenge [230]. Recently, using nanostructured devices and nanomaterials to deliver medications against cancer is the most advanced method for treating cancer [231]. Metal nanocomposites are being investigated for theranostics, and various functional groups are being incorporated to modify metal/metal oxide nanocomposites [232]. Recently, ZnO-TiO_2-chitosan-amygdalin nanoparticles have gained much interest as potent anticancer agents. MOLT-4 (T-lymphoblast malignant cells) were treated with nanocomposite (ZnO-TiO_2-chitosan-amygdalin) to evaluate its cytotoxic effect on these cells. The results showed increased cytotoxicity, mitochondrial membrane depolarization, caspase activation, and ROS generation in leukemia cells [233].

6.11 Oral cancer

Oral Squamous Cell Carcinoma (OSCC) is characterized by local hypoxia and tumoral necrosis spreading on a large area, which is the cause of drug resistance and low chemotherapeutic response [234]. Immune suppression is also a factor that limits the therapeutic response and poor prognosis [235]. The primary therapy is surgical resection for OSCC, while radiotherapy and chemotherapy are additional treatment options [236]. However, with all the present treatment options, the five-year survival rate is still 60%, which severely damages the life quality [237]. Photodynamic theory utilizing nanoparticles as photosensitizers has gained much attention for OSCC cure and prevention [238, 239]. TiO_2 NPs have widely investigated nanoparticles as photosensitizers in photodynamic therapy since their photocatalytic activity was discovered in 1972 [240–242]. Metal polypoidal complexes have attracted scientists as photosensitizers. Ru(II) complex TLD-1433 photosensitizers have been used in clinical trials for bladder cancer (non-muscle invasive bladder cancer) in Canada [243, 244]. TLD-1433 can potentially cause DNA damage under hypoxic conditions [243, 245]. Based on this phenomenon, TiO_2@Ru@siRNA nanocomposite comprised SiRNA-loaded TiO_2 NPs modified with ruthenium-based photosensitizers. This nanocomposite shows photodynamic effects upon irradiation with visible light. It can cause lysosomal damage, HIF-1α gene silencing, production of type I and type II ROS, and eradication of OSCC cells efficiently. In addition, it also reduces the expression of immunosuppressive factors and elevates the antitumor immune response. The PDX and oral rat carcinoma model significantly improved antitumor immunity and inhibited tumor progression and growth [246]. Pure TiO_2 and TiO_2 nanoparticles modified with ginger, garlic, and turmeric were used for anticancer activity against KB oral cell line by Maheshwari et al. They found that modified TiO_2 showed better anticancer activity against oral cancer cells than pure TiO_2 [247].

7. Conclusion

In summary, the nano titania application in cancer therapy and diagnosis is highly favorable due to its biocompatible and porous nature, surface modification, and ROS generation properties. The TiO_2 surface can be coated with polymeric and metallic nanostructures to enhance drug loading ability and target desired tissue viz. tumor. Due to their inert nature, nano titania is commonly implemented as food additives and cosmetic products. However, UV light application limits its photoactivation, which is inconsistent with WHO recommended therapeutic window (600–1000 nm). Indeed, their surface coating or nanocomposite formation can shift its absorption from UV to NIR range, which holds promising effects in anticancer therapy and diagnosis via bioimaging. Their photodynamic or photothermal therapy effect suits topical and body cavity cancer resection. Employing titanium nanoparticles as drug carriers for anticancer therapy might help improve therapeutic effects and avoid undesirable side effects. Combining titanium NPs with other nanoparticles also holds great therapeutic potential in cancer. The applications of nano titania and their conjugates discussed in this chapter can be utilized to improve cancer theranostics.

Acknowledgements

AAM acknowledges HEC Pakistan and World Bank's Grand Challenge Fund (GCF)-543 and National Research Program for Universities. ZC acknowledges Southeast University Postdoctoral Foundation (1107032211).

Conflict of interest

The authors declare no conflict of interest.

Author details

Rida e Maria Qazi[1], Zahra Sajid[1], Chunqiu Zhao[2], Fawad Ur Rehman[1*] and Afsar Ali Mian[1*]

1 Centre for Regenerative Medicine and Stem Cells Research, The Aga Khan University, Karachi, Sindh, Pakistan

2 State Key Lab of Bioelectronics, Southeast University, Nanjing Jiangsu, China

*Address all correspondence to: rehman.fawad@aku.edu and afsar.mian@aku.edu

References

[1] Katz E, Willner I. Integrated nanoparticle–biomolecule hybrid systems: Synthesis, properties, and applications. Angewandte Chemie International Edition. 2004;**43**(45):6042-6108. DOI: 10.1002/anie.200400651

[2] Jafari S, Mahyad B, Hashemzadeh H, Janfaza S, Gholikhani T, Tayebi L. Biomedical applications of TiO2 nanostructures: Recent advances. International Journal of Nanomedicine. 2020;**15**:3447. DOI: 10.2147/IJN.S249441

[3] Hashimoto K, Irie H, Fujishima A. TiO_2 photocatalysis: A historical overview and future prospects. Japanese Journal of Applied Physics. 2005;**44**(12R):8269. DOI: 10.1143/JJAP.44.8269

[4] Fernández-Garcia M, RODGRIGUEZ J. Metal Oxide Nanoparticles. Upton, NY (United States): Brookhaven National Lab.(BNL); 2007

[5] Grover T, Pandey A, Kumari ST, Awasthi A, Singh B, Dixit P, et al. Role of titanium in bio implants and additive manufacturing: An overview. Materials Today: Proceedings. 2020;**26**:3071-3080. DOI: 10.1016/j.matpr.2020.02.636

[6] Huang B-H, Lu Y-J, Lan W-C, Ruslin M, Lin H-Y, Ou K-L, et al. Surface properties and biocompatibility of anodized titanium with a potential pretreatment for biomedical applications. Metals. 2021;**11**(7):1090. DOI: 10.3390/met11071090

[7] Gojznikar J, Zdravković B, Vidak M, Leskošek B, Ferk P. TiO2 nanoparticles and their effects on eukaryotic cells: A double-edged sword. International Journal of Molecular Sciences. 2022;**23**(20):12353. DOI: 10.3390/ijms232012353

[8] Hou J, Wang L, Wang C, Zhang S, Liu H, Li S, et al. Toxicity and mechanisms of action of titanium dioxide nanoparticles in living organisms. Journal of Environmental Sciences. 2019;**75**:40-53. DOI: 10.1016/j.jes.2018.06.010

[9] Diebold U. The surface science of titanium dioxide. Surface Science Reports. 2003;**48**(5-8):53-229. DOI: 10.1016/S0167-5729(02)00100-0

[10] Sun S, Song P, Cui J, Liang S. Amorphous TiO 2 nanostructures: Synthesis, fundamental properties and photocatalytic applications. Catalysis Science & Technology. 2019;**9**(16):4198-4215. DOI: 10.1039/C9CY01020C

[11] Musial J, Krakowiak R, Mlynarczyk DT, Goslinski T, Stanisz BJ. Titanium dioxide nanoparticles in food and personal care products—What do we know about their safety? Nanomaterials. 2020;**10**(6):1110. DOI: 10.3390/nano10061110

[12] Fiordaliso F, Bigini P, Salmona M, Diomede L. Toxicological impact of titanium dioxide nanoparticles and food-grade titanium dioxide (E171) on human and environmental health. Environmental Science: Nano. 2022;**2022**:1199-1211. DOI: 10.1039/D1EN00833A

[13] Bischoff NS, de Kok TM, Sijm DT, van Breda SG, Briedé JJ, Castenmiller JJ, et al. Possible adverse effects of food additive E171 (titanium dioxide) related to particle specific human toxicity, including the immune system. International Journal of Molecular Sciences. 2020;**22**(1):207. DOI: 10.3390/ijms22010207

[14] Lu N, Chen Z, Song J, Weng Y, Yang G, Liu Q, et al. Size effect of TiO2

nanoparticles as food additive and potential toxicity. Food Biophysics. 2022;**17**(1):75-83. DOI: 10.1007/s11483-021-09695-7

[15] Vance ME, Kuiken T, Vejerano EP, McGinnis SP, Hochella MF Jr, Rejeski D, et al. Nanotechnology in the real world: Redeveloping the nanomaterial consumer products inventory. Beilstein Journal of Nanotechnology. 2015;**6**(1):1769-1780. DOI: 10.3762/bjnano.6.181

[16] Lagopati N, Evangelou K, Falaras P, Tsilibary E-PC, Vasileiou PV, Havaki S, et al. Nanomedicine: Photo-activated nanostructured titanium dioxide, as a promising anticancer agent. Pharmacology & Therapeutics. 2021;**222**:107795. DOI: 10.1016/j.pharmthera.2020.107795

[17] Vamvakas I, Lagopati N, Andreou M, Sotiropoulos M, Gatzis A, Limouris G, et al. Patient specific computer automated dosimetry calculations during therapy with 111In octreotide. European Journal of Radiography. 2009;**1**(4):180-183. DOI: 10.1016/j.ejradi.2010.08.001

[18] Cai R, Hashimoto K, Itoh K, Kubota Y, Fujishima A. Photokilling of malignant cells with ultrafine TiO2 powder. Bulletin of the Chemical Society of Japan. 1991;**64**(4):1268-1273. DOI: 10.1246/bcsj.64.1268

[19] Mahendran D, Kavi Kishor P, Geetha N, Manish T, Sahi S, Venkatachalam P. Efficient antibacterial/biofilm, anti-cancer and photocatalytic potential of titanium dioxide nanocatalysts green synthesised using Gloriosa superba rhizome extract. Journal of Experimental Nanoscience. 2021;**16**(1):11-30. DOI: 10.1080/17458080.2021.1872781

[20] Xing Y, Yi R, Yang H, Xu Q, Huang R, Tang J, et al. Antifungal effect of chitosan/Nano-TiO2 composite coatings against Colletotrichum gloeosporioides, Cladosporium oxysporum and Penicillium steckii. Molecules. 2021;**26**(15):4401. DOI: 10.3390/molecules26154401

[21] Maneerat C, Hayata Y. Antifungal activity of TiO2 photocatalysis against Penicillium expansum in vitro and in fruit tests. International Journal of Food Microbiology. 2006;**107**(2):99-103. DOI: 10.1016/j.ijfoodmicro.2005.08.018

[22] Yan P, Liu L-H, Wang P. Sonodynamic therapy (SDT) for cancer treatment: Advanced sensitizers by ultrasound activation to injury tumor. ACS Applied Bio Materials. 2020;**3**(6):3456-3475. DOI: 10.1021/acsabm.0c00156

[23] Costley D, Mc Ewan C, Fowley C, McHale AP, Atchison J, Nomikou N, et al. Treating cancer with sonodynamic therapy: A review. International Journal of Hyperthermia. 2015;**31**(2):107-117. DOI: 10.3109/02656736.2014.992484

[24] McEwan C, Nesbitt H, Nicholas D, Kavanagh ON, McKenna K, Loan P, et al. Comparing the efficacy of photodynamic and sonodynamic therapy in non-melanoma and melanoma skin cancer. Bioorganic & Medicinal Chemistry. 2016;**24**(13):3023-3028. DOI: 10.1016/j.bmc.2016.05.015

[25] McHale AP, Callan JF, Nomikou N, Fowley C, Callan B. Sonodynamic therapy: Concept, mechanism and application to cancer treatment. Therapeutic. Ultrasound. 2016;**2016**: 429-450. DOI: 10.1007/978-3-319-22536-4_22

[26] Pan X, Wang H, Wang S, Sun X, Wang L, Wang W, et al. Sonodynamic therapy (SDT): A novel strategy for cancer nanotheranostics. Science China

Life Sciences. 2018;**61**(4):415-426. DOI: 10.1007/s11427-017-9262-x

[27] Shimizu N, Ogino C, Dadjour MF, Murata T. Sonocatalytic degradation of methylene blue with TiO2 pellets in water. Ultrasonics Sonochemistry. 2007;**14**(2):184-190. DOI: 10.1016/j.ultsonch.2006.04.002

[28] Nejad SM, Takahashi H, Hosseini H, Watanabe A, Endo H, Narihira K, et al. Acute effects of sono-activated photocatalytic titanium dioxide nanoparticles on oral squamous cell carcinoma. Ultrasonics Sonochemistry. 2016;**32**:95-101. DOI: 10.1016/j.ultsonch.2016.02.026

[29] Harada Y, Ogawa K, Irie Y, Endo H, Feril LB Jr, Uemura T, et al. Ultrasound activation of TiO2 in melanoma tumors. Journal of Controlled Release. 2011;**149**(2):190-195. DOI: 10.1016/j.jconrel.2010.10.012

[30] Aksel M, Kesmez Ö, Yavaş A, Bilgin MD. Titaniumdioxide mediated sonophotodynamic therapy against prostate cancer. Journal of Photochemistry and Photobiology B: Biology. 2021;**225**:112333. DOI: 10.1016/j.jphotobiol.2021.112333

[31] Li W. Nanoparticles for photodynamic therapy. In: Handbook of Biophotonics. Vol. 2. Weinheim, Germany: Wiley Online Library; 2013

[32] Yang G, Xu L, Xu J, Zhang R, Song G, Chao Y, et al. Smart nanoreactors for pH-responsive tumor homing, mitochondria-targeting, and enhanced photodynamic-immunotherapy of cancer. Nano Letters. 2018;**18**(4):2475-2484. DOI: 10.1021/acs.nanolett.8b00040

[33] Shi J, Chen Z, Wang B, Wang L, Lu T, Zhang Z. Reactive oxygen species-manipulated drug release from a smart envelope-type mesoporous titanium nanovehicle for tumor sonodynamic-chemotherapy. ACS Applied Materials & Interfaces. 2015;**7**(51):28554-28565. DOI: 10.1021/acsami.5b09937

[34] Yang C-C, Sun Y-J, Chung P-H, Chen W-Y, Swieszkowski W, Tian W, et al. Development of Ce-doped TiO2 activated by X-ray irradiation for alternative cancer treatment. Ceramics International. 2017;**43**(15):12675-12683. DOI: 10.1016/j.ceramint.2017.06.149

[35] Kanpittaya K, Teerakapong A, Morales NP, Hormdee D, Priprem A, Weera-Archakul W, et al. Inhibitory effects of erythrosine/curcumin derivatives/nano-titanium dioxide-mediated photodynamic therapy on Candida albicans. Molecules. 2021;**26**(9):2405. DOI: 10.3390/molecules26092405

[36] Kang X, Liu S, Dai Z, He Y, Song X, Tan Z. Titanium dioxide: From engineering to applications. Catalysts. 2019;**9**(2):191. DOI: 10.3390/catal9020191

[37] Linsebigler AL, Lu G, Yates JT Jr. Photocatalysis on TiO2 surfaces: Principles, mechanisms, and selected results. Chemical Reviews. 1995;**95**(3):735-758. DOI: 10.1021/cr00035a013

[38] Shang H, Han D, Ma M, Li S, Xue W, Zhang A. Enhancement of the photokilling effect of TiO2 in photodynamic therapy by conjugating with reduced graphene oxide and its mechanism exploration. Journal of Photochemistry and Photobiology B: Biology. 2017;**177**:112-123. DOI: 10.1016/j.jphotobiol.2017.10.016

[39] Li Z, Pan X, Wang T, Wang P-N, Chen J-Y, Mi L. Comparison of the killing effects between nitrogen-doped and pure TiO 2 on HeLa cells with

visible light irradiation. Nanoscale Research Letters. 2013;**8**(1):1-7. DOI: 10.1186/1556-276X-8-96

[40] Lagopati N, Tsilibary E-P, Falaras P, Papazafiri P, Pavlatou EA, Kotsopoulou E, et al. Effect of nanostructured TiO2 crystal phase on photoinduced apoptosis of breast cancer epithelial cells. International Journal of Nanomedicine. 2014;**9**:3219. DOI: 10.2147/IJN.S62972

[41] Moosavi MA, Sharifi M, Ghafary SM, Mohammadalipour Z, Khataee A, Rahmati M, et al. Photodynamic N-TiO2 nanoparticle treatment induces controlled ROS-mediated autophagy and terminal differentiation of leukemia cells. Scientific Reports. 2016;**6**(1):1-16. DOI: 10.1038/srep34413

[42] Wang Y, Cui H, Zhou J, Li F, Wang J, Chen M, et al. Cytotoxicity, DNA damage, and apoptosis induced by titanium dioxide nanoparticles in human non-small cell lung cancer A549 cells. Environmental Science and Pollution Research. 2015;**22**(7):5519-5530. DOI: 10.1007/s11356-014-3717-7

[43] Ghosh M, Bandyopadhyay M, Mukherjee A. Genotoxicity of titanium dioxide (TiO2) nanoparticles at two trophic levels: Plant and human lymphocytes. Chemosphere. 2010;**81**(10):1253-1262. DOI: 10.1016/j.chemosphere.2010.09.022

[44] Lagopati N, Kitsiou P, Kontos A, Venieratos P, Kotsopoulou E, Kontos A, et al. Photo-induced treatment of breast epithelial cancer cells using nanostructured titanium dioxide solution. Journal of Photochemistry and Photobiology A: Chemistry. 2010;**214**(2-3):215-223. DOI: 10.1016/j.jphotochem.2010.06.031

[45] Ramachandran P, Khor B-K, Lee CY, Doong R-A, Oon CE, Thanh NTK, et al. N-doped graphene quantum dots/titanium dioxide nanocomposites: A study of ROS-forming mechanisms, cytotoxicity and photodynamic therapy. Biomedicine. 2022;**10**(2):421. DOI: 10.3390/biomedicines10020421

[46] Wang Q, Jian-Ying H, Li H-Q, Chen Z, Zhao AZ-J, Wang Y, et al. TiO 2 nanotube platforms for smart drug delivery: A review. International Journal of Nanomedicine. 2016;**11**:4819. DOI: 10.2147/IJN.S108847

[47] Motiei Pour M, Moghbeli MR, Larijani B, Akbari JH. pH-sensitive mesoporous bisphosphonate-based TiO2 nanoparticles utilized for controlled drug delivery of dexamethasone. Chemical Papers. 2022;**76**(1):439-451. DOI: 10.1007/s11696-021-01870-x

[48] Levina AS, Repkova MN, Ismagilov ZR, Shikina NV, Malygin EG, Mazurkova NA, et al. High-performance method for specific effect on nucleic acids in cells using TiO2~ DNA nanocomposites. Scientific Reports. 2012;**2**(1):1-6. DOI: 10.1038/srep00756

[49] Salahuddin N, Abdelwahab M, Gaber M, Elneanaey S. Synthesis and Design of Norfloxacin drug delivery system based on PLA/TiO2 nanocomposites: Antibacterial and antitumor activities. Materials Science and Engineering: C. 2020;**108**:110337. DOI: 10.1016/j.msec.2019.110337

[50] Gulen B, Demircivi P. Synthesis and characterization of montmorillonite/ciprofloxacin/TiO2 porous structure for controlled drug release of ciprofloxacin tablet with oral administration. Applied Clay Science. 2020;**197**:105768. DOI: 10.1016/j.clay.2020.105768

[51] Rosu M-C, Bratu I. Promising psyllium-based composite containing TiO2 nanoparticles as aspirin-carrier

matrix. Progress in Natural Science: Materials International. 2014;**24**(3):205-209. DOI: 10.1016/j.pnsc.2014.05.007

[52] Rehman FU, Rauf MA, Ullah S, Shaikh S, Qambrani A, Muhammad P, et al. Ultrasound-activated nano-TiO2 loaded with temozolomide paves the way for resection of chemoresistant glioblastoma multiforme. Cancer Nanotechnology. 2021;**12**(1):1-17. DOI: 10.1186/s12645-021-00088-6

[53] Kim S, Im S, Park E-Y, Lee J, Kim C, Kim T-i, et al. Drug-loaded titanium dioxide nanoparticle coated with tumor targeting polymer as a sonodynamic chemotherapeutic agent for anti-cancer therapy. Nanomedicine: Nanotechnology, Biology and Medicine. 2020;**24**:102110. DOI: 10.1016/j.nano.2019.102110

[54] Faria HAM, de Queiroz AAA. A novel drug delivery of 5-fluorouracil device based on TiO2/ZnS nanotubes. Materials Science and Engineering: C. 2015;**56**:260-268. DOI: 10.1016/j.msec.2015.06.008

[55] Khoee MH, Khoee S, Lotfi M. Synthesis of titanium dioxide nanotubes with liposomal covers for carrying and extended release of 5-FU as anticancer drug in the treatment of HeLa cells. Analytical Biochemistry. 2019;**572**:16-24. DOI: 10.1016/j.ab.2019.02.027

[56] Wang Y, Yuan L, Yao C, Fang J, Wu M. Cytotoxicity evaluation of pH-controlled antitumor drug release system of titanium dioxide nanotubes. Journal of Nanoscience and Nanotechnology. 2015;**15**(6):4143-4148. DOI: 10.1166/jnn.2015.9792

[57] Venkatasubbu GD, Ramasamy S, Ramakrishnan V, Kumar J. Folate targeted PEGylated titanium dioxide nanoparticles as a nanocarrier for targeted paclitaxel drug delivery. Advanced Powder Technology. 2013;**24**(6):947-954. DOI: 10.1016/j.apt.2013.01.008

[58] Zhao C, Ur Rehman F, Yang Y, Li X, Zhang D, Jiang H, et al. Bio-imaging and photodynamic therapy with tetra sulphonatophenyl porphyrin (TSPP)-TiO2 nanowhiskers: New approaches in rheumatoid arthritis theranostics. Scientific Reports. 2015;**5**(1):1-11. DOI: 10.1038/srep11518

[59] Rehman FU, Zhao C, Wu C, Li X, Jiang H, Selke M, et al. Synergy and translation of allogenic bone marrow stem cells after photodynamic treatment of rheumatoid arthritis with tetra sulfonatophenyl porphyrin and TiO2 nanowhiskers. Nano Research. 2016;**9**(11):3305-3321. DOI: 10.1007/s12274-016-1208-5

[60] Rehman FU, Zhao C, Jiang H, Selke M, Wang X. Protective effect of TiO2 nanowhiskers on tetra sulphonatophenyl porphyrin (TSPP) complexes induced oxidative stress during photodynamic therapy. Photodiagnosis and Photodynamic Therapy. 2016;**13**:267-275. DOI: 10.1016/j.pdpdt.2015.08.005

[61] Cleary AS, Leonard TL, Gestl SA, Gunther EJ. Tumour cell heterogeneity maintained by cooperating subclones in Wnt-driven mammary cancers. Nature. 2014;**508**(7494):113-117. DOI: 10.1038/nature13187

[62] Counihan JL, Grossman EA, Nomura DK. Cancer metabolism: Current understanding and therapies. Chemical Reviews. 2018;**118**(14):6893-6923. DOI: 10.1021/acs.chemrev.7b00775

[63] Anand P, Kunnumakara AB, Sundaram C, Harikumar KB, Tharakan ST, Lai OS, et al. Cancer is a preventable disease that requires major lifestyle

changes. Pharmaceutical Research. 2008;**25**(9):2097-2116. DOI: 10.1007/s11095-008-9661-9

[64] Trachootham D, Alexandre J, Huang P. Targeting cancer cells by ROS-mediated mechanisms: A radical therapeutic approach? Nature Reviews Drug Discovery. 2009;**8**(7):579-591. DOI: 10.1038/nrd2803

[65] Yang B, Chen Y, Shi J. Reactive oxygen species (ROS)-based nanomedicine. Chemical Reviews. 2019;**119**(8):4881-4985. DOI: 10.1021/acs.chemrev.8b00626

[66] Zhang Z, Bragg LM, Servos MR, Liu J. Gold nanoparticles as dehydrogenase mimicking nanozymes for estradiol degradation. Chinese Chemical Letters. 2019;**30**(9):1655-1658. DOI: 10.1016/j.cclet.2019.05.062

[67] Yang S, Zhou L, Su Y, Zhang R, Dong C-M. One-pot photoreduction to prepare NIR-absorbing plasmonic gold nanoparticles tethered by amphiphilic polypeptide copolymer for synergistic photothermal-chemotherapy. Chinese Chemical Letters. 2019;**30**(1):187-191. DOI: 10.1016/j.cclet.2018.02.015

[68] He H, Liu L, Morin EE, Liu M, Schwendeman A. Survey of clinical translation of cancer nanomedicines—Lessons learned from successes and failures. Accounts of Chemical Research. 2019;**52**(9):2445-2461. DOI: 10.1021/acs.accounts.9b00228

[69] Sun H, Dong Y, Feijen J, Zhong Z. Peptide-decorated polymeric nanomedicines for precision cancer therapy. Journal of Controlled Release. 2018;**290**:11-27. DOI: 10.1016/j.jconrel.2018.09.029

[70] Shi J, Kantoff PW, Wooster R, Farokhzad OC. Cancer nanomedicine: Progress, challenges and opportunities. Nature reviews Cancer. 2017;**17**(1):20-37. DOI: 10.1038/nrc.2016.108

[71] Riley RS, June CH, Langer R, Mitchell MJ. Delivery technologies for cancer immunotherapy. Nature reviews Drug Discovery. 2019;**18**(3):175-196. DOI: 10.1038/s41573-018-0006-z

[72] Milling L, Zhang Y, Irvine DJ. Delivering safer immunotherapies for cancer. Advanced Drug Delivery Reviews. 2017;**114**:79-101. DOI: 10.1016/j.addr.2017.05.011

[73] Zhou J, Rao L, Yu G, Cook TR, Chen X, Huang F. Supramolecular cancer nanotheranostics. Chemical Society Reviews. 2021;**50**(4):2839-2891. DOI: 10.1039/d0cs00011f

[74] Ulbrich K, Hola K, Subr V, Bakandritsos A, Tucek J, Zboril R. Targeted drug delivery with polymers and magnetic nanoparticles: Covalent and noncovalent approaches, release control, and clinical studies. Chemical Reviews. 2016;**116**(9):5338-5431. DOI: 10.1021/acs.chemrev.5b00589

[75] Stewart MP, Sharei A, Ding X, Sahay G, Langer R, Jensen KF. In vitro and ex vivo strategies for intracellular delivery. Nature. 2016;**538**(7624):183-192. DOI: 10.1038/nature19764

[76] Sabri T, Pawelek PD, Capobianco JA. Dual activity of rose Bengal functionalized to albumin-coated lanthanide-doped upconverting nanoparticles: Targeting and photodynamic therapy. ACS Applied Materials & Interfaces. 2018;**10**(32):26947-26953. DOI: 10.1021/acsami.8b08919

[77] Gu T, Cheng L, Gong F, Xu J, Li X, Han G, et al. Upconversion composite nanoparticles for tumor hypoxia

modulation and enhanced near-infrared-triggered photodynamic therapy. ACS Applied Materials & Interfaces. 2018;**10**(18):15494-15503. DOI: 10.1021/acsami.8b03238

[78] Cheng L, Wang C, Feng L, Yang K, Liu Z. Functional nanomaterials for phototherapies of cancer. Chemical Reviews. 2014;**114**(21):10869-10939. DOI: 10.1021/cr400532z

[79] Dong X, Liang J, Yang A, Qian Z, Kong D, Lv F. Fluorescence imaging guided CpG nanoparticles-loaded IR820-hydrogel for synergistic photothermal immunotherapy. Biomaterials. 2019;**209**:111-125. DOI: 10.1016/j.biomaterials.2019.04.024

[80] Liang R, Li Y, Huo M, Lin H, Chen Y. Triggering sequential catalytic Fenton reaction on 2D MXenes for hyperthermia-augmented synergistic nanocatalytic cancer therapy. ACS Applied Materials & Interfaces. 2019;**11**(46):42917-42931. DOI: 10.1021/acsami.9b13598

[81] Guo Q, Wang D, Yang G. Photoacoustic imaging guided photothermal and chemodynamic combined therapy for cancer using. Journal of Biomedical Nanotechnology. 2019;**15**(10):2090-2099. DOI: 10.1166/jbn.2019.2832

[82] Liu S, Wang L, Lin M, Wang D, Song Z, Li S, et al. Cu (II)-doped polydopamine-coated gold nanorods for tumor theranostics. ACS Applied Materials & Interfaces. 2017;**9**(51):44293-44306. DOI: 10.1021/acsami.7b13643

[83] An L, Cao M, Zhang X, Lin J, Tian Q, Yang S. pH and glutathione synergistically triggered release and self-assembly of Au nanospheres for tumor theranostics. ACS Applied Materials & Interfaces. 2020;**12**(7):8050-8061. DOI: 10.1021/acsami.0c00302

[84] Soares S, Sousa J, Pais A, Vitorino C. Nanomedicine: Principles, properties, and regulatory issues. Frontiers in Chemistry. 2018;**6**:360. DOI: 10.3389/fchem.2018.00360

[85] Dong P, Rakesh K, Manukumar H, YHE M, Karthik C, Sumathi S, et al. Innovative nano-carriers in anticancer drug delivery-a comprehensive review. Bioorganic Chemistry. 2019;**85**:325-326. DOI: 10.1016/j.bioorg.2019.01.019

[86] Zhang J, Wang Q, Liu J, Guo Z, Yang J, Li Q, et al. Saponin-based near-infrared nanoparticles with aggregation-induced emission behavior: Enhancing cell compatibility and permeability. ACS Applied Bio Materials. 2019;**2**(2):943-951. DOI: 10.1021/acsabm.8b00812

[87] Ji X, Wang C, Tang M, Guo D, Peng F, Zhong Y, et al. Biocompatible protamine sulfate@ silicon nanoparticle-based gene nanocarriers featuring strong and stable fluorescence. Nanoscale. 2018;**10**(30):14455-14463. DOI: 10.1039/c8nr03107j

[88] Racca L, Canta M, Dumontel B, Ancona A, Limongi T. Zinc oxide nanostructures in biomedicine. Smart Nanoparticles Biomedicine. 2018;**10**:B978. DOI: 10.1016/B978-0-12-814156-4.00012-4

[89] De Matteis V, Cascione M, Toma CC, Leporatti S. Silver nanoparticles: Synthetic routes, in vitro toxicity and theranostic applications for cancer disease. Nanomaterials. 2018;**8**(5):319. DOI: 10.3390/nano8050319

[90] Limongi T, Canta M, Racca L, Ancona A, Tritta S, Vighetto V, et al. Improving dispersal of therapeutic

nanoparticles in the human body. Nanomedicine. 2019;**14**(7):797-801. DOI: 10.2217/nnm-2019-0070

[91] Youn YS, Bae YH. Perspectives on the past, present, and future of cancer nanomedicine. Advanced Drug Delivery Reviews. 2018;**130**:3-11. DOI: 10.1016/j.addr.2018.05.008

[92] Liang L, Yue Z, Du W, Li Y, Tao H, Wang D, et al. Molecular imaging of inducible VEGF expression and tumor progression in a breast cancer model. Cellular Physiology and Biochemistry. 2017;**42**(1):407-415. DOI: 10.1159/000477485

[93] Siegel RL, Miller KD, Jemal A. Cancer statistics, 2018. CA: A Cancer Journal for Clinicians. 2018;**68**(1):7-30. DOI: 10.3322/caac.21442

[94] Baneshi M, Dadfarnia S, Shabani AMH, Sabbagh SK, Haghgoo S, Bardania H. A novel theranostic system of AS1411 aptamer-functionalized albumin nanoparticles loaded on iron oxide and gold nanoparticles for doxorubicin delivery. International Journal of Pharmaceutics. 2019;**564**:145-152. DOI: 10.1016/j.ijpharm.2019.04.025

[95] Dhankhar R, Vyas SP, Jain AK, Arora S, Rath G, Goyal AK. Advances in novel drug delivery strategies for breast cancer therapy. Artificial Cells, Blood Substitutes, and Biotechnology. 2010;**38**(5):230-249. DOI: 10.3109/10731199.2010.494578

[96] Gao S, Li X, Ding X, Qi W, Yang Q. Cepharanthine induces autophagy, apoptosis and cell cycle arrest in breast cancer cells. Cellular Physiology and Biochemistry. 2017;**41**(4):1633-1648. DOI: 10.1159/000471234

[97] Qin N, Lu S, Chen N, Chen C, Xie Q, Wei X, et al. Yulangsan polysaccharide inhibits 4T1 breast cancer cell proliferation and induces apoptosis in vitro and in vivo. International Journal of Biological Macromolecules. 2019;**121**:971-980. DOI: 10.1016/j.ijbiomac.2018.10.082

[98] Kwapisz D. Pembrolizumab and atezolizumab in triple-negative breast cancer. Cancer Immunology, Immunotherapy. 2021;**70**(3):607-617. DOI: 10.1007/s00262-020-02736-z

[99] Menaa F. When pharma meets nano or the emerging era of nanopharmaceuticals. Pharm Anal Acta. 2013;**4**:223. DOI: 10.4172/2153-2435.1000223

[100] Batool A, Menaa F, Uzair B, Khan BA, Menaa B. Progress and prospects in translating nanobiotechnology in medical theranostics. Current Nanoscience. 2020;**16**(5):685-707. DOI: 10.2174/1573413715666191126093258

[101] de Melo GD, Buzaid AC, Perez-Garcia J, Cortes J. Immunotherapy in breast cancer: Current practice and clinical challenges. BioDrugs. 2020;**34**(5):611-623. DOI: 10.1007/s40259-020-00436-9

[102] von Roemeling C, Jiang W, Chan CK, Weissman IL, Kim BY. Breaking down the barriers to precision cancer nanomedicine. Trends in Biotechnology. 2017;**35**(2):159-171. DOI: 10.1016/j.tibtech.2016.07.006

[103] Chowdhury D, Paul A, Chattopadhyay A. Photocatalytic polypyrrole– TiO2– nanoparticles composite thin film generated at the air– water interface. Langmuir. 2005;**21**(9):4123-4128. DOI: 10.1021/la0475425

[104] Kim H, Jeon D, Oh S, Nam K, Son S, Gye MC, et al. Titanium dioxide nanoparticles induce apoptosis by

interfering with EGFR signaling in human breast cancer cells. Environmental Research. 2019;**175**:117-123. DOI: 10.1016/j.envres.2019.05.001

[105] Zheng J, Lee HCM. Bin Sattar MM, Huang Y, Bian J-S. Cardioprotective effects of epigallocatechin-3-gallate against doxorubicin-induced cardiomyocyte injury. European Journal of Pharmacology. 2011;**652**(1-3):82-88. DOI: 10.1016/j.ejphar.2010.10.082

[106] Ibsen S, Zahavy E, Wrasdilo W, Berns M, Chan M, Esener S. A novel doxorubicin prodrug with controllable photolysis activation for cancer chemotherapy. Pharmaceutical Research. 2010;**27**(9):1848-1860. DOI: 10.1007/s11095-010-0183-x

[107] Iqbal H, Razzaq A, Uzair B, Ul Ain N, Sajjad S, Althobaiti NA, et al. Breast cancer inhibition by biosynthesized titanium dioxide nanoparticles is comparable to free doxorubicin but appeared safer in balb/c mice. Materials. 2021;**14**(12):3155. DOI: 10.3390/ma14123155

[108] Siegel RL, Miller KD, Fedewa SA, Ahnen DJ, Meester RG, Barzi A, et al. Colorectal cancer statistics, 2017. CA: A cancer journal for clinicians. 2017;**67**(3):177-193. DOI: DOI, 10.3322/caac.21395

[109] Mizrahi JD, Surana R, Valle JW, Shroff RT. Pancreatic cancer. The Lancet. 2020;**395**(10242):2008-2020. DOI: 10.1016/S0140-6736(20)30974-0

[110] Hayashi A, Hong J, Iacobuzio-Donahue CA. The pancreatic cancer genome revisited. Nature Reviews Gastroenterology & Hepatology. 2021;**18**(7):469-481. DOI: 10.1038/s41575-021-00463-z

[111] Souchek JJ, Baine MJ, Lin C, Rachagani S, Gupta S, Kaur S, et al. Unbiased analysis of pancreatic cancer radiation resistance reveals cholesterol biosynthesis as a novel target for radiosensitisation. British Journal of Cancer. 2014;**111**(6):1139-1149. DOI: 10.1038/bjc.2014.385

[112] Strobel O, Neoptolemos J, Jäger D, Büchler MW. Optimizing the outcomes of pancreatic cancer surgery. Nature reviews Clinical Oncology. 2019;**16**(1):11-26. DOI: 10.1038/s41571-018-0112-1

[113] Natale CA, Li J, Pitarresi JR, Norgard RJ, Dentchev T, Capell BC, et al. Pharmacologic activation of the g protein–coupled estrogen receptor inhibits pancreatic ductal adenocarcinoma. Cellular and Molecular Gastroenterology and Hepatology. 2020;**10**(4):868-880. DOI: 10.1016/j.jcmgh.2020.04.016

[114] Pereira SP, Oldfield L, Ney A, Hart PA, Keane MG, Pandol SJ, et al. Early detection of pancreatic cancer. The Lancet Gastroenterology & Hepatology. 2020;**5**(7):698-710. DOI: 10.1016/S2468-1253(19)30416-9

[115] Zhang H, Dai Z, Wu W, Wang Z, Zhang N, Zhang L, et al. Regulatory mechanisms of immune checkpoints PD-L1 and CTLA-4 in cancer. Journal of Experimental & Clinical Cancer Research. 2021;**40**(1):1-22. DOI: 10.1186/s13046-021-01987-7

[116] Lisi L, Lacal PM, Martire M, Navarra P, Graziani G. Clinical experience with CTLA-4 blockade for cancer immunotherapy: From the monospecific monoclonal antibody ipilimumab to probodies and bispecific molecules targeting the tumor microenvironment. Pharmacological Research. 2022;**175**:105997. DOI: 10.1016/j.phrs.2021.105997

[117] Esmaily M, Masjedi A, Hallaj S, Afjadi MN, Malakotikhah F, Ghani S,

et al. Blockade of CTLA-4 increases anti-tumor response inducing potential of dendritic cell vaccine. Journal of Controlled Release. 2020;**326**:63-74. DOI: 10.1016/j.jconrel.2020.06.017

[118] Zhu G, Lynn GM, Jacobson O, Chen K, Liu Y, Zhang H, et al. Albumin/vaccine nanocomplexes that assemble in vivo for combination cancer immunotherapy. Nature Communications. 2017;**8**(1):1-15. DOI: 10.1038/s41467-017-02191-y

[119] Morrison AH, Byrne KT, Vonderheide RH. Immunotherapy and prevention of pancreatic cancer. Trends in Cancer. 2018;**4**(6):418-428. DOI: 10.1016/j.trecan.2018.04.001

[120] Winograd R, Simeone DM, Bar-Sagi D. A novel target for combination immunotherapy in pancreatic cancer: IL-1β mediates immunosuppression in the tumour microenvironment. British Journal of Cancer. 2021;**124**(11):1754-1756. DOI: 10.1038/s41416-021-01303-2

[121] Shang J, Han X, Zha H, Tao H, Li X, Yuan F, et al. Systemic immune-inflammation index and changes of neutrophil-lymphocyte ratio as prognostic biomarkers for patients with pancreatic cancer treated with immune checkpoint blockade. Frontiers in Oncology. 2021;**11**:585271. DOI: 10.3389/fonc.2021.585271

[122] Asadzadeh Z, Safarzadeh E, Safaei S, Baradaran A, Mohammadi A, Hajiasgharzadeh K, et al. Current approaches for combination therapy of cancer: The role of immunogenic cell death. Cancers. 2020;**12**(4):1047. DOI: 10.3390/cancers12041047

[123] Wang J, Meng J, Ran W, Lee RJ, Teng L, Zhang P, et al. Hepatocellular carcinoma growth retardation and PD-1 blockade therapy potentiation with synthetic high-density lipoprotein. Nano Letters. 2019;**19**(8):5266-5276. DOI: 10.1021/acs.nanolett.9b01717

[124] Lamberti MJ, Nigro A, Mentucci FM, Rumie Vittar NB, Casolaro V, Dal CJ. Dendritic cells and immunogenic cancer cell death: A combination for improving antitumor immunity. Pharmaceutics. 2020;**12**(3):256. DOI: 10.3390/pharmaceutics12030256

[125] Shao S, Wang S, Ren L, Wang J, Chen X, Pi H, et al. Layer-by-layer assembly of lipid nanobubbles on microneedles for ultrasound-assisted transdermal drug delivery. ACS Applied Bio Materials. 2022;**5**(2):562-569. DOI: 10.1021/acsabm.1c01049

[126] Dong F, An J, Zhang J, Yin J, Guo W, Wang D, et al. Blinking acoustic nanodroplets enable fast super-resolution ultrasound imaging. ACS Nano. 2021;**15**(10):16913-16923. DOI: 10.1021/acsnano.1c07896

[127] Bhargava N, Mor RS, Kumar K, Sharanagat VS. Advances in application of ultrasound in food processing: A review. Ultrasonics Sonochemistry. 2021;**70**:105293. DOI: 10.1016/j.ultsonch.2020.105293

[128] Jeanjean P, El Hamrani D, Genevois C, Quesson B, Couillaud F. Combination of MRI-guided high-intensity focused ultrasound and bioluminescent biological systems to assess thermal therapies for tumor and tumor microenvironment. Advanced Materials Technologies. 2022;**2022**:2101258. DOI: 10.1002/admt.202101258

[129] Wang X, Shang M, Sun X, Guo L, Xiao S, Shi D, et al. Dual-responsive nanodroplets combined with ultrasound-targeted microbubble destruction

suppress tumor growth and metastasis via autophagy blockade. Journal of Controlled Release. 2022;**343**:66-77. DOI: 10.1016/j.jconrel.2022.01.009

[130] Ren J, Zhou J, Liu H, Jiao X, Cao Y, Xu Z, et al. Ultrasound (US)-activated redox dyshomeostasis therapy reinforced by immunogenic cell death (ICD) through a mitochondrial targeting liposomal nanosystem. Theranostics. 2021;**11**(19):9470. DOI: 10.7150/thno.62984

[131] Nowak KM, Schwartz MR, Breza VR, Price RJ. Sonodynamic therapy: Rapid progress and new opportunities for non-invasive tumor cell killing with sound. Cancer Letters. 2022;**2022**:215592. DOI: 10.1016/j.canlet.2022.215592

[132] Zhan G, Xu Q, Zhang Z, Wei Z, Yong T, Bie N, et al. Biomimetic sonodynamic therapy-nanovaccine integration platform potentiates Anti-PD-1 therapy in hypoxic tumors. Nano Today. 2021;**38**:101195. DOI: 10.1016/j.nantod.2021.101195

[133] Duo Y, Luo G, Li Z, Chen Z, Li X, Jiang Z, et al. Photothermal and enhanced photocatalytic therapies conduce to synergistic anticancer phototherapy with biodegradable titanium diselenide nanosheets. Small. 2021;**17**(40):2103239. DOI: 10.1002/smll.202103239

[134] Chen L, Xue W, Cao J, Zhang S, Zeng Y, Ma L, et al. TiSe2-mediated sonodynamic and checkpoint blockade combined immunotherapy in hypoxic pancreatic cancer. Journal of Nanobiotechnology. 2022;**20**(1):1-14. DOI: 10.1186/s12951-022-01659-4

[135] Zhu Y, Sui B, Liu X, Sun J. The reversal of drug resistance by two-dimensional titanium carbide Ti2C (2D Ti2C) in non-small-cell lung cancer via the depletion of intracellular antioxidant reserves. Thoracic Cancer. 2021;**12**(24):3340-3355. DOI: 10.1111/1759-7714.14208

[136] Mei L, Zhu S, Yin W, Chen C, Nie G, Gu Z, et al. Two-dimensional nanomaterials beyond graphene for antibacterial applications: Current progress and future perspectives. Theranostics. 2020;**10**(2):757. DOI: 10.7150/thno.39701

[137] Balachandran K, Mageswari S, Preethi A. Photocatalytic decomposition of A549-lung cancer cancer cells by TiO2 nanoparticles. Materials Today: Proceedings. 2021;**37**:1071-1074. DOI: 10.1016/j.matpr.2020.06.297

[138] Kukia NR, Rasmi Y, Abbasi A, Koshoridze N, Shirpoor A, Burjanadze G, et al. Bio-effects of TiO2 nanoparticles on human colorectal cancer and umbilical vein endothelial cell lines. Asian Pacific Journal of Cancer Prevention: APJCP. 2018;**19**(10):2821. DOI: 10.22034/APJCP.2018.19.10.2821

[139] Wei X, Liu Y, El-kott A, Ahmed AE, Khames A. Calendula officinalis-based green synthesis of titanium nanoparticle: Fabrication, characterization, and evaluation of human colorectal carcinoma. Journal of Saudi Chemical Society. 2021;**25**(11):101343. DOI: 10.1016/j.jscs.2021.101343

[140] Vigneshwaran R, Ezhilarasan D, Rajeshkumar S. Inorganic titanium dioxide nanoparticles induces cytotoxicity in colon cancer cells. Inorganic Chemistry Communications. 2021;**133**:108920. DOI: 10.1016/j.inoche.2021.108920

[141] Sung H, Ferlay J, Siegel RL, Laversanne M, Soerjomataram I, Jemal A, et al. Global cancer statistics

2020: GLOBOCAN estimates of incidence and mortality worldwide for 36 cancers in 185 countries. CA: A Cancer Journal for Clinicians. 2021;**71**(3):209-249. DOI: 10.3322/caac.21660

[142] Burke WM, Orr J, Leitao M, Salom E, Gehrig P, Olawaiye AB, et al. Endometrial cancer: A review and current management strategies: Part II. Gynecologic Oncology. 2014;**134**(2):393-402. DOI: 10.1016/j.ygyno.2014.06.003

[143] Guillotin D, Martin SA. Exploiting DNA mismatch repair deficiency as a therapeutic strategy. Experimental Cell Research. 2014;**329**(1):110-115. DOI: 10.1016/j.yexcr.2014.07.004

[144] Goorani S, Shariatifar N, Seydi N, Zangeneh A, Moradi R, Tari B, et al. The aqueous extract of Allium saralicum RM Fritsch effectively treat induced anemia: Experimental study on Wistar rats. Oriental Pharmacy and Experimental Medicine. 2019;**19**(4):403-413. DOI: 10.1007/s13596-019-00361-5

[145] Prasad KS, Shivamallu C, Shruthi G, Prasad M. A novel and one-pot green synthesis of vanadium oxide nanorods using a phytomolecule isolated from *Phyllanthus amarus*. ChemistrySelect. 2018;**3**(13):3860-3865. DOI: 10.1002/slct.201800653

[146] Moradi R, Hajialiani M, Salmani S, Almasi M, Zangeneh A, Zangeneh MM. Effect of aqueous extract of Allium saralicum RM Fritsch on fatty liver induced by high-fat diet in Wistar rats. Comparative Clinical Pathology. 2019;**28**(5):1205-1211. DOI: 10.1007/s00580-018-2834-y

[147] Su YH, Yin ZF, Xin HL, Zhang HQ, Sheng JY, Yang YL, et al. Optimized antimicrobial and antiproliferative activities of titanate nanofibers containing silver. International Journal of Nanomedicine. 2011;**6**:1579. DOI: 10.2147/IJN.S18765

[148] Seo JH, Jeon WI, Dembereldorj U, Lee SY, Joo S-W. Cytotoxicity of serum protein-adsorbed visible-light photocatalytic Ag/AgBr/TiO2 nanoparticles. Journal of Hazardous Materials. 2011;**198**:347-355. DOI: 10.1016/j.jhazmat.2011.10.059

[149] Flak D, Yate L, Nowaczyk G, Jurga S. Hybrid ZnPc@ TiO2 nanostructures for targeted photodynamic therapy, bioimaging and doxorubicin delivery. Materials Science and Engineering: C. 2017;**78**:1072-1085. DOI: 10.1016/j.msec.2017.04.107

[150] Yurt F, Ocakoglu K, Ince M, Colak SG, Er O, Soylu HM, et al. Photodynamic therapy and nuclear imaging activities of zinc phthalocyanine-integrated TiO2 nanoparticles in breast and cervical tumors. Chemical Biology & Drug Design. 2018;**91**(3):789-796. DOI: 10.1111/cbdd.13144

[151] Pandurangan M, Enkhtaivan G, Young JA, Hoon HJ, Lee H, Lee S, et al. In vitro therapeutic potential of Tio2 nanoparticles against human cervical carcinoma cells. Biological Trace Element Research. 2016;**171**(2):293-300

[152] Hodson R. The brain. Nature. 2019;**571**(7766):S1. DOI: 10.1038/d41586-019-02206-2

[153] Kim S-S, Harford JB, Pirollo KF, Chang EH. Effective treatment of glioblastoma requires crossing the blood–brain barrier and targeting tumors including cancer stem cells: The promise of nanomedicine. Biochemical and Biophysical Research Communications. 2015;468(3):485-489. DOI: 10.1016/j.bbrc.2015.06.137

[154] Adamson C, Kanu OO, Mehta AI, Di C, Lin N, Mattox AK, et al.

Glioblastoma multiforme: A review of where we have been and where we are going. Expert Opinion on Investigational Drugs. 2009;**18**(8):1061-1083. DOI: 10.1517/13543780903052764

[155] Harilal S, Jose J, Kumar R, Unnikrishnan MK, Uddin MS, Mathew GE, et al. Revisiting the blood-brain barrier: A hard nut to crack in the transportation of drug molecules. Brain Research Bulletin. 2020;**160**:121-140. DOI: 10.1016/j.brainresbull.2020.03.018

[156] Feng M. Assessment of blood-brain barrier penetration: In silico, in vitro and in vivo. Current Drug Metabolism. 2002;**3**(6):647-657. DOI: 10.2174/1389200023337063

[157] Ciura K, Dziomba S. Application of separation methods for in vitro prediction of blood–brain barrier permeability—The state of the art. Journal of Pharmaceutical and Biomedical Analysis. 2020;**177**:112891. DOI: 10.1016/j.jpba.2019.112891

[158] Upadhyay RK. Drug delivery systems, CNS protection, and the blood brain barrier. BioMed Research International. 2014;**2014**:1-37. DOI: 10.1155/2014/869269

[159] Rehman FU, Liu Y, Zheng M, Shi B. Exosomes based strategies for brain drug delivery. Biomaterials. 2022;**2002**:121949. DOI: 10.1016/j.biomaterials.2022.121949

[160] Rehman FU, Liu Y, Yang Q, Yang H, Liu R, Zhang D, et al. Heme Oxygenase-1 targeting exosomes for temozolomide resistant glioblastoma synergistic therapy. Journal of Controlled Release. 2022;**345**:696-708. DOI: 10.1016/j.jconrel.2022.03.036

[161] Del Prado-Audelo ML, Caballero-Florán IH, Meza-Toledo JA, Mendoza-Muñoz N, González-Torres M, Florán B, et al. Formulations of curcumin nanoparticles for brain diseases. Biomolecules. 2019;**9**(2):56. DOI: 10.3390/biom9020056

[162] Mulvihill JJ, Cunnane EM, Ross AM, Duskey JT, Tosi G, Grabrucker AM. Drug delivery across the blood–brain barrier: Recent advances in the use of nanocarriers. Nanomedicine. 2020;**15**(2):205-214. DOI: 10.2217/nnm-2019-0367

[163] Saeedi M, Eslamifar M, Khezri K, Dizaj SM. Applications of nanotechnology in drug delivery to the central nervous system. Biomedicine & Pharmacotherapy. 2019;**111**:666-675. DOI: 10.1016/j.biopha.2018.12.133

[164] Boyes WK, van Thriel C. Neurotoxicology of nanomaterials. Chemical Research in Toxicology. 2020;**33**(5):1121-1144. DOI: 10.1021/acs.chemrestox.0c00050

[165] Jia J, Wang Z, Yue T, Su G, Teng C, Yan B. Crossing biological barriers by engineered nanoparticles. Chemical Research in Toxicology. 2020;**33**(5):1055-1060. DOI: 10.1021/acs.chemrestox.9b00483

[166] Lee J, Jeong J-S, Kim SY, Park M-K, Choi S-D, Kim U-J, et al. Titanium dioxide nanoparticles oral exposure to pregnant rats and its distribution. Particle and Fibre Toxicology. 2019;**16**(1):1-12. DOI: 10.1186/s12989-019-0313-5

[167] Kao Y-Y, Cheng T-J, Yang D-M, Wang C-T, Chiung Y-M, Liu P-S. Demonstration of an olfactory bulb–brain translocation pathway for ZnO nanoparticles in rodent cells in vitro and in vivo. Journal of Molecular Neuroscience. 2012;**48**(2):464-471. DOI: 10.1007/s12031-012-9756-y

[168] Fuster E, Candela H, Estévez J, Vilanova E, Sogorb MA. Titanium dioxide,

but not zinc oxide, nanoparticles cause severe transcriptomic alterations in T98G human glioblastoma cells. International Journal of Molecular Sciences. 2021;**22**(4):2084. DOI: 10.3390/ijms22042084

[169] Marino A, Almici E, Migliorin S, Tapeinos C, Battaglini M, Cappello V, et al. Piezoelectric barium titanate nanostimulators for the treatment of glioblastoma multiforme. Journal of Colloid and Interface Science. 2019;**538**:449-461. DOI: 10.1016/j.jcis.2018.12.014

[170] Lin H-P, Lin C-Y, Liu C-C, Su L-C, Huo C, Kuo Y-Y, et al. Caffeic acid phenethyl ester as a potential treatment for advanced prostate cancer targeting akt signaling. International Journal of Molecular Sciences. 2013;**14**(3):5264-5283. DOI: 10.3390/ijms14035264

[171] Szliszka E, Krol W. Soy isoflavones augment the effect of TRAIL-mediated apoptotic death in prostate cancer cells. Oncology Reports. 2011;**26**(3):533-541. DOI: 10.3892/or.2011.1332

[172] Cimino S, Sortino G, Favilla V, Castelli T, Madonia M, Sansalone S, et al. Polyphenols: Key issues involved in chemoprevention of prostate cancer. Oxidative Medicine and Cellular Longevity. 2012;**2012**:1-8. DOI: 10.1155/2012/632959

[173] Ayca T, Cakmak NK, Agbektas T, Silig Y. Cytotoxic activity of zinc oxide/titanium dioxide nanoparticles on prostate cancer cells. International Journal of Chemistry and Technology. 2019;**3**(2):113-120. DOI: 10.32571/ijct.613536

[174] Fan X, Zhao X, Qu X, Fang J. pH sensitive polymeric complex of cisplatin with hyaluronic acid exhibits tumor-targeted delivery and improved in vivo antitumor effect. International Journal of Pharmaceutics. 2015;**496**(2):644-653. DOI: 10.1016/j.ijpharm.2015.10.066

[175] Zhao Q, Liu J, Zhu W, Sun C, Di D, Zhang Y, et al. Dual-stimuli responsive hyaluronic acid-conjugated mesoporous silica for targeted delivery to CD44-overexpressing cancer cells. Acta Biomaterialia. 2015;**23**:147-156. DOI: 10.1016/j.actbio.2015.05.010

[176] Hou Z, Zhang Y, Deng K, Chen Y, Li X, Deng X, et al. UV-emitting upconversion-based TiO2 photosensitizing nanoplatform: Near-infrared light mediated in vivo photodynamic therapy via mitochondria-involved apoptosis pathway. ACS Nano. 2015;**9**(3):2584-2599. DOI: 10.1021/nn506107c

[177] Lucky SS, Muhammad Idris N, Li Z, Huang K, Soo KC, Zhang Y. Titania coated upconversion nanoparticles for near-infrared light triggered photodynamic therapy. ACS Nano. 2015;**9**(3):191-205. DOI: 10.1021/nn503450t

[178] Ninomiya K, Ogino C, Oshima S, Sonoke S. Kuroda S-i, Shimizu N. targeted sonodynamic therapy using protein-modified TiO2 nanoparticles. Ultrasonics Sonochemistry. 2012;**19**(3):607-614. DOI: 10.1016/j.ultsonch.2011.09.009

[179] Ninomiya K, Fukuda A, Ogino C, Shimizu N. Targeted sonocatalytic cancer cell injury using avidin-conjugated titanium dioxide nanoparticles. Ultrasonics Sonochemistry. 2014;**21**(5):1624-1628. DOI: 10.1016/j.ultsonch.2014.03.010

[180] Ninomiya K, Noda K, Ogino C, Kuroda S, Shimizu N. Enhanced OH radical generation by dual-frequency ultrasound with TiO2 nanoparticles: Its application to targeted sonodynamic therapy. Ultrasonics Sonochemistry. 2014;**21**(1):289-294. DOI: 10.1016/j.ultsonch.2013.05.005

[181] Yuan P, Song D. MRI tracing non-invasive TiO2-based nanoparticles activated by ultrasound for multi-mechanism therapy of prostatic cancer*. Nanotechnology. 2018;**29**(12):125101. DOI: 10.1088/1361-6528/aaa92a

[182] Lenis AT, Lec PM, Chamie K. Bladder cancer: A review. Journal of the American Medical Association. 2020;**324**(19):1980-1991

[183] Burger M, Oosterlinck W, Konety B, Chang S, Gudjonsson S, Pruthi R, et al. ICUD-EAU international consultation on bladder Cancer 2012: Non–muscle-invasive urothelial carcinoma of the bladder. European Urology. 2013;**63**(1):36-44. DOI: 10.1016/j.eururo.2012.08.061

[184] Schrier BP, Hollander MP, van Rhijn BW, Kiemeney LA, Witjes JA. Prognosis of muscle-invasive bladder cancer: Difference between primary and progressive tumours and implications for therapy. European Urology. 2004;**45**(3):292-296. DOI: 10.1016/j.eururo.2003.10.006

[185] Bhindi B, Kool R, Kulkarni GS, Siemens DR, Aprikian AG, Breau RH, et al. Canadian Urological Association guideline on the management of non-muscle-invasive bladder cancer–full-text. Canadian Urological Association Journal. 2021;**15**(8):E424. DOI: 10.5489/cuaj.7367

[186] Jain P, Kathuria H, Momin M. Clinical therapies and nano drug delivery systems for urinary bladder cancer. Pharmacology & Therapeutics. 2021;**226**:107871. DOI: 10.1016/j.pharmthera.2021.107871

[187] Hong JK, Bang JY, Xu G, Lee J-H, Kim Y-J, Lee H-J, et al. Thickness-controllable electrospun fibers promote tubular structure formation by endothelial progenitor cells. International Journal of Nanomedicine. 2015;**10**:1189. DOI: 10.2147/IJN.S73096

[188] Nogueira AE, Ribeiro LS, Gorup LF, Silva GT, Silva FF, Ribeiro C, et al. New approach of the oxidant peroxo method (OPM) route to obtain Ti (OH) 4 nanoparticles with high photocatalytic activity under visible radiation. International Journal of Photoenergy. 2018;**2018**:1-10. DOI: 10.1155/2018/6098302

[189] Ribeiro LS, Nogueira AE, Aquino JM, Camargo ER. A new strategy to obtain nano-scale particles of lithium titanate (Li4Ti5O12) by the oxidant peroxo method (OPM). Ceramics International. 2019;**45**(18):23917-23923. DOI: 10.1016/j.ceramint.2019.07.274

[190] Robeldo T, Ribeiro LS, Manrique L, Kubo AM, Longo E, Camargo ER, et al. Modified titanium dioxide as a potential visible-light-activated photosensitizer for bladder Cancer treatment. ACS Omega. 2022;**2022**:17563-17574. DOI: 10.1021/acsomega.1c07046

[191] Ni W, Li M, Cui J, Xing Z, Li Z, Wu X, et al. 808 nm light triggered black TiO2 nanoparticles for killing of bladder cancer cells. Materials Science and Engineering: C. 2017;**81**:252-260. DOI: 10.1016/j.msec.2017.08.020

[192] Lim D-J, Vines JB, Park H, Lee S-H. Microneedles: A versatile strategy for transdermal delivery of biological molecules. International Journal of Biological Macromolecules. 2018;**110**:30-38. DOI: 10.1016/j.ijbiomac.2017.12.027

[193] Mishra H, Mishra PK, Ekielski A, Jaggi M, Iqbal Z, Talegaonkar S. Melanoma treatment: From conventional to nanotechnology. Journal of Cancer Research and Clinical Oncology. 2018;**144**(12):2283-2302. DOI: 10.1007/s00432-018-2726-1

[194] Garrubba C, Donkers K. Skin cancer. Journal of the American Academy of PAs. 2020;**33**(2):49-50. DOI: 10.1097/01.JAA.0000651756.15106.3e

[195] Cichorek M, Wachulska M, Stasiewicz A, Tymińska A. Skin melanocytes: Biology and development. Advances in Dermatology and Allergology/Postępy Dermatologii i Alergologii. 2013;**30**(1):30-41. DOI: 10.5114/pdia.2013.33376

[196] Nataraj AJ, Trent JC 2nd, Ananthaswamy HN. p53 gene mutations and photocarcinogenesis. Photochemistry and Photobiology. 1995;**62**(2):218-230. DOI: 10.1111/j.1751-1097.1995.tb05262.x

[197] Soehnge H, Ouhtit A, Ananthaswamy HN. Mechanisms of induction of skin cancer by UV radiation. Frontiers in Bioscience-Landmark. 1997;**2**(4):538-551. DOI: 10.2741/a211

[198] Gulla S, Lomada D, Srikanth VV, Shankar MV, Reddy KR, Soni S, et al. Recent advances in nanoparticles-based strategies for cancer therapeutics and antibacterial applications. Methods in Microbiology. 2019;**46**:255-293. DOI: 10.1016/bs.mim.2019.03.003

[199] Subbiah R, Veerapandian M. Nanoparticles: Functionalization and multifunctional applications in biomedical sciences. Current Medicinal Chemistry. 2010;**17**(36):4559-4577. DOI: 10.2174/092986710794183024

[200] Ashree J, Wang Q, Chao Y. Glyco-functionalised quantum dots and their progress in cancer diagnosis and treatment. Frontiers of Chemical Science and Engineering. 2020;**14**(3):365-377. DOI: 10.1007/s11705-019-1863-7

[201] Hu C-MJ, Zhang L. Therapeutic nanoparticles to combat cancer drug resistance. Current Drug Metabolism. 2009;**10**(8):836-841. DOI: 10.2174/138920009790274540

[202] Dadwal A, Baldi A, Kumar Narang R. Nanoparticles as carriers for drug delivery in cancer. Artificial Cells, Nanomedicine, and Biotechnology. 2018;**46**(Sup 2):295-305. DOI: 10.1080/21691401.2018.1457039

[203] Zhou W, Liu H, Boughton RI, Du G, Lin J, Wang J, et al. One-dimensional single-crystalline Ti–O based nanostructures: Properties, synthesis, modifications and applications. Journal of Materials Chemistry. 2010;**20**(29):5993-6008. DOI: 10.1039/B927224K

[204] Latha TS, Reddy MC, Durbaka PV, Muthukonda SV, Lomada D. Immunomodulatory properties of titanium dioxide nanostructural materials. Indian Journal of Pharmacology. 2017;**49**(6):458. DOI: 10.4103/ijp.IJP_536_16

[205] Gulla S, Lomada D, Araveti PB, Srivastava A, Murikinati MK, Reddy KR, et al. Titanium dioxide nanotubes conjugated with quercetin function as an effective anticancer agent by inducing apoptosis in melanoma cells. Journal of Nanostructure in Chemistry. 2021;**11**(4):721-734. DOI: 10.1007/s40097-021-00396-8

[206] Gulla S, Reddy VC, Araveti PB, Lomada D, Srivastava A, Reddy MC, et al. Synthesis of titanium dioxide nanotubes (TNT) conjugated with quercetin and its in vivo antitumor activity against skin cancer. Journal of Molecular Structure. 2022;**1249**:131556. DOI: 10.1016/j.molstruc.2021.131556

[207] Wang Y, Yao C, Ding L, Li C, Wang J, Wu M, et al. Enhancement of the immune function by titanium dioxide

nanorods and their application in cancer immunotherapy. Journal of Biomedical Nanotechnology. 2017;**13**(4):367-380. DOI: 10.1166/jbn.2017.2323

[208] Annu AA, Ahmed S. Green synthesis of metal, metal oxide nanoparticles, and their various applications. Handbook of Ecomaterials. 2018;**2018**:1-45. DOI: 10.1007/978-3-319-48281-1_115-1

[209] Jiang S, Liu S, Feng W. PVA hydrogel properties for biomedical application. Journal of the Mechanical Behavior of Biomedical Materials. 2011;**4**(7):1228-1233. DOI: 10.1016/j.jmbbm.2011.04.005

[210] Morgado PI, Miguel SP, Correia IJ, Aguiar-Ricardo A. Ibuprofen loaded PVA/chitosan membranes: A highly efficient strategy towards an improved skin wound healing. Carbohydrate Polymers. 2017;**159**:136-145. DOI: 10.1016/j.carbpol.2016.12.029

[211] Kheradmandi M, Vasheghani-Farahani E, Ghiaseddin A, Ganji F. Skeletal muscle regeneration via engineered tissue culture over electrospun nanofibrous chitosan/PVA scaffold. Journal of Biomedical Materials Research Part A. 2016;**104**(7):1720-1727. DOI: 10.1002/jbm.a.35702

[212] Kayal S, Ramanujan R. Doxorubicin loaded PVA coated iron oxide nanoparticles for targeted drug delivery. Materials Science and Engineering: C. 2010;**30**(3):484-490. DOI: 10.1016/j.msec.2010.01.006

[213] Ekambaram R, Saravanan S, Selvam N, Dharmalingam S. Statistical optimization of novel acemannan polysaccharides assisted TiO2 nanorods based nanofibers for skin cancer application. Carbohydrate Polymer Technologies and Applications. 2021;**2**:100048. DOI: 10.1016/j.carpta.2021.100048

[214] Rahman S, Mansour MR. The role of noncoding mutations in blood cancers. Disease Models & Mechanisms. 2019;**12**(11):dmm041988. DOI: 10.1242/dmm.041988

[215] Pulte D, Jansen L, Brenner H. Changes in long term survival after diagnosis with common hematologic malignancies in the early 21st century. Blood Cancer Journal. 2020;**10**(5):1-8. DOI: 10.1038/s41408-020-0323-4

[216] Kushibiki T, Tu Y, Abu-Yousif AO, Hasan T. Photodynamic activation as a molecular switch to promote osteoblast cell differentiation via AP-1 activation. Scientific Reports. 2015;**5**(1):1-11. DOI: 10.1038/srep13114

[217] Tada DB, Baptista MS. Photosensitizing nanoparticles and the modulation of ROS generation. Frontiers in Chemistry. 2015;**3**:33. DOI: 10.3389/fchem.2015.00033

[218] Agostinis P, Berg K, Cengel KA, Foster TH, Girotti AW, Gollnick SO, et al. Photodynamic therapy of cancer: An update. CA: A Cancer Journal for Clinicians. 2011;**61**(4):250-281. DOI: 10.3322/caac.20114

[219] Cai R, Kubota Y, Shuin T, Sakai H, Hashimoto K, Fujishima A. Induction of cytotoxicity by photoexcited TiO2 particles. Cancer Research. 1992;**52**(8):2346-2348

[220] Huang N-P, Min-hua X, Yuan C-W, Rui-rong Y. The study of the photokilling effect and mechanism of ultrafine TiO2 particles on U937 cells. Journal of Photochemistry and Photobiology A: Chemistry. 1997;**108**(2-3):229-233. DOI: 10.1016/S1010-6030(97)00093-2

[221] Manke A, Wang L, Rojanasakul Y. Mechanisms of nanoparticle-induced oxidative stress and toxicity. BioMed

Research International. 2013;**2013**:1-15. DOI: 10.1155/2013/942916

[222] Khataee A, Fathinia M, Aber S, Zarei M. Optimization of photocatalytic treatment of dye solution on supported TiO2 nanoparticles by central composite design: Intermediates identification. Journal of Hazardous Materials. 2010;**181**(1-3):886-897. DOI: 10.1016/j.jhazmat.2010.05.096

[223] Flak D, Coy E, Nowaczyk G, Yate L, Jurga S. Tuning the photodynamic efficiency of TiO 2 nanotubes against HeLa cancer cells by Fe-doping. RSC Advances. 2015;**5**(103):85139-85152. DOI: 10.1039/c5ra17430a

[224] Townley HE, Rapa E, Wakefield G, Dobson PJ. Nanoparticle augmented radiation treatment decreases cancer cell proliferation. Nanomedicine: Nanotechnology, Biology and Medicine. 2012;**8**(4):526-536. DOI: 10.1016/j.nano.2011.08.003

[225] Li Z, Mi L, Wang P-N, Chen J-Y. Study on the visible-light-induced photokilling effect of nitrogen-doped TiO 2 nanoparticles on cancer cells. Nanoscale Research Letters. 2011;**6**(1):1-7. DOI: 10.1186/1556-276X-6-356

[226] Pan X, Xie J, Li Z, Chen M, Wang M, Wang P-N, et al. Enhancement of the photokilling effect of aluminum phthalocyanine in photodynamic therapy by conjugating with nitrogen-doped TiO2 nanoparticles. Colloids and Surfaces B: Biointerfaces. 2015;**130**:292-298. DOI: 10.1016/j.colsurfb.2015.04.028

[227] Terwilliger T, Abdul-Hay M. Acute lymphoblastic leukemia: A comprehensive review and 2017 update. Blood Cancer Journal. 2017;**7**(6):e577. DOI: 10.1038/bcj.2017.53

[228] Chang JHC, Poppe MM, Hua CH, Marcus KJ, Esiashvili N. Acute lymphoblastic leukemia. Pediatric Blood & Cancer. 2021;**68**:e28371. DOI: 10.1002/pbc.28371

[229] Kato M, Manabe A. Treatment and biology of pediatric acute lymphoblastic leukemia. Pediatrics International. 2018;**60**(1):4-12. DOI: 10.1111/ped.13457

[230] Balsat M, Cacheux V, Carre M, Tavernier-Tardy E, Thomas X. Treatment and outcome of Philadelphia chromosome-positive acute lymphoblastic leukemia in adults after relapse. Expert Review of Anticancer Therapy. 2020;**20**(10):879-891. DOI: 10.1080/14737140.2020.1832890

[231] Javed R, Rais F, Fatima H, et al. Chitosan encapsulated ZnO nanocomposites: Fabrication, characterization, and functionalization of bio-dental approaches. Materials Science and Engineering: C. 2020;**116**:111184. DOI: 10.1016/j.msec.2020.111184

[232] Kustiningsih I, Ridwan A, Abriyani D, Syairazy M, Kurniawan T, Dhena RB. Development of chitosan-TiO2 nanocomposite for packaging film and its ability to inactive *Staphylococcus aureus*. Oriental Journal of Chemistry. 2019;**35**(3):1132. DOI: 10.13005/ojc/350329

[233] Elderdery AY, Alzahrani B, Alanazi F, Hamza SM, Elkhalifa AM, Alhamidi AH, et al. Amelioration of human acute lymphoblastic leukemia (ALL) cells by ZnO-TiO2-chitosan-amygdalin nanocomposites. Arabian Journal of Chemistry. 2022;**15**(8):103999. DOI: 10.1016/j.arabjc.2022.103999

[234] Pérez-Sayáns M, Suárez-Peñaranda JM, Pilar G-D, Barros-Angueira F, Gándara-Rey JM, García-García A. Hypoxia-inducible factors in OSCC. Cancer Letters.

2011;**313**(1):1-8. DOI: 10.1016/j.canlet.2011.08.017

[235] Eckert AW, Wickenhauser C, Salins PC, Kappler M, Bukur J, Seliger B. Clinical relevance of the tumor microenvironment and immune escape of oral squamous cell carcinoma. Journal of Translational Medicine. 2016;**14**(1):1-13. DOI: 10.1186/s12967-016-0828-6

[236] Neville BW, Day TA. Oral cancer and precancerous lesions. CA: A Cancer Journal for Clinicians. 2002;**52**(4):195-215. DOI: 10.3322/canjclin.52.4.195

[237] Chinn SB, Myers JN. Oral cavity carcinoma: Current management, controversies, and future directions. Journal of Clinical Oncology. 2015;**33**(29):3269. DOI: 10.1200/JCO.2015.61.2929

[238] H-y F, Zhu Z-l, W-l Z, Y-j Y, Y-l T, Liang X-h, et al. Light stimulus responsive nanomedicine in the treatment of oral squamous cell carcinoma. European Journal of Medicinal Chemistry. 2020;**199**:112394. DOI: 10.1016/j.ejmech.2020.112394

[239] Saini R, Lee NV, Liu KY, Poh CF. Prospects in the application of photodynamic therapy in oral cancer and premalignant lesions. Cancers. 2016;**8**(9):83. DOI: 10.3390/cancers8090083

[240] Fujishima A, Honda K. Electrochemical photolysis of water at a semiconductor electrode. Nature. 1972;**238**(5358):37-38. DOI: 10.1038/238037a0

[241] Rajh T, Dimitrijevic NM, Bissonnette M, Koritarov T, Konda V. Titanium dioxide in the service of the biomedical revolution. Chemical Reviews. 2014;**114**(19):10177-10216. DOI: 10.1021/cr500029g

[242] Rehman F, Zhao C, Jiang H, Wang X. Biomedical applications of nano-titania in theranostics and photodynamic therapy. Biomaterials Science. 2016;**4**(1):40-54. DOI: 10.1039/c5bm00332f

[243] Monro S, Colon KL, Yin H, Roque J III, Konda P, Gujar S, et al. Transition metal complexes and photodynamic therapy from a tumor-centered approach: Challenges, opportunities, and highlights from the development of TLD1433. Chemical Reviews. 2018;**119**(2):797-828. DOI: 10.1021/acs.chemrev.8b00211

[244] Karges J. Clinical development of metal complexes as photosensitizers for photodynamic therapy of cancer. Angewandte Chemie International Edition. 2022;**61**(5):e202112236

[245] McFarland SA, Mandel A, Dumoulin-White R, Gasser G. Metal-based photosensitizers for photodynamic therapy: The future of multimodal oncology? Current Opinion in Chemical Biology. 2020;**56**:23-27. DOI: 10.1016/j.cbpa.2019.10.004

[246] Zhou J-Y, Wang W-J, Zhang C-Y, Ling Y-Y, Hong X-J, Su Q, et al. Ru (II)-modified TiO2 nanoparticles for hypoxia-adaptive photo-immunotherapy of oral squamous cell carcinoma. Biomaterials. 2022;**289**:121757. DOI: 10.1016/j.biomaterials.2022.121757

[247] Maheswari P, Ponnusamy S, Harish S, Ganesh M, Hayakawa Y. Hydrothermal synthesis of pure and bio modified TiO2: Characterization, evaluation of antibacterial activity against gram positive and gram negative bacteria and anticancer activity against KB Oral cancer cell line. Arabian Journal of Chemistry. 2020;**13**(1):3484-3497. DOI: 10.1016/j.arabjc.2018.11.020